Computer Security Management

Karen A. Forcht
James Madison University

boyd & fraser publishing company

DEDICATION

To my daughters, Nicole and Michelle . . .
The reasons for it all.

Executive Editor: James H. Edwards
Project Manager: Christopher T. Doran
Production Editor: Barbara Worth
Compositor and Production Services: Custom Editorial Productions, Inc.
Interior Design: Jeff Davidson
Cover Design: Karen Gourley Lehman
Cover Photos: Courtesy of Hewlett-Packard Company (left)
        Courtesy of Kmart Corporation (middle)
        Courtesy of Finger Matrix Corporation (right)
Manufacturing Coordinator: Tracy Megison

© 1994 by boyd & fraser publishing company
One Corporate Place • Ferncroft Village
Danvers, Massachusetts 01923

International Thomson Publishing
boyd & fraser publishing company is an ITP company.
The ITP trademark is used under license.

This book is printed on recycled, acid-free paper that meets
Environmental Protection Agency standards.

Manufactured in the United States of America

Library of Congress Cataloging-in-Publication Data

Forcht, Karen Anne, 1944-
    Computer security management / Karen A. Forcht.
        p.   cm.
    Includes bibliographical references and index.
    ISBN 0-87835-881-1
    1. Computer security.   I. Title.
  QA76.9.A25F64   1994
  658.4'78--dc20

                                                        94-698
                                                          CIP

1 2 3 4 5 6 7 8 9 10 D 8 7 6 5 4

BRIEF CONTENTS

CONTENTS

V MANAGERIAL ISSUES

PREFACE

ABOUT THIS BOOK

As its title implies, this text discusses computer systems management and introduces the security issues that result from automation. *Computer Security Management* addresses today's heightened concerns regarding confidentiality, privacy, and volatility in our increasingly computerized society.

Computers impact each one of us directly, and learning to manage them properly and efficiently is a valuable skill. This book presents the basic principles of computer system security, providing a strong platform of knowledge for managers at all levels. *Computer Security Management* reinforces the basic tenet that planning up front is preferable to reorganization and punishment after the fact through a series of pertinent articles at the beginning of each chapter that are right out of today's headlines.

SUMMARY OF CONTENTS

This book is divided into five parts:

Part I, Introduction to Computer Security, includes an overview of computer security and a look at physical protection.

Part II, Systems Security and Control, includes chapters on hardware security controls, software controls, and encryption techniques.

Part III, Special Considerations, covers database security, network and telecommunications security, microcomputer security, and viruses.

Part IV, Legal and Ethical Issues, includes coverage of legal issues, current legislation, and ethical use of computers.

Part V, Managerial Issues, includes chapters on managerial issues, disaster recovery and contingency planning, and new technologies and future trends.

Each chapter concludes with exercises and case studies, allowing readers to apply the topics presented in the chapter. Computer security is not a "fixed" discipline; it must be applied to the issues at hand. The "moving target" that computers represent is, in most cases, situational, and careful discussion of the unique circumstances is vital.

ACKNOWLEDGMENTS

Many people were very involved in the development and publication of this book, and I would like to thank each one for the active part they played. The author is only one player in the orchestra—the rest of the band is most valuable in producing the song in unison. Thanks especially to the reviewers of this text: Jan Cook, Illinois State University; John Grillo, Bentley College; and David A. Nuesse, University of Wisconsin at Eau Claire.

At boyd & fraser publishing company, my special thanks go to: Thomas Walker, Publisher, for encouraging the project from its inception; James Edwards, Executive Editor, for the initial idea for publishing this book; and to the people who read the manuscript to keep the project moving forward: Barbara Worth, Production Editor; Christopher Doran, Project Manager; and Daphne Snow, Editorial Assistant.

To Leslie Kauffman, Project Editor, Custom Editorial Productions, Cincinnati, go my special thanks for the invaluable editing that brought the project to completion.

INTRODUCTION TO COMPUTER SECURITY

OVERVIEW OF COMPUTER SECURITY

LEARNING OBJECTIVES

After studying this chapter, you will be able to:

1. Identify the various security vulnerabilities in computer systems.
2. Explain the effect of security breaches on hardware, software, data, and people.
3. Consider the role of ethical awareness in computer security programs.
4. Define the unavoidable threats to security involving natural disasters, malfunctions, and other uncontrollable events.
5. Explain some of the measures necessary to preserve the security and operating integrity of information systems.
6. Explain the information access problems caused by unauthorized users.
7. Describe the various security measures currently being used.
8. Define the following terms.

TERMS

- Software Piracy
- Configuration Management
- Software Alteration
- Hacker
- Data Manipulation
- Ethics
- Natural Disaster
- Contingency Planning
- Backup
- Sabotage
- Fraud
- Theft
- Espionage

- Operating Integrity
- Access Logs
- Exception Report
- Verification
- Editing
- Reasonableness Check
- Totals Check
- Check Digits
- Trapdoor
- Trojan Horse
- Salami Attack
- Virus
- Worm

TERMS (continued)

- Password
- Encryption
- Dial-Back Device
- Progress Review
- Acceptance Test

- Post-Installation Review
- Cost-Benefit Analysis
- Audit
- Performance Monitor
- Nondisclosure Agreement

ATTACK OF THE KILLER VIRUS

Natural disasters and direct attacks by humans are not the only threats to computer system security. "Infectious" programs called viruses are much more insidious foes, because they infect the computer and cause damage without the user's knowledge. Often, an innocuous screen message such as "Gotcha!" is the user's only clue that something has gone awry with the computer. And the potential damage can be as serious as loss of the entire contents of the hard disk.

How are these viruses transmitted? Typically, a user obtains data and programs from other users via diskettes or directly via electronic bulletin boards. When the user inserts an infected disk or downloads an infected bulletin board program, the virus as well as the data is transmitted to the user's disk. The virus might then be triggered at some planned date or time. For example, in 1988 some users of a pre-release version of a popular commercial drawing program learned about viruses the hard way: On March 2 their screens displayed a universal message of peace, then the contents of the machines' hard disks were scrambled. It is likely that this virus was activated by the computers' internal clocks.

A computer virus is the latest refinement of a type of rogue program that has been present for at least 20 years. Known also as a Trojan Horse or a Logic Bomb, this type of program damages or destroys data or programs. Computer viruses have two insidious characteristics: First, they appear to be introduced into an organization's computing

Continued

ATTACK . . .

From page 3

activities from outside. Second, they are designed to replicate themselves many times. Often, each time the virus is activated, it copies itself repeatedly to numerous memory locations. Potentially, each successive virus copy can be activated independently and reproduce itself, thus furthering the chain of infection.

Experts are still assessing the potential danger to information system security posed by viruses, but it is clear that organizations and individual users should take some basic steps to protect their computers:

1. Restrict downloading of files and software from bulletin boards.

2. Limit the use of software—commercial programs as well as freeware and shareware—that has not been independently verified to be virus-free.

3. Design interfaces to network gateways and other telecommunications ports that prevent direct access to disk contents from outside sources.

4. Protect the memory of mainframe-supported systems by segmenting storage device interconnections and the program that manages them. This step will reduce the system's vulnerability to infection.

As reported in "The Computer Virus Danger Grows," by Belden Menkus, *Modern Office Technology*, February 1989, pp. 38–40.

WHY WORRY ABOUT COMPUTER SECURITY?

Before going to bed at night, do you leave your front door unlocked? When parking your car, do you leave the door unlocked and the keys in the ignition? The answer to both questions is probably "no."

You automatically take basic precautions to secure your valuables. Information is a valuable asset for many companies. In this age of computing, information is increasingly valuable; it can even become the life blood of a company. With advanced technology, information becomes readily available to the user, and it can be obtained at ever faster rates.

Because the information stored in computers is both essential and valuable, not securing data with as many safeguards as you would use to protect your home or other assets is ludicrous. Unfortunately, computer security is a fairly new idea; it lags behind the proliferation of new techniques and equipment. All too often, security measures are either minimal or completely ignored.

In this chapter, we will discuss the security issue; areas of vulnerability in systems; the people involved in computer crime; methods of trespassing; and ways to counteract intrusion. The purpose of this chapter is to introduce the concept of computer security and to provide a brief overview of security problems and ways to deal with them.

The Need for Security

Although all the assets of an organization are subject to loss, damage, or destruction from various causes, information systems tend to be particularly susceptible to these dangers for several reasons. First, their components are comparatively fragile. Computer hardware can be damaged more easily than, say, the tools on an automobile assembly line, and data files are extremely fragile compared with most other organizational assets. Second, computer systems are likely to be the target of disgruntled workers, protestors, and even criminals. Finally, the decentralization of facilities and the use of distributed processing have increased the difficulty of protecting information and computers.

Areas of Vulnerability

Hardware. Hardware includes the physical devices, which are the most visible part of the computing system. Because of its accessibility, hardware can be subject to common mishaps such as coffee spills, crumbs getting into keyboards, and dust. Intentional attacks against hardware include its removal from the premises by computer thieves; kicking, slapping, and punching peripherals; and sabotage of terminals with water and fire. Pens and pencils have been used to short circuit boards to limit the availability of the computer. Most of these acts can be prevented simply by placing locks on machine room doors or installing a surveillance system (equipped with cameras and a guard). The advent of the microcomputer and local area networks, however, further complicates the security issue by creating a "system" on almost every desk. If the organization is serious about and committed to security, these computers can be locked or even placed in cabinets specially designed to keep the computers secure and out of sight when not in use.

Software. Software is another integral part of the computing system. Without the operating system, utility programs, and application programs, the hardware would be just machinery. Attacks on software occur in three ways: piracy, deletion, and alteration of software.

Software piracy. **Software piracy**—the illegal copying and distribution of software—is a serious problem. Although most people think nothing of making a copy of software and giving it to a friend or making extra copies for personal use, these actions violate copyright laws and are punishable by law. The creator of the software should be compensated for his work just as a musician is paid royalties for an original score of music.

Deletion of software. Deletion of software is a common occurrence. Have you ever accidentally deleted a file, a program, or perhaps your whole directory? This is quite easy to do. This mishap can be controlled by a process called configuration management. With **configuration management** (done either by human or computer), every change made to a program is controlled and recorded. The goal of this process is to ensure the availability of the correct version of the program. With uncontrolled modifications by numerous individuals, the original version of a program will most likely be lost.

Software alteration. **Software alteration** is another form of sabotage that is difficult to detect. Changing a few lines of code in a program that is thousands of lines long is difficult to detect. Nevertheless, these few lines can alter the program and have destructive results. (See vignette.)

Data. Because data has essentially no intrinsic value, it is difficult to put a monetary value on it, even though it is crucial to an organization and its livelihood. Reconstructing lost data can be very time-consuming and, at the least, causes measurable costs to the organization—the opportunity cost of lost computing time.

An actual exposure of data may aid the competition, leak important inside information, or damage lives if personal data (for example, police files) are revealed. Authorizing users' access only to limited areas of data, shredding or disintegrating sensitive data after its use, and protecting data until it loses its value are a few of the commonly used safeguards.

People. People can cause a great deal of damage to any computer. For different reasons, people attack computers or, more often, the information they contain.

Intruders. Disgruntled employees can seek revenge against an organization by planting a logic bomb or any other destructive program. Many times such intruders are easy to identify; managers should keep an eye on unhappy employees or employees with personal problems, such as illness in the family or drug abuse. These people are prime candidates for computer crimes because of their need for fast cash.

Hackers. Hackers are a different kind of computer intruder. According to a large glossary file maintained at MIT and Stanford, the term "hacker" is defined as follows:

> HACKER (originally someone who makes furniture with an axe) 1. A person who enjoys learning the details of programming systems and how to stretch their capabilities, as opposed to most *users,* who prefer to learn only the minimum necessary. 2. One who programs enthusiastically, or who enjoys programming rather than just theorizing about programming.[1]

Hackers are often teenagers who are passionate devotees of computers. They may prefer beating, crashing, or penetrating a system to meals, dating, or even earning good grades.[2] (See Figure 1.1.) Despite the abundance of hackers at such institutions as MIT, Stanford, and Berkeley, not all hackers are technical experts. In many cases, all the hacker needs to "hack" a system is a personal computer, modem, and a fair amount of computer literacy. One infamous "hack" involved a group of teenagers dubbed the 414 Gang. Named for their Milwaukee area code,

FIGURE 1.1

Profile of a Typical
Computer Crime
Perpetrator

- ☑ Relatively young (under 30 years of age)
- ☑ Highly motivated
- ☑ Intelligent
- ☑ Personable
- ☑ A good worker
- ☑ Happy with job
- ☑ Employed for several years
- ☑ No history of job problems
- ☑ Trusted
- ☑ Possibly overqualified for current position
- ☑ Sees self as a "borrower"

this group met as Explorer Scouts and cultivated their interest in computers by breaking into more than 60 systems across the nation. Fortunately, this group was discovered before it did any serious harm, but the 414 Gang is one example of how a computer, a little curiosity, and a lot of extra time can lead to disastrous results. Often, companies whose systems have been penetrated by hackers fail to prosecute them because of public embarrassment. How would you feel if your bank's system had been broken into and had millions of dollars (including some of yours) diverted to a Swiss bank account belonging to an unethical employee? Your confidence in that bank would plummet, and you would most likely take your business elsewhere.

Hackers can do a great deal of harm, but they generally are not malicious—their goal is to challenge the system and discover its vulnerabilities. Instead of prosecuting hackers, some companies hire them to penetrate the system in a controlled environment and to identify weaknesses in the system.

Computer Criminals. Computer crime is big business. In a survey conducted by the American Bar Association, about half of the 283 business and governmental institutions surveyed had been victims of computer crime in the preceding year.[20] It is estimated that the computer crime industry generates revenues of $100 million to $5 billion a year, but the exact figures from this computer crime will probably never be known.

Some areas of computer crime include the following:

- **Theft of computer time.** This is a common practice that ranges from employees borrowing the computer at work to figure out personal finances to running a business for profit on the side at someone else's expense. Theft of computer time can also include the time it takes to repair the damage done by a virus, bomb, or other destructive program.

- **Theft of data.** This category can involve physically removing data from trash receptacles or from files stored on the computer.

- **Manipulation of data/computer programs.** Changing or inserting one line of code in a program can alter the purpose of a program. One example of **data manipulation** is a college student changing his or her grades or number of credit hours in the computer files containing transcripts so that he or she can graduate.

- **Software piracy.** This is the illegal copying of software ranging anywhere from games to word processing and spreadsheet packages. Most of the time the pirated software is traded among users, but some do charge a price for the illegal copies. Even the basic elements of a program, such as source code and flowcharts, are deemed copyrightable by the Copyright Act of 1976.

These are some major areas of computer crime that are crippling the computer industry today. The people committing these crimes are not typical criminals. They are intelligent, white-collar professionals who are taking huge risks for the possibility of astronomical profits. It is a "clean" crime; no blood or violence are involved, and millions of dollars are just a few keystrokes away.

ETHICAL CONSIDERATIONS

The discussion of users, hackers, and computer criminals brings us to an extremely important topic in computing: ethics. Making a copy of a handy word processing program for a friend may seem like a courtesy, except for one thing—you break the law. It is estimated that for each legal copy of software, there are ten bootleg copies. People try to justify illegal copying because "the software prices are so high." Software prices, in reality, are high because the programmers and vendors raise prices to compensate for illegal copies and their high development costs.

In other instances, the line between what is ethical and what isn't may appear to be fuzzy. For example, students getting together to work on their programs produce programs that look extremely similar. They can argue that they worked together; therefore, their programs should look relatively similar and produce similar output. Is this ethical? A statement of ethical standards in computing is becoming common in colleges and universities. These statements include guidelines relating to programming and other student assignments, the appropriate use of school equipment, prohibiting the use of computers to harrass (for example, misuse of e-mail), and other items pertaining to the individual school.

Criminals used to get away with computer crimes because there simply were no laws governing them. Legislation now deals with this special issue; the laws are growing in sophistication as different degrees and types of computer crime occur.

THREATS TO SECURITY

Natural Disasters

Some threats to computer systems are obvious; they vary little from threats to other assets. **Natural disasters** fall into this category. Fires, floods, windstorms, lightning, and earthquakes can occur without warning and destroy or damage hardware, software, and data.

Although we can do little to *prevent* natural disasters, we can do much to minimize their effects. Hardware, like other valuable assets, should be insured. It can be replaced if lost. The disruption to operations that results from the loss of hardware can be minimized if contingency planning has identified available backup facilities—perhaps from the manufacturer, a computer service center, or the user's organization.

Software and data losses are more serious. They, too, can be covered by insurance, but unlike hardware, their replacements can not be bought. Therefore, disaster planning must include provisions for backup data and software as well as backup hardware. Many companies maintain off-site, fireproof vaults for storing backup material. This approach is particularly efficient in large, multisite organizations in which the various sites back each other up. Smaller or single-site firms often turn to service centers for backup. Sunguard, a subsidiary of the Sun Oil Company, is a service center that offers its customers both computer backup and vault storage for data and software. Complete backup service such as this typically costs $3,000 to $5,000 per year.

Malfunctions

Natural disasters can be catastrophic, but fortunately they are rare. Malfunctions usually cause much less damage, but they occur with greater frequency. Even worse, managers are highly aware when disaster strikes, but they may be totally unaware that a malfunction is taking place. Malfunctions can occur in hardware as a result of power surges or failures, stray electrical forces (static electricity or the output of nearby electronic devices), the introduction of dirt or other foreign matter into equipment, mechanical wear, operator error, or the failure of electronic components. Software can malfunction, usually because of logic errors, but also because of contamination of the recording medium or stray electrical charges. Although we generally do not speak of "data malfunctions," the proper functioning of computer centers depends heavily on accurate data; any inaccuracies in data may have the same negative effect as hardware or software malfunctions.

Hardware reliability is at least partially a function of design and production quality control; as such, it is beyond the control of the user organization. Users can, however, be sure that routine maintenance is performed on schedule, that the machine room is kept relatively free from impurities, that primary power is regulated against surges, that operators are properly trained, and that backup power is available. (Many computers now contain sufficient emergency battery power to preserve data until backup power can be applied.)

Testing and debugging are supposed to ensure software reliability, but anyone who has ever written a computer program knows how an unanticipated combination of events can produce inaccurate results. Liability for coding errors is a matter of growing concern for personnel and has created an expanding market for insurance companies. Three Midwest banks recently sued their data processing service company over software errors that caused the overpayment of interest to the banks' customers. Insurance firms are experiencing dramatic growth in the sales of policies insuring against programming errors. Malpractice insurance, which until recently was available only to doctors, lawyers, architects, and other professionals, is now available to computer programmers. If software reliability cannot be absolutely *assured*, at least it can now be *insured*. Insurance does not completely solve the problem, of course. Incidental losses, such as not being able to collect accounts receivable if a file is inadvertently destroyed, are still uninsurable. Insurance does not relieve managers of their responsibility to eliminate software errors, but it does somewhat soften the blow that can result from such errors.

Criminal Acts

In the world of computing, victims may be doubly exposed to criminal activity. Crimes can be, and are, committed both *against* and *with* computers. Crimes against computers include malicious acts such as **sabotage** and mischievous acts such as "data diddling." Crimes committed with the aid of a computer include fraud, theft, and espionage.

Crimes Against Computers. In the late 1960s and early 1970s, a few extremists expressed their protest against the war in Vietnam by sabotaging the computers of an arms manufacturer and destroying the punched-card records of a local draft office. Although these acts do not seem to have altered the course of the war to any measurable degree, they did point out the vulnerability of computers to acts of violence. The results of these acts can be as devastating as natural disasters.

Defenses Against Computer Crime. Although you cannot lock out a flood, you can lock out, or at least make entry very difficult for, a would-be saboteur.

Hardware can be protected by passive measures, such as by locating the machine room away from normal traffic flows and by protecting the site against fire and minor explosions. Active measures include the use of security guards and controlled access to hardware locations via keys, coded locks, or sign-in stations.

Protecting data and software from sabotage, particularly in distributed systems where there are many points of access to databases, is more difficult. Data can be protected by the use of passwords, a "read-only" access mode, limited access to specific applications, and the continuous screening of employees who are given access to the system. The last measure is especially important because many acts of sabotage against data are committed by those who have authorized access. For example, an employee who is given a one-week termination notice should have his or her access revoked at the time of notification. Revoking access privileges should trigger a change of passwords. In any event, passwords should be changed periodically.

Less serious, but still of major concern, is the threat of **data diddling**—the alteration of data. Numerous cases of data diddling have been traced to teenage whiz-kids who gain access to databases through communications networks and their home or school computers. Data diddlers rarely have any malicious intent; they do their mischief for the challenge involved, perhaps without fully understanding the consequences of their acts. The benign intentions of data diddlers are small consolation to the database administrator who must sort garbage out of the files!

Crimes with Computers. Those who sabotage hardware may gain some perverse satisfaction from their acts, but those with the ability to sabotage software and data stand to gain much more. The ability to tap into databases or to modify applications software can be used to defraud, steal, or engage in espionage.

Crimes committed with computers are much more difficult to detect than those committed against computers. Where sabotage leaves ruined equipment or garbled data in its wake, the clever computer-aided thief can use computer technology to cover his or her tracks as well as to perpetrate the crime.

Documented cases of computer crime include the now-famous case of the bank employee who diverted fractional cents (the result of rounding) to his own account, a team of racetrack employees in Florida who printed winning parimutuel tickets on the backup computer, and the California programmer who diverted millions of dollars through electronic funds transfer channels to his numbered Swiss bank account. Less dramatic cases involve a government employee who used his agency's computer to maintain records of his stock market transactions (thereby "stealing" valuable computer time) and a marketing organization that gained access to its service center's files to obtain the mailing lists of rival firms (a simple case of industrial espionage).

No one knows the extent of such crime because even those crimes that are detected—which may be less than 10 percent of the total—are not always reported. Companies may not report computer crime out of fear of appearing incompetent and to avoid documenting how the crime was committed and thereby making a how-to guide for computer crime.

Some data need to be protected from illegal access even though their loss or damage might cause no financial harm. For example, personnel files are subject to privacy legislation that prohibits the disclosure of information not only to outsiders but also to users in other functional areas who may otherwise have unrestricted access. Personnel files are also unique in that old backup or "grandfather" files that no longer have value must be given the same protection afforded current files. This is an unusual requirement, one that is easily overlooked if security is lax.

Operating Integrity

Even if natural disasters, malfunctions, and criminal acts were not threats to computers, systems managers would still need to take precautions to safeguard data. The very nature of processing—changing, adding, and deleting data—raises the possibility for error. Changes may be posted improperly, incorrect data may be added, or data that should have been saved may be deleted. **Operating integrity** deals with elimination of these errors in processing.

Operating integrity can be compromised in a variety of ways, the most common being by errors committed during data entry. The old phrase "garbage in, garbage out" may seem a little worn now, but it is still true. Even if data were perfect, computers are still vulnerable to logic errors and rare, but possible, internal processing errors. Any one or a combination of these errors can result in direct financial losses caused, for example, by overpaying suppliers or underbilling customers. Errors also can lead to indirect losses by providing faulty information to decision makers. A logic error in the computation of a net present value used in financial planning may ultimately be far more costly to an organization than one that, for example, bills a customer a few hundred dollars less than the correct amount.

Some security measures, such as the use of access logs or the maintenance of backup files, also help preserve operating integrity, but other techniques, such as editing or the use of check digits, are strictly error-detection methods. We will look briefly at some methods in each category to get a better idea of how systems managers can maintain operating integrity.

Backup Files

If you have an item of extrinsic value—a rare coin, for example—you can protect it by keeping it in a safe-deposit box at your bank. But even when an item has no intrinsic value—the telephone number of a friend back home, for example—you probably keep a duplicate copy somewhere.

Data fall into the second category: they have no intrinsic value. Data are represented by magnetic impressions on metal oxide or holes in a card. Those physical representations, along with the digital numbers they stand for, also have no intrinsic value. Put them in the context of a formula or a report, which does have extrinsic value, and the need to keep the data secure becomes much stronger. The need for security increases as data change from day to day. Thus, data can have both intrinsic and extrinsic value; they may also be volatile. Backing data up gives some help to those affected by volatile data. Data, however, are unlike your friend's telephone number, which does not change from day to day. Data change constantly, thus complicating the procedures for maintaining backup files. The specific approach to backing up data depends on the mode of processing.

Backup in Batch Processing. In batch processing, the old master and transaction files are maintained for at least one cycle to facilitate reconstruction of the current master file should it be damaged—perhaps as it is being used as the old master file in the next cycle. As an extra precaution, the transaction documents are also saved to permit reconstruction of the old transaction file, should it be damaged as well. As shown in Figure 1.2, this combination of old master file, old transaction file, and transaction documents will cover all backup requirements in a batch system. Of course, the backup files and documents must be stored in a location and a manner that preclude a simultaneous loss of both primary and backup data. A fireproof vault some distance from the processing area would satisfy these conditions.

Backup in Transactional Processing. Transactional processing poses a somewhat more difficult problem because the master file is updated continuously and there is no clear distinction between "old" and "new" files. In this case, one can define "cycles" arbitrarily by making periodic copies of the master file. If transaction documents are saved for the most recent "cycle," a new master file can be created at any time from the last copy and the intervening transaction documents. Again, the backup data must be stored separately from the primary data to achieve a reliable backup capability.

Figure 1.3 shows a transactional system in which the master file is copied periodically onto tape (the medium of the backup file is not important). Of course,

FIGURE 1.2
Backup in Batch Processing

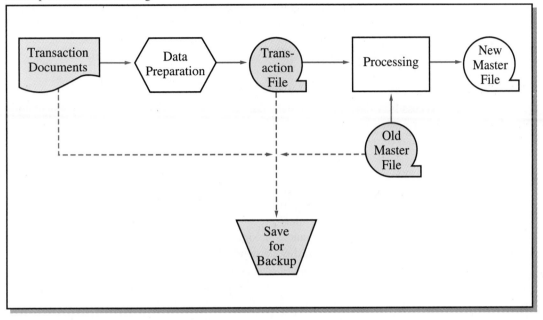

FIGURE 1.3
Backup in Transactional Processing

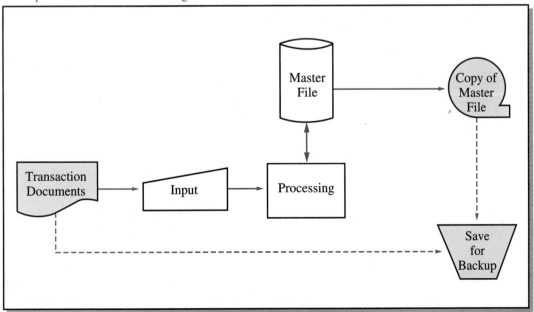

if the backup master file is on a sequential access medium such as tape, it is necessary to copy it back onto a direct access medium before resuming processing. Because it is rare actually to use a backup file (just as it is rare for a homeowner to use a fire insurance policy), the slight inconvenience of recopying the file is more than offset by the use of a less expensive medium for backup. Utility programs are available in system software that copy from disk to tape or tape to disk.

Access Logs

Access logs show the time, date, duration, and user identification of access to computer operations. Increased frequency of access or unusually long access times on the part of a user are cause for suspicion. **Exception reports** can be prepared to show deviations from past individual usage or from group norms. In most cases, there will be valid reasons for such deviations; in some instances, the deviations may point to a user who is a real or potential threat to operating integrity.

Access logs can be either manual or automated. Physical access—to the machine room or the terminal area—is usually recorded manually on sign-in and sign-out sheets at the entrance. Automated logs, maintained by service programs in the system software, record the user identification, the sign-on and sign-off times, and the application software or files accessed.

Originally, access logs were intended to thwart unauthorized access to computer areas. Currently, the popularity of distributed processing, the proliferation of terminals and other data entry devices in functional work areas, and the increased use of data communications have made it almost impossible to keep out unauthorized personnel. Unauthorized users are not likely to use their correct identification—if they even have one—to help security personnel locate them. Access logs may tell managers when unauthorized access has occurred, but they are of little help in apprehending intruders. As a result, access logs are now used primarily for accounting purposes and for identifying legitimate users who may be inadvertently threatening computer operating integrity. Once identified, these persons can be educated in the proper use of the system.

Error Detection Measures

Errors can be detected at various stages of processing. **Verification** is a means of detecting errors in input. Other input errors can be detected by **editing**. For example, a preprocessing edit of the transaction file for personnel applications might check social security numbers for nine numerical characters. The edit would not ensure that the nine digits used are the correct ones, of course, but it would screen out gross errors, such as the presence of alphabetic characters or an eight-digit number.

A **reasonableness check** makes sure that the data are within certain reasonable limits, say, $3.50 to $20.00 for the hourly wage rate in a payroll system. In an interactive system, unreasonable values—those outside the limits—cause a program interruption. They must be verified or corrected by the operator before the program can continue. In other systems, values failing the reasonableness check may be printed on an error listing for correction before a second attempt is made to process them. Reasonableness checks can be employed at any stage in the system—

during input, processing, or output. Like editing, reasonableness checks do not ensure accuracy, but they can eliminate gross errors, such as the $10,000 or $15,000 monthly residential electricity bill we read about occasionally.

A **totals check** compares beginning and ending totals—of inventory, for example—and shows the net change created by all the transactions. Any difference indicates an error, probably the loss of one or more transactions during the update process. This is the same principle that we apply to balancing our checkbooks: We add the outstanding checks and the unrecorded deposits, then compare the result with the difference between the bank statement and the amount shown in our checkbook. The bank also uses this technique, but in a grander, computer-supported version, to check its daily calculations. As these examples suggest, totals checks are most appropriate where the data are numerical.

Finally, **check digits** guard against transmission errors that might occur as data are moved internally within the CPU and externally between the CPU and storage devices. A check digit is an extra digit determined by an arithmetic manipulation of numerical data, say, the result of alternately adding and subtracting each digit in a number. In this example, the number 32,754 would have a 7 added to it (3-2+7-5+4=7) and be transmitted as 327547. Before it is used in processing, the computation of the check digit is repeated and compared with the transmitted check digit. Any difference is an indication that the number was somehow altered during transmission. Of course, the check digit is not part of the number for processing purposes and must be recomputed if the updating process changes the original number.

Information Access Problems

Programs, written by authorized users, serve as vehicles for accessing and manipulating data to produce the desired output. Because the programs can access data (e.g., payroll data, company accounts), they are perfect targets for computer criminals. If you can access the program and modify it, then you can access the data. The following are a few vehicles commonly used to access data.

Trapdoors. Trapdoors are secret entry points into a program inserted by the programmer during the program's development to provide a way back into the program for modification after implementation. Trapdoors are useful in tracking the flow of operations in a program, but they become dangerous when they are used for unauthorized access. The programmer might later exploit the trapdoor as a covert means of access after the program has been implemented in the system, or an intruder may stumble upon the opening after exhaustive attempts to enter the system.

Trojan Horses. Like the Trojan Horse from Greek mythology, a program containing a **Trojan Horse** can be very deceiving. Although the program may seem productive and beneficial, it hides something within its code that can be fatal. For example, a program designed to list all the users enrolled in a certain class may be quite useful, but concealed subroutines may also determine the passwords for the users' accounts. These subroutines or commands can be scattered throughout the program or even scrambled to make detection very difficult.

Salami Attacks. A **Salami Attack**, named for the way salami can be sliced very thin, is likely to hit a financial program dealing with interest computations or manipulation of funds. Suppose an unethical bank employee altered the interest computation on all checking accounts so that one extra cent was generated from each account and diverted to his or her account. One cent alone is not much, but it would add up if the computation was not detected for five years. It is highly unlikely that a customer would detect such a small difference in his or her balance. Even if customers found the error, they would probably attribute it to a mistake in rounding or balancing their checkbook.

Infectious Programs. Programs can be used to affect computer systems as well as data. These programs are termed "infectious" because they infect the system (just as a virus would infect a human body) and make it unavailable for use. **Viruses** are programs that pass along their "infection" to any other programs they contact, thus eventually infecting the whole system. For example, an infected program in a network setting containing commands to delete user directories can do a large amount of damage in very little time. A **worm** is similar to a virus in that it is a malicious code embedded in useful programs. It performs its specific task, then exits, looking for other programs to penetrate. A worm program can continue execution without control and it will eventually deny access to authorized users. "Vaccine" programs are available that can detect and eliminate viruses in your programs, but even these must be used with care—viruses have been known to exist in these anti-virus programs. When that happens, the program the user expects will prevent damage actually doubles it. (See Figure 1.4.)

SECURITY MEASURES

There are many ways to prevent access to a system—from physical security involving locks and guards to measures embedded in the system itself. The user is a source of vulnerability; every user represents a way to access the system. Therefore, many security measures begin with users.

Passwords

Passwords, the most common means of user authentication, are words that the user and system have chosen to allow the user access. Passwords are generally used in this sequence: The user enters a user ID, then the system asks for a password. If the password the user enters matches the one assigned to the user, access is allowed. If not, access is denied and the user may try again. Sounds easy? If the password is easy for the user to remember, it is likely that an intruder could guess the password just by knowing a little about the user. To make this process more secure, a few rules should be followed when choosing passwords:

- Don't leave your password out in the open. Don't write it down and tape it to the terminal or another obvious place.
- Choose a password with at least six letters. The time it takes to exhaust a list of probable passwords increases exponentially as the number of

FIGURE 1.4
Virus Infection: Is Your Computer Next?

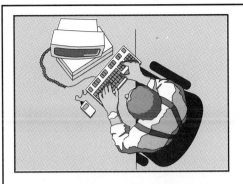

A "virus" is a small computer program that contains malicious or damaging codes or instructions.

Viruses are usually transmitted into a computer by an infected disk that contains the malicious codes.

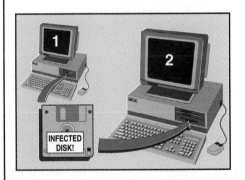

The disk is then used on other machines, or copies are shared and the infection spreads. In some cases, the computer itself becomes infected so that anyone communicating with that computer (e.g., via a bulletin board, e-mail) becomes infected.

letters increases. Choose a password that isn't an actual word in the dictionary; exhausting all the words in a dictionary doesn't take very long on a computer!

■ Don't choose a password that is obvious. Obvious choices include first and last names, the name of your boyfriend or girlfriend, or even such words as "enter." Anyone who knows you could guess these passwords after a few tries.

Encryption

Encryption, another form of concealing or protecting data, is the translation of data into secret codes so that an intruder cannot read the data even if he or she were to gain access to it. The process requires an encryption device that actually converts the data into the encrypted form and a decryption device that returns the data to its original form. The federal Data Encryption Standard (DES), created by IBM and several government agencies, is currently the most commonly used encryption process. It is endorsed by the American National Standards Institute (ANSI).

Dial-Back Devices

Dial-back devices are used to verify that users at remote terminals are indeed authorized users. Figure 1.5 illustrates how a dial-back device works.

Control

Control begins during planning and is exercised to some extent in every phase of the information systems life cycle, just as control is included in each phase of the management cycle. During planning, control standards are established as objectives of the system, and periodic evaluations are scheduled. One type of evaluation, the **progress review**, occurs during the development phase. Another, the **acceptance test**, takes place during implementation. Shortly after implementation, early in the operating phase, a **post-installation review** is conducted. Periodic **audits** are administered at regular intervals, usually six months or one year, for the remaining life of the system.

Progress Reviews. It would be very unusual for a project to proceed precisely on schedule in exact conformance to a master plan. More likely, some activities will be delayed, others will proceed more rapidly than planned, new or revised information requirements will precipitate design modifications, and environmental changes will force a restatement of objectives. The purpose of a progress review is to bring these changes to light, to revise the master plan if necessary, and to redirect development efforts when that is required.

Progress is measured by a comparison of the amount of work completed with the amount scheduled for completion using a milestone schedule or Gantt chart. Discrepancies are resolved by modification in the schedules or reallocations of development resources. Also, the expenditure of funds is compared with the budget, and any necessary adjustments are made. Any changes in either the schedule or the budget must be conveyed to all affected parties.

Deviation from schedules is to be expected and, unless major discrepancies arise, is not a cause for concern. More worrisome are modifications caused by changing information requirements. The project management team should establish change policies very early and enforce them strictly during development and operation. Once the logical design is completed, any changes desired must be fully justified by the originator and approved by the project management team. Changes are usually judged more harshly than original proposals because changes may render some completed work useless and may result in disproportionate expenses. At

FIGURE 1.5
Dial-Back System

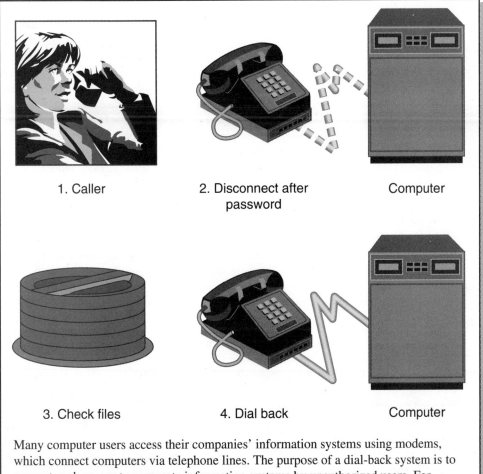

1. Caller   2. Disconnect after   Computer
                      password

3. Check files   4. Dial back   Computer

Many computer users access their companies' information systems using modems, which connect computers via telephone lines. The purpose of a dial-back system is to prevent such access to corporate information systems by unauthorized users. For example, (1) the user dials a computer by touch-tone phone. (2) The computer asks the user for a password, which the user gives. The computer disconnects and (3) checks its files to see if the password matches an authorized phone number. The computer then dials that number and allows the user to obtain access to the computer. If the computer finds that the call came from an unauthorized location, the computer dials the authorized number (4) and warns the user that there has been an attempt to use the password.

some point, the project management team must draw the line on changes and approve only those brought about by environmental conditions beyond organizational control. One exception to this policy occurs in the development of decision support systems, where changes are expected more frequently than in management information systems or transaction processing systems.

Acceptance Test. The acceptance test is the final activity before conversion to the new system. In a modular design, each module will be tested as it is introduced, but then the total system must be tested in its entirety.

The acceptance test is a systems test that includes user personnel and that must assure the project management team that the module is ready for implementation. In contrast to testing and debugging, which detect errors in programming only, the acceptance test evaluates user procedures, personnel training, operator procedures, the analysis and design effort, data communications, and every other aspect of the system, including a recheck of the program testing done earlier.

The acceptance test may be designed by personnel, but it should be conducted by an agency with less personal involvement. The project management team may form an ad hoc acceptance test team expressly for this purpose. As in all evaluations, clear, quantitative standards for acceptance (or rejection) must be determined prior to the test. Error rates, turnaround times, and measures of accuracy provide unambiguous standards for comparison.

Post-Installation Review. When operation of the new system is technically satisfactory—that is, when personnel are achieving the expected speed and accuracy in entering data, all programming errors are corrected, and users are accustomed to the new output—it is time to evaluate the system in operational terms. In brief, is the system accomplishing what it was intended to do? This operational test is called the post-installation review and is evaluated according to several different criteria.

Cost-Benefit Analysis. The standard for this comparison is the **cost-benefit analysis** conducted in the economic feasibility analysis, as modified by changes encountered during development. Because it is frequently difficult to place a dollar value on the benefits, some organizations treat the computer operation as a *profit center* and "charge" customers (users) for services. Realistic charges can be developed from a comparison with the fees charged by computer service centers for similar work. In effect, this approach allows the user to "buy" services, doing without some or all of them, or even going to an outside agency to obtain computer service. In this concept, "income" is equated with benefits, and the cost-benefit analysis is greatly simplified. This method obviously works best in those organizations already using the profit center concept for other services. Just as obviously, an organization with a fully integrated system cannot really give users the option of not participating in the system. It is inconceivable, for example, that the production department would not participate in the model.

Attitude Surveys. The cost-benefit analysis validates the economic feasibility study; however, validating behavioral feasibility is just as important. One way to do this is to survey the attitude of the users toward the computer application. The design of survey questionnaires is itself a highly specialized skill and may require the assistance of consultants. The questions typically address changes in the volume, the quality, the difficulty, and the enjoyment of work as a result of the new system. A well-designed questionnaire will include items that can be validated from second sources, so that differences between actual impact and perceived impact can be determined.

Measurement. Much of the impact of the new computer applications can be measured against and compared with project standards or past performance. For example, if inventory stock-outs had averaged 3.7 per week prior to implementation and the objective was to reduce this number to 2.5, a new rate of 2.3 represents both an improvement and the meeting (or exceeding) of system objectives. Other measures may be more subtle and more difficult to trace. Certainly, one expects better information to result in better decisions and ultimately in an improvement in an organization-wide measure such as return on investment. It is difficult, however, to isolate the contribution of such an improvement.

Audits. Historically, *financial audits* were conducted to provide an independent validation of an organization's financial affairs. The auditing concept was expanded after World War II to include *operational audits* of other, nonfinancial activities. More recently, *management audits* have been directed specifically at management practices. The **audit** of an information system includes some features of each type of audit.

Information systems make an impact on the finances of an organization in two ways: They represent between 5 and 10 percent of the cost of operating the organization, and they are used to maintain the organization's financial records. It is not surprising, therefore, that auditors have had to become skilled in the working of information systems and that those skills are frequently employed to audit the operation and management of information systems as well.

Auditors may come from internal or external sources. The source is not important, as long as the auditor is able to maintain an objective point of view. In a small organization, where the only auditors are the accounting staff and where the audit must evaluate the performance of the auditors' peers and superiors, objectivity may be lost; external auditors are then called for. Larger organizations may be able to maintain a full-time auditing group that is, for all practical purposes, external to the activities it audits. The financial auditing practice of using external auditors is so well established, however, that many firms that could maintain their own auditing staff still use accounting firms or management consulting groups to audit their information systems function.

Performance Monitors. The auditing of operations may be facilitated by the use of **performance monitors**. There are two kinds of performance monitors: *hardware monitors* and *software monitors*. Hardware monitors are "black boxes" connected to the computer that keep track of the active and idle time of the various components. For example, a hardware monitor might show that a printer is active 90 percent of the time, whereas the CPU is active only 20 percent of the time. In this situation, hardware efficiency may be improved by the addition of a second printer.

Software monitors are computer programs that record the processing times of various activities. Software monitors are used to locate inefficiencies in programming. They provide more specific information than hardware monitors, but they take up memory space, which then introduces certain inefficiencies. Hardware monitors do not interfere with normal processing, but they cannot report on individual programs; because contemporary computers process a number of programs at once under multiprocessing or multiprogramming, the hardware monitor cannot tell which program is running at any given moment.

Contingency Planning

Contingency planning is another area that is crucial to a successful computer security system. This back-up plan is based on a contingent event—an event that may or may not occur. In the event that it does occur, a plan to get the system up and running as soon as possible will be put into action. The computer center manager must include a few vital functions in his or her plan, such as:

- A prioritized list of functions and applications that are to be restored when the disaster hits. It is not possible to restore everything at once so the most vital functions are dealt with first.

- A statement of responsibility that establishes the duties of the users in the event of a disaster.

- Documentation of the contingency procedures for reference during and after a disaster.

- Periodical run-throughs of the contingency plan to keep it up-to-date with users and to reflect changing requirements.

Although the probability of a disaster may be small, contingency planning is "insurance" against such an occurrence. Like our reasoning for having car insurance, we have contingency plans because of the liability and financial responsibility we would incur if we encountered a disaster, not because we plan to have an accident every day.

Hiring Practices

Employees are key ingredients in the success of a computer security system. Their attitudes, ethics, and work habits have a direct impact on whether a security system is successful. Therefore, the hiring and recruiting of employees should include thorough screening and user education from the start. Initial security checks should include references, educational records, law enforcement records, and military records. If the company has a standard of ethics, it should be included in the employee orientation, training, and handbook so that each new employee becomes aware of the organization's ethical standards. An employee may be asked to sign a **nondisclosure agreement** in an attempt to ensure that the employee will not use confidential company information for his or her own benefit. These types of agreements are difficult to enforce unless an outright sale of information can be traced back to the company and the employee. Nevertheless, emphasis on computer ethics from the beginning may prevent confusion or ignorance about ethical standards and company policies later.

Management's Role in Security

Jay BloomBecker, Director of the National Center for Computer Crime Data in Santa Cruz, California, states that the approach to security by obscurity (based on the traditional military model) must now be augmented with:

1. Technology
2. Laws and legal actions
3. Good management techniques

The rise in the importance of computer security has generated a need for management specializing in security. Whether or not a company actually has an information security administrator or computer security manager, security awareness must start with management. A security policy means nothing if the top-level managers don't practice what they preach. The management must be able to sell the benefits of security and maintain an awareness of security throughout the organization. This is no small feat. It takes strong communication skills and knowledge of the industry and all security-related functions. Security is not the sole responsibility of security personnel, but the responsibility of the whole organization. This concept of shared security responsibility must be promoted by all levels of management in order to have a successful security system.

Bennett Asbury, Eastern Regional Manager of Corporate Information Security for EDS Corporation in Herndon, Virginia, sums up the challenge facing information systems security today by stating that a good security program must include:

- Management
- Security organization
- Security activities
- Data ownership
- Users
- Data processing operations
- Applications development
- Disaster recovery

SUMMARY

Although the assets of an organization are subject to loss, damage, or destruction from various causes, information systems tend to be particularly susceptible to these dangers. Because IS components are comparatively fragile, computer hardware can be damaged more easily than other tools, and IS data files are highly tenuous compared with most other organizational assets. Information systems may also be the object of hostility on the part of disgruntled workers, protestors, and even criminals. Trends in the decentralization of IS facilities and distributed processing have increased the difficulty of protecting IS.

Areas of vulnerability are hardware, software, data, and people. Some of the areas of computer crime are theft of computer time, theft of data, **manipulation of data**/computer programs, and **software piracy**.

Ethical considerations, as well as adherence to current legislation, should be adopted to ensure that computer systems are protected fully.

Natural disasters, malfunctions, and criminal acts all should be considered when building an effective security policy for an organization. In addition, systems managers need to take precautions to safeguard data so that **operating integrity** is maintained. Some of the methods used to ensure the integrity of the system are **backup** files, **access logs**, and error detection measures (such as **reasonableness checks**, **totals checks**, and **check digits**).

Several information access routines that are used today are **trapdoors**, **Trojan Horses, Salami Atacks**, and infectious programs (such as **viruses** or **worms**). To thwart some of these infections, **password** protection routines, **encryption**, and **dial-back devices** are used.

Control of IS begins during planning and is exercised to some extent in every phase of the IS life cycle by using **progress reviews**, **acceptance testing**, **post-installation reviews**, **cost-benefit analysis**, attitude surveys, measurement, **audits**, **performance monitors**, and **contingency planning**.

Management's role in security is a vital one. It should include well-defined hiring practices, **nondisclosure agreements**, and full support of security programs.

REFERENCES

1. Alexander, Michael, "Newsman Faces Tamper Charge," *Computerworld*, February 13, 1989, p. 4.

2. Alexander, Michael, "Prison Term for First U.S. Hacker-Law Convict," *Computerworld*, February 20, 1989, p. 1.

3. American Bar Association, *Report on Computer Crime*, Washington, D.C., 1984.

4. Baskerville, Richard, *Designing Information Systems Security* (New York: John Wiley and Sons, 1988).

5. BloomBecker, Jay J., *Computer Crime, Computer Security, Computer Ethics*, National Center for Computer Crime Data, Los Angeles, 1986.

6. *Computer Viruses*, Proceedings of an Invitational Symposium, Deloitte, Haskins, and Sells, cosponsored with Information Systems Security Association, October 10-11, 1988.

7. Fisher, Royal P., *Information Systems Security* (Englewood Cliffs, NJ: Prentice-Hall, 1984).

8. Fites, Philip E., Martin P. J. Kratz, and Alan F. Brebner, *Control and Security of Computer Information Systems* (Rockville, MD: Computer Science Press, 1989).

9. Forcht, Karen A., "Data Security Concerns in the Microcomputer Environment," *The Journal of Data Education*, Summer 1985, pp. 14-16.

10. Forcht, Karen, and J.K. Pierson, "Developing Computer Security Awareness," *Information Resources Management Bulletin*, Vol. 1, No. 3, November 1987, p. 2.

11. Forcht, Karen A., "The Path to Network Security," *Security Management*, Vol. 31, No. 9, September 1987, p. 152.

12. Forcht, Karen A., "The Special Considerations of Computer Security in the Microcomputer Environment," *Computer Crime Digest*, January 1987, pp. 1-5.

13. Lewis, Barry, "Things to consider for an Ideal Security Match," *Software Magazine*, February 1989, p. 43.

14. Lickson, Charles P., "Those Nasty Little Viruses," *Tech Exec*, March 1989, pp. 46-48.

15. McAfee, John D., "Managing the Virus Threat," *Computerworld*, November 14, 1989, pp. 89-96.

16. McAfee, John D., "The Virus Cure," *Datamation*, February 15, 1989, pp. 29-40.

17. Mandell, Steven L., *Computers, Data Processing, and the Law* (St. Paul, MN: West Publishing Company, 1984), pp. 1-1081.

18. Manning, George, and Kent Curtis, *Ethics At Work. . . Fire in a Dark World* (Cincinnati, OH: South-Western Publishing Company, 1988).

19. Moates, William, and Karen A. Forcht, "Computer Security Education: Are Business Schools Lagging Behind?" *Data Management*, March 1986, p. 54.

20. O'Leary, T.J., and Brian K. Williams, *Computers and Information Systems*, Benjamin Cummings Publishing Company, 1989, p. 529.

21. Parker, Donn B., *Computer Abuse Assessment* (Menlo Park, CA: Stanford Research Institute, 1975).

22. Parker, Donn B., *Computer Security Management* (Reston, VA: 1981), pp.1-308.

23. Pfleeger, Charles P., *Security in Computing* (Englewood Cliffs, NJ: Prentice-Hall, 1989), pp. 1-538.

24. Rose, Frank, "Joy of Hacking," *Science 82*, November 1982, pp. 59-66.

25. Thomas, Daphyne, and Karen A. Forcht, "Unseen Liabilities in Computer Use," *Security Management*, February 1987, pp. 87-90.

26. Wood, Charles Cresson, William N. Banks, Sergio B. Buarro, Abel A. Barcia, Viktor E. Hampel, and Henry P. Sartorio, *Computer Security. . . A Comprehensive Controls Checklist* (New York: John Wiley and Sons, 1987).

REVIEW QUESTIONS

1. Briefly describe how a virus enters a computer system.

2. Explain why hardware is considered the most visible device in a computer system and therefore one of the most vulnerable.

3. Briefly discuss how configuration management operates.

4. Why are hackers considered very dangerous to the security of a computer system?

5. What are the four major areas of computer crime that are crippling the computer industry today?

6. Briefly explain some of the error detection measures that can be used in various stages of processing.

7. What are some of the natural disasters that can pose a threat to computer systems, and why?

8. What is the difference between sabotage and fraud as applied to a computer crime?

9. What is the basic purpose of backup files?

10. Explain how password protection can be used to thwart unauthorized entry into a computer system.

DISCUSSION QUESTIONS

1. It has been asserted that IS are more likely to be targets of sabotage and other intentional acts of violence than other physical assets. Do you agree? If so, why do you suppose this is the case?

2. If you were responsible for the IS activities in an organization that produced chemicals, would you be more concerned about crimes committed *against* computers or crimes committed *with* computers? Why? Would your answer

change if the organization were an insurance firm instead of a chemical manufacturer? Explain your reasons for answering the way you did.

3. Access logs give records of entry and exit to computer facilities. Similar records can be maintained to show who (which password account) has accessed data or used applications software. Is this procedure likely to stop a person from sabotage? Is it likely to stop a person from theft or fraud? Explain your answers.

4. A progress review reveals that problems in the cost-accounting module have necessitated a major redesign of the input for the application, and it is about three weeks behind schedule. What are the implications of these findings? What additional actions might be necessary as a result of the changes made?

5. Many critics of end-user computing argue that as more and more people are introduced to the computer, the instances of computer abuse will increase dramatically. Computer security experts say that the key to curbing abuse is to teach employees about ethical use of computers—thereby making certain that employees and end-users use computers in a responsible manner. What topics should be covered in a computer ethics awareness program?

EXERCISES

1. Select an organization or a company that has a computer installation. Interview one of its employees to learn computer security measures that have been instituted in the organization or to evaluate how well-informed that employee is regarding the company's published computer security program.

2. Choose a computer location on your campus and list any vulnerabilities or security violations that may be evident to you.

3. Locate and review a computer security-related article in a current periodical.

4. Discuss the following statement: "Hackers are not a danger to computer operations, because they point out flaws in the system."

5. Prepare a list of the various ways passwords can be protected.

PROBLEM-SOLVING EXERCISES

1. Many corporations today require employees to sign an Ethics Oath before beginning employment. Is this policy a violation of the privacy and individual rights of the employee? Why?

2. Which of the following computer files hold a high degree of confidentiality and should be protected at all costs?

   a. Psychiatrist's patient files

   b. Overdue library books at a campus library

   c. Doctor's health records of individual patients

   d. Electric company's billing statements

   e. College students' transcript files in the Records Office

    f. Bank records of an individual's checking account

    g. Christmas card mailing list

3. Privacy and confidentiality of records are of prime concern today. What measures do organizations take to protect records from disclosure?

4. What level of security and backup should be attached to the following computer output?

    a. Payroll files

    b. Personnel absence reports

    c. Sales reports

    d. New product development reports

    e. Accounts receivable

    f. Accounts payable

    g. New employee physical examination reports

    h. Long-range strategic planning reports

    i. Daily manufacturing output logs

    j. Profit-and-loss statements

5. Many computer experts predict that computer security threats will take on a global perspective in the near future. What are some of the international aspects of and inherent dangers to, computer systems in this age of the shrinking globe?

CASES

Marcum State University

The personnel department at Marcum State University has recently purchased PCs for individual offices in the various departments so that they can keep their own records. In the past, all records were stored on the university's mainframe computer. By decentralizing this computer function, each department will have better control over its individual records, and the security of these records will be easier to manage. Most of the departments would like to transfer personnel records of faculty, staff, and student employees in their departments to their PCs from the mainframe.

    Dr. John Gould, Chairperson of the Accounting Department, would like to use the personnel data regarding the people in his department to generate some statistics concerning salaries, vacation days used, and absences. Rather than code the files or use social security numbers, Dr. Gould would like to keep the names of the individuals with the information that is recorded about them.

1. What are some of the security considerations in this conversion?

2. What are some of the ethical and legal ramifications of keeping files with a person's name attached?

3. What might be a better way of extracting (and storing) this information to ensure maximum security and control?

Commonwealth Bank

Commonwealth Bank is the largest bank in the city of Alta, Virginia. The home office of Commonwealth is located in Richmond, Virginia, with banks situated throughout the state. Recently, Mrs. Runyon, the Alta Bank manager, discovered that the Commonwealth system had been breached and nearly $5,000 was diverted to a "dummy" account that has yet to be accessed. Mrs. Runyon immediately closed out this account before the funds could be withdrawn; thus, the bank suffered no cash loss. Mrs. Runyon has now called in a team of auditors to trace the original source of this diversion. She suspects that someone within the bank has diverted the funds from several dormant or inactive accounts that are not being closely watched by the depositors. At this point, Mrs. Runyon has not traced the source of the money or the responsible party.

1. Does Mrs. Runyon have an obligation to inform the public of the existence of this account?

2. Once the source is found, should Mrs. Runyon notify the account holder of the problem or quietly transfer the money back into the account?

3. If the responsible party is discovered, what action should be taken with this individual?

4. What measures can the bank employ to be sure that this does not happen in the future?

PHYSICAL PROTECTION

LEARNING OBJECTIVES

After studying this chapter, you will be able to:

1. Identify the natural disasters that threaten systems.
2. Determine damage assessment and reconstruction techniques.
3. Select and organize the physical location of a computer center.
4. Measure the air conditioning and power supply sources for the computer center environment.
5. Understand the three document destruction techniques.
6. Describe the various access control mechanisms.
7. Describe ideal configurations for the areas related to the computer center.
8. Explain the mechanisms involved in smart cards and biometric devices.
9. Define the following terms.

TERMS

- Smart Card
- Access Key
- Tokens
- PIN Number
- Natural Disaster
- Artificial Disaster
- Halon Fire Extinguisher
- Backup
- Data Safe
- Off-Site Location
- Physical Security
- UPS
- Document Destruction Program
- Shredder
- Disintegrator

- Incinerator
- Annunciation Panel
- Mantrap
- Guard Station
- Internal Communications
- Telephone System
- Access Controls
- Biometric Device
- Retina Pattern Matching
- Fingerprint Matching
- Hand Geometry
- Voice Pattern Matching
- Signature Dynamics
- Keystroke Dynamics

SECURITY TOKENS: PASSPORTS TO AUTHENTICATION

James Bond and other fictional spies are renowned for their ability to circumvent intricate security systems to gain entry to tightly protected facilities. Similarly, IS managers need to protect their information systems from high-tech "spies" who attempt to access the data stored in corporate databases. Increasingly, IS managers are turning to security tokens, which act as a third level of user authentication—adding a safety net to the first level (user identification) and the second level (passwords). A *token* is a device or representation that passes security data from one device to another.

Two types of tokens—access keys and smart cards—generate constantly changing, nonrepeatable codes that help verify a user's identity. An *access key* is a device that allows entry to a room, building, or computer. A *smart card* is a plastic card that contains a magnetic strip, much like a credit card. The strip holds security data that can be read by a terminal and transmitted to a central computer for verification.

Although smartcard tokens have been around for more than 10 years, they have recently become more popular due to the contrasting cumbersome nature of dial-back systems and the increasing mobility of microcomputer users, largely thanks to the proliferation of notebook and laptop computers. Industry experts estimate that approximately 200,000 Americans now carry some kind of security token.

One type of token resembles a credit card with a little window in the corner that displays a number. The number, called a *passcode*, changes every 60 seconds, in sync with the hardware at the data center. Each user's passcode is on file in the protected computer. A remote user accesses the secured computer from a terminal or laptop keyboard using the passcode flashing on the card and his or her *personal identification number*, or PIN. A PIN is a unique number that is assigned to and identifies a user of a computer system, just like your bank's passcode and PIN identifies you as an authorized user of your PIN identifies you as an authorized user of your bank's passcode and confirms its accuracy through an algorithm based on the current time. After the passcode is confirmed, the system software then asks the caller for a conventional user ID and password.

Because the user must both know something—the PIN—and possess something—the token—to gain access to a secured host, this type of system is doubly secure.

As reported in "Smart Cards: Smarter Than Passwords," by Lamont Wood, Datamation, July 15, 1991, pp. 69–70.

The military's basic premise, "Secure the perimeter," also applies to the computer battlefield. To assess the dangers and vulnerabilities that can plague a computer facility, we will discuss natural disasters, fire and water damage, vandalism, and intruders. We will also look into protective devices and procedures, including badges and access controls, and disposal of sensitive media. Once the hard, physical protections are in place, the soft issues of software, operating systems, personnel, management, and policies should then be implemented.

NATURAL DISASTERS

A **natural disaster** can be defined as any event that is an act of God or the result of environmental or natural causes that are not avoidable or predictable. While it is virtually impossible to prevent natural disasters from threatening computer centers, measures should be taken to assess the potential risk, just as we do with homes, automobiles, and office buildings. A computer center in California is more prone to earthquakes than centers in other parts of the country. Floods are prominent in areas near bodies of water or near sewage lines. Hurricanes and their aftermath, which often includes flooding, tend to cluster around coastal areas. (See Figure 2.1.) Computers are extremely sensitive to environmental and temperature conditions. Excessive heat, humidity, or cold, or inadequate power are devastating to the equipment and software.

FIGURE 2.1
Disasters
Threatening
Computer Centers

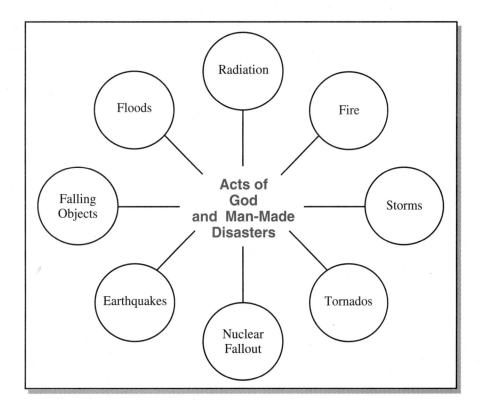

By proper planning, damage and destruction can be lessened. Some of the disasters to be considered include floods, water, fire, power loss, power surges, heat, and humidity.

Floods

Floods generally result from natural causes such as rainstorms, tides, and waves. **Artificial disasters** are man-made. Floods resulting from broken water pipes, sewer pipes, or human error are artificial. Regardless of the source of the water, the damage to the information systems may be the maximum possible. (See Figure 2.2.)

Rising Water. In areas prone to flooding, the computer center should be located well above ground level. In hilly areas, the computer center should always be located at the highest end of the slope and well above ground level. In any flood-prone area, computer centers should never be placed underground.

Falling Water. If sprinkler systems or overhead pipes are present in computer centers, large plastic sheets should be stored in an accessible location so that employees can quickly cover all equipment in the event of a sudden cascade of water. Many computer centers follow the rule that all equipment (including microcomputers) are to be covered with made-to-fit plastic covers when not in use. (See Figure 2.3.)

FIGURE 2.2
Water Damage

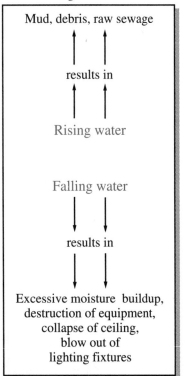

Mud, debris, raw sewage

↑ ↑

results in

↑ ↑

Rising water

Falling water

↓ ↓

results in

↓ ↓

Excessive moisture buildup, destruction of equipment, collapse of ceiling, blow out of lighting fixtures

FIGURE 2.3
Water Protection Guidelines

Threats from Water
✔ Flooding may be caused by rain, melting ice/snow, toilet/sink/pipes overflowing, or sprinklers used to suppress fire or mounted in ceilings.
✔ Once a computer is wet, it cannot be turned on unless it is thoroughly dry.
✔ Water sensors should be installed on the floor (or beneath the floor in raised flooring).

Earthquakes. The computer industry has made great strides in the past few years in protecting its assets against earthquake damage as much as possible. Consider that the epicenter of the earthquake that hit Northern California in October 1989 was located just 70 miles south of downtown San Francisco. Financial institutions, which are at the heart of the area's economy, suffered nominal losses and were able to resume processing in a relatively short time. Many data centers were operating the next day, and most resumed normal operations by the end of the week. The biggest problem was the power outage of nearly 36 hours following the quake. The power company was able to restore services to private homes before it could restore power to the business community.

Typically data center disruptions from natural causes represent approximately 10 percent of the total—in 1989, with the San Francisco earthquake and Hurricane Hugo, the figure was closer to 25 percent.[35] (See Figure 2.4.)

FIGURE 2.4
Earthquake
Protection
Guidelines

Protecting Against Earthquakes
☑ Locate computers away from windows.
☑ In high-risk areas, place computer near floor, not in raised/high areas.
☑ Avoid placing objects near computer (e.g., plants, bookcases, large pictures, filing cabinets).

Hurricanes. In 1989, Hurricane Hugo did a tremendous amount of damage in southern coastal areas. Many of the roads through Charleston, South Carolina, were impassable after Hugo hit. "We were unable to assess the damage for two days until the roads were clear," said Jim McDaniel, Vice President of MIS for Heritage Trust Federal Credit Union. McDaniel said that a massive oak desk in the president's office was overturned and a PC was blown across the room. Although the central computer system was not damaged, a power outage rendered the center useless for processing for four days; they were then able to call a commercial recovery service in Atlanta.[35]

Fire

Fire can present a much more serious threat than water because fire's spontaneous origin gives people little time to react. Human lives, as well as computer center facilities, are often at stake so the danger presented by fire is much more immediate and threatening.

In 1984, a $138 million fire occurred at Tinker Air Force Base in Oklahoma City, Oklahoma. The fire broke out in the facility used to repair jet aircraft, burned for two days, and destroyed the entire facility. A large computing center was located in an adjacent building, and much of the computer center equipment and information storage media had to be removed. There was a great concern that the burning building would collapse as a result of structural damage.

As the fire spread, a safe shutdown of the computer system was undertaken and vital equipment was removed. There was no time, however, to retrieve and move the entire massive library of tapes. Many of the employees—working under stress in dangerous conditions—could not identify which tapes were most vital to the ongoing operation of the computing facility. A system of color coding or labeling with priority codes, plus better employee training, would have speeded up the process a great deal.

Plans and Drills. A well-planned disaster recovery plan should always be a part of every computing center's procedural policy. Periodic "fire drills" should be practiced to ensure that the plan is up to date and efficient.

Placement. A natural defense against fire is careful placement of the computing facility. A windowless location with fire-resistant access doors and nonflammable walls can prevent a fire from spreading into the computer room from adjacent areas. With a fire- and smoke-resistant facility, personnel are then able to shut down the system, remove vital media and equipment, and resume normal operations once the crisis has passed. (See Figure 2.5.)

FIGURE 2.5
Fire and Smoke
Protection
Guidelines

Protecting Against Fire and Smoke
☑ Use smoke detectors liberally and check often.
☑ Locate fire extinguishers in/near computer room, in visible/well-marked places.
☑ Enforce no smoking policy in computer areas.
☑ Use Halon extinguishers rather than water-based. Halon activation warnings should be visible. Slight time delay should be a standard procedure to avoid health risks when Halon activates.

Halon Fire Extinguishers. Several years ago, a major issue in computer centers was how to control a fire if one were to break out around the sensitive equipment. E.I. DuPont de Nemours & Co., Inc. accidentally came across a substance it called Halon that fought fires rapidly and left little or no damage to the facilities. **Halon fire extinguishers** contain a white bromine powder produced in the manufacture of chlorofluorocarbon compounds (CFCs) such as styrofoam egg cartons. It is very effective in extinguishing fires.

Halon can be loaded into a dump system. When a fire breaks out, the Halon can be *dumped* onto the fire. This seems to be a safe fire extinguishing method for electronic equipment because it avoids water damage to the sensitive machinery. Unfortunately, it has been discovered that if an insufficient amount of Halon is dumped on a fire, the Halon will break down into several corrosive compounds that can destroy computer logic circuits and equipment.

Recently, it has been discovered that Halon is destroying the earth's ozone layer because of the large emission of bromine into the atmosphere. This fact has attracted the attention of many government officials and corporations. Because of this finding and government regulations, DuPont will stop production of Halon by the year 2000.

Fire experts now suggest that fire systems not be totally Halon-based. The computer center and rooms with high concentrations of sensitive equipment should use Halon dump systems, but other rooms should use standard water sprinkler systems. Experts say until a new chemical is found to replace Halon, minimizing Halon's use will help preserve the ozone layer.

Halon must mix with the air, have a minimum concentration of 5 percent of total air mixture, and be maintained for at least ten minutes to work effectively. Most Halon failures occur because Halon concentration never reaches 5 percent in the space where the fire takes place. This problem is sometimes caused by doors that do not automatically close to seal off the room or by a ventilation system that draws the Halon gas out of the room.

By using a combination of Halon and standard water sprinkling systems, some of these problems can be reduced. Because water and computers do not mix well, the computers should be turned off before water is applied to the fire. Automatic shut-offs can be installed to work in conjunction with smoke alarms or another alarm device.

The most sensible approach is a two-step one: apply Halon gas to the fire area first; if the Halon system fails to extinguish the fire, then activate the water sprinkler system.

Backup Storage

Backups are copies of all data and information considered to be critical to the ongoing operation of the computer facility. If making backup copies of important data files is the way to prepare for a fire, then these backups are critical. Backups are stored within 25 feet of the system in most computer centers today. Because these backups will be used to restore a system after a major disaster, they should be placed somewhere other than in the same environment as the system. If the system hard drives are destroyed by a fire, more than likely the magnetic tapes or cartridges used for backups will also be destroyed.

A storage system that is available to almost any size computer center is a data safe. A **data safe** is a fire- and water-resistant safe used to store magnetic tapes and cartridges. A data safe usually has an outer shell of steel. It is lined with such things as gypsum, concrete, dry chemicals, and sometimes wood. These safes are usually designed to withstand temperatures of up to 1,770° F for an hour or more.

Although a fire or flood does not always destroy a disk, it does not take much to warp a disk. Once a disk is warped, the information is usually lost forever. By using a data storage unit such as a data safe, a company can cut down on the number of backups and data files that are lost through warping.

Some large corporations have too many backup tapes and cartridges to use a data safe. In these instances, it would be best to use a storage location that is not near the computer center. By shipping the data and backups to an off-site location, the risk of having the original and the backup destroyed is lessened. An **off-site location** is a separate facility used exclusively for storage of backups, media, and equipment. Although using off-site storage can be very inconvenient at times, it is still preferable to having backup data near the system.

Property protection can range from a very low level to a high level in which redundant fire suppression, alarm, and structural systems are used to minimize the probability of significant loss. Minimum protection levels can be used when the cost of the computer is so low that complete protection can be more expensive than replacing the computer hardware. In many instances, protected backup data, duplicated programs, and easily replaceable computers provide a reasonable level of protection, especially if the actual data processing operations could be easily conducted at an acceptable, alternate facility.[5] Some low-level alternatives for protection include the following strategies:

- Eliminate sprinkler protection within the computer room enclosure and rely on a Halon extinguishing system and combustion detection products.
- Eliminate the Halon extinguishing system but maintain automatic sprinkler protection and combustion detection products in the computer room.
- Eliminate Halon and combusion detection products and rely solely on automatic sprinkler protection in the computer room.

The highest level of protection is generally provided for on-line systems considered to have a major impact on life, production, profitability, or business continuity. Some high-level alternatives for protection include the following:

- Structural fire separation of the computer center from other fire-prone areas of the building
- Automatic wet-pipe sprinkler protection to extinguish fires on either sides of the fire separation walls
- Automatic Halon 1301 extinguishing systems in the computer center (Full power-down and air control must be integrated into this system for it to be effective.)
- Current backups of critical programs with appropriate file updates
- An uninterruptible power supply system in case of power loss

PHYSICAL FACILITIES

In the early days of computing, protecting the system was quite simple because all equipment was housed in a single area and kept locked. Only persons having keys or access cards were permitted entry into the computer room. Today, with computers becoming cheaper, smaller, and more accessible to almost everyone, protecting the physical environment is much more difficult. Traditional security systems today must be augmented with special considerations for emerging technology and applications. Physical security, however, remains a very basic component of an organization's total security plan. **Physical security** involves tangible measures that are instituted to protect the facility, equipment, and information from theft, tampering, careless misuse, and disclosure (both accidental and intentional). In many instances, physical security is the most obvious type of security. It is both visible and constantly reassuring. Because physical security is a tangible sign to both employees and outsiders, the corporation's commitment to good security practices can be conveyed to everyone.

Buildings, computer room facilities, computers and related equipment, and media libraries provide an important outer, physical perimeter of security. (See Figure 2.6.) Once this physical perimeter is adequately secured, access controls and other types of security provide additional layers of protection of information.

FIGURE 2.6
Protection of
Physical Facilities

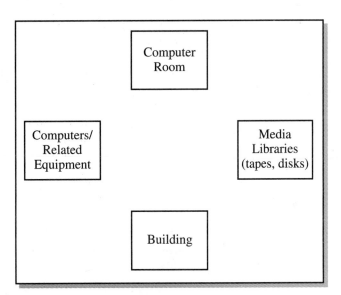

Selecting and Preparing the Physical Location

Most basic physical requirements for security can be met by careful planning prior to installation of the computer or as a refinement or adjustment to an already established facility. Physical security considerations should include the following:

- Locate the computer away from main building traffic areas.
- Avoid a location with outside walls and windows.
- Provide secure door locks and related access control devices.
- Install power sources and air conditioning systems that can function independent of regular utilities during an emergency.

Many computer equipment manufacturers offer site selection and preparation guidelines. To ensure full adherence to construction and building codes, many organizations require that the computer facility plan be reviewed by an architect, construction engineer, building planner, and fire protection engineer. Special considerations must be given to the unique requirements of computer room facilities because temperature and humidity controls are critical to the ongoing operation. For example, air conditioning system vent closures installed in the tape library room may be designed to shut automatically when the temperature exceeds 150° F. In a fire, stored tapes and disks will be seriously damaged long before the 150° cut-off. In addition, computer area power sources and air conditioning units may not be isolated from those used by the rest of the building.

If air conditioning is automatically cut off because of power load fluctuations or a fire, the humidity level of the computer facility will become dangerously high in a short period of time; this can cause extreme damage to equipment and storage media.

Building. The building housing the computer center should meet the basic requirements for a secure location and should follow these guidelines:

- Building should be situated away from obvious hazards, such as gas stations, airports, garages, paint storage areas, chemical plants, fire walls, heating systems, and electrical generation substations.

- Building should be not located next to a factory area used for machining or other basic production activities that will create airborne dust.

- Building should be located in an area free of water problems caused by flooding, creeks or springs, drainage, or runoff areas.

- Building should be located well away from main traffic areas and exterior building entrances and walls.

- Building access should be limited to only those users or personnel who need access to equipment.

- Computer room should be made as inconspicuous as possible by removing door and direction signs that identify the computer center's location.

- Walls and ceilings of the computer center should be made water-tight to prevent possible damage due to plumbing lines that pass near the room or through the walls.

- Raised floors of computer centers should be equipped with a drain opening to eliminate any fluid build-up quickly.

Air Conditioning Equipment. To ensure that adequate levels of air conditioning and humidity control are properly installed, these guidelines should be followed:

- The computer facility should have an air conditioning system independent of the rest of the building.

- The air conditioning should be capable of cooling adjacent service areas and office space in the event of an emergency so that ongoing operations can be maintained.

- The computer center air conditioning system should be connected to the fire detection and extinguishing system.
- Supplemental smoke exhaust facilities should be installed for emergency use.
- An audible alarm should allow sufficient time for a system halt while evacuation is taking place.

Power Supply and Electricity. To ensure a continuous power supply in the event of emergencies, as well as daily use, the following strategies should be considered:

- Install an uninterruptible power source (**UPS**) that will isolate the computer center from fluctuations in the primary power supply load. A UPS should be capable of sustaining the required load for 15 to 18 minutes, allowing for an alternate generator to take over and automatic restart or shutdown of operations.
- Install a line filter on the computer's main power supply.
- Install a special electrical circuit with an isolated ground.
- Install an emergency electrical power supply to illuminate support work areas, hallways, stairwells, and at least one elevator, and to operate work area heating and air conditioning, and a pump of sufficient capability to supply water to strategically located lavatories and drinking fountains.
- Place reserve battery packs in selected computer room fluorescent light fixtures, allowing for a minimum of 90 minutes of sustained illumination during emergency operations.
- Install antistatic carpeting with special filaments that minimize static electricity.
- In the event of lightning, turn off the computer and unplug all electric units or install surge protectors designed to absorb surges.
- Protect backup tapes from the magnetic field created by lightning striking the building. Store tapes and disks as far as possible from the building's steel support.

Document Destruction

The disposal of sensitive media must be considered as part of the overall security plan. A **document destruction program** involves the destruction of confidential reports, research and development data, duplicate copies of documents, and disks and ribbons.

The most common types of document destruction devices are the following:

- Shredders
- Disintegrators
- Incinerators

Shredders. **Shredders**, used for destruction of paper, floppy disks, printer ribbons, and tapes, convert their input into thin strips or pulp so fine that piecing the strands back together again is impossible. Some organizations burn the shreds for added protection. (See Figures 2.7, 2.8, 2.9, and 2.10.)

FIGURE 2.7

Materials to Consider for Shredding

Advertising Schedules	Meeting Notes
Assembly Data/Bills of Material	Obsolete Forms such as Checks or Purchase Orders
"Bad" Copies	Obsolete Reports
Bids and Quotations	Old Applications for Employment
Blueprints	
Business Plans	Old Files
Computer Printouts	Old Purchase Orders
Confidential Report Overruns	P&L Reports
Correspondence	Paper Documents after Microfilming
Customer Lists	
Design Ideas and Concepts	Patent Application Material
Drafts and Interim Copies	Payroll Data
Engineering Documentation	Personnel Documents
Engineering Notes	Price List Data
Expired Documents in Personnel Files	Production Schedules
	Resumes
Government Classified Materials	Salary Information
	Status Reports—Periodic and Year to Date
Labor Negotiation Data	
Marketing Plans	Unneeded "Extra Copies"
Meeting Minutes	Vendor Files and Quotes

Disintegrators. Disintegrators grind and pulverize documents under water pressure or by the use of chemicals. Items to be considered for disintegration could be spoiled impact printer output, continuous form carbon paper accumulation, used machine tabulation cards, tapes, disks, and computer-generated waste products.

Incinerators. Incinerators burn computer-generated output. The facility must meet environmental standards before an incinerator is installed.

Waste products should be processed through one of these devices before they are placed in the building trash collection site. Failure to provide this extra protection before disposal could allow snoopers, industrial spies, or unauthorized insiders to compromise confidentiality.

FIGURE 2.8
Types of Shredders

Strip/Ribbon Cut

• Paper is cut into 1/4" strips
• Most common type due to low cost
• Strip width can be altered with some units
• Has high capacity
• Simple to use

Crosscut/Particle Cut

• 1/4" x 1 1/2" strips or particles result
• High security is provided
• Bulk is reduced so use of shredder bags is minimized

Top Security Crosscut

• Meets most government high security requirements
• Recommended for classified materials
• Reduces bulk to a minimum

Additional Considerations and Measures

Each computer center has unique assets, which may vary a great deal from those of other computer center operations. A bank may have a greater need for control than a small retail business. The extent to which a business may or may not adopt different types of control, security measures, mechanisms, and procedures depends on its size, its economic strength, the sensitivity of its data, and the regulations imposed by government agencies.

FIGURE 2.9
Shredding Techniques

Stream Feeding
- ☑ Allows for continuous stream of papers (e.g., computer printouts)
- ☑ Minimal operator time required

Batch Feeding
- ☑ Allows for multiple sheets to be fed
- ☑ Fastest method for large volume of material

Combination Feeding
- ☑ Allows for operator to add batched material while stream feeding is taking place
- ☑ Provides greatest efficiency of shredder types

FIGURE 2.10
Selecting Shredders and Suppliers

Needs
- ☑ Number of users
- ☑ Type of material to be shredded
- ☑ Volume of shredding required

Use
- ☑ What is classification of shredded materials (confidential, top secret, general)?
- ☑ Are small personal units needed in addition to large unit?
- ☑ Can shredder be shared or open to multiple users?
- ☑ Should shredder be placed near copier, fax, blueprint machines?
- ☑ Is electrical circuitry sufficient?

Supplies
- ☑ Are vendors available locally?
- ☑ Is pricing competitive?
- ☑ What is the reputation of the vendor?
- ☑ Is service available, as well as needed parts?

Some other physical measures that may be installed are:

- Annunciation panels
- Mantraps
- Guard services
- Internal communications
- Telephone service

Annunciation Panels. **Annunciation panels** are used to signal abnormal conditions, including fluctuations in electrical power, water, low fuel levels in power generators, status of coolant pumps, and unauthorized entry and intrusion.

Mantraps. **Mantraps** are usually sequential-entry double doors or turnstiles at computer room entrances activated by security guards inside the guard station or by keycard. Frequently, they include keycard door locks, audible alarms, closed-circuit TV surveillance, and metal detectors. A mantrap generally assists the security guards in detaining a person attempting to enter or leave the computer room until the guards are satisfied that the person is authorized to be there and presents no threat to the center.

Guard Stations. A **guard station** is a specially constructed and designed enclosure that is usually connected to, or part of, the mantrap. Often, these stations are manned 24 hours per day, 7 days per week. They are equipped to monitor the security of the data center through the following:

- TV monitors
- Public address speakers
- Direct manual alarms to police
- Private security service
- Direct alarms to fire departments
- Intercoms with the data center
- TV surveillance
- Automatic photographing of persons entering the facility
- Radio police scanners tuned to emergency channels
- Walkie-talkies for emergency communications

Sometimes a wide array of automatic shutoff switches is used to reduce harm to the equipment if an abnormal function is detected within the center. Guard stations are often constructed with bulletproof walls, doors, and windows, depending on the nature of the perceived threat to the data center and required protection capabilities.

Internal Communications. **Internal communications** include intercom systems linking guard stations and all areas concerned with the daily operation of the data center. Generally, the systems provide the guards with override capability of all stations, conference-calling, and busy-line indicators. Direct communications lines with police and fire departments are often provided, as well as walkie-talkies and public address systems to all data center areas.

Telephone Service. The telephone system installed must be reasonably secure against willful or accidental damage. Consequently, the wiring of the system is often installed under the raised floor, encased in fire protective materials, and equipped with smoke and heat detectors. Generally, several lines are used to minimize disruption if one becomes inoperable. The telephone wire terminal closets are locked and within the secure perimeter.

ACCESS CONTROLS

Access controls often include card-key locks, automatic door closing, fingerprint or photo identification and other means of logging in, and cameras trained on entrances, hallways, loading docks, elevator doors, outside building entries, and potentially vulnerable public access areas above, below, and around the data center. These controls also might include mirrors to eliminate blind spots in these areas and emergency lighting units to be turned on if the regular lighting system fails.

Access to the computer facility must be closely controlled on a continuing basis to reduce vulnerability to arson, vandalism, and unauthorized visitors.

Security of Areas Related to the Computer Center

In addition to restricting access to computer center facilities, strict procedures must be established and maintained for adjoining areas. (See Figure 2.11.)

FIGURE 2.11
Computer-Related
Areas to Be
Restricted

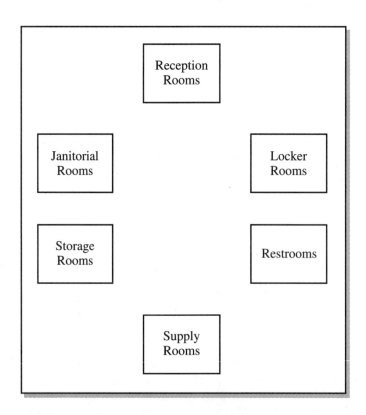

Reception Rooms. Reception rooms for visitors and users should be located outside the data center. Stations for delivery or pickup of materials should be adjacent to other data center rooms or critical areas. If an elevator is necessary, its use should be restricted to data center personnel or authorized personnel; moreover, when the elevator stops at the data center level, its monitoring system should signal a guard station. Operators' lounges are frequently provided, particularly in installations that have 24-hour operations. Preferably, they are not accessible except through the data center.

Janitorial Rooms. Janitorial rooms should have central access to the data center so that adequate cleaning and maintenance may be efficiently performed. Because these rooms store various cleaning supplies and are generally equipped with deep sinks, they should be protected against fire and water damage as well as the extension of that damage to the data center. Accordingly, janitorial rooms must be constructed in much the same way as the computer room (except for the raised floors) and have the same kinds of protective devices to monitor against potential damage.

Locker Rooms and Restrooms. Locker rooms and restrooms should not be located adjacent to the computer room or any critical mechanical and electrical equipment rooms or facilities. Nevertheless, because these rooms obviously must be reasonably close to the data center, they should have a public address system and protective devices against fire and water damage. For the same reasons, these rooms should have no windows or other means of access that would present vulnerability to the data center.

Storage and Supply Rooms. Storage and supply rooms should not be located next to the computer room or mechanical and electrical equipment rooms because the materials they contain are often combustible. For this reason, these rooms should have intercom stations, fire extinguishers, water systems that are independent of all other data center areas, floor drains with backwater valves, water alarm connections to annunciation panels, and intrusion alarm systems that are monitored by the guard station.

Access Control Plan

A successful access control plan is based on this instruction:

Limit entry to employees who have a need to be in the computer center.

Management should approve a formal policy; *all* employees should be briefed on the policy. The policy should include the following tenets:

- Access to the computer center will be on a need-to-enter basis. Company employees are *not* to be granted automatic admission to the facility.
- Where access must be monitored closely, a personal access control booth will be installed, consisting of two sets of doors separated by a short corridor. The corridor will be equipped with a closed circuit television camera or a one-way window to permit observation of individuals attempting to secure access. A standard badge reader access control system will also be installed to aug-

ment the control booth. A one-way revolving door or turnstile may be installed in place of the corridor or double doors.

■ Vendor representatives, auditors, cleaning crews, maintenance employees, janitorial staff, and other visitors will be required to :

1. Provide positive identification before admittance.

2. Record the date, time, and purpose of their visit.

3. Wear a visible visitor's badge when in the computer center.

4. Be escorted by an assigned member of the computer center staff.

■ Messengers and vendor representatives will be informed that deliveries are *not* to be made directly to the computer center. All items should be delivered to a central receiving area. Messengers should deliver materials to the mail room or a reception area for routing to the appropriate party.

■ All employees will be required to wear identification cards at all times.

A few simple but effective measures can be instituted; they include the following:

■ Eliminate or block any doors in the perimeter not essential to safe emergency evacuation.

■ Remove outside door knob hardware on all doors in the perimeter that are not to be used as entrance points. (Panic door opening hardware should be installed on all interior doors in place of knobs.)

■ Attach audible exit alarms to all emergency doors.

Figure 2.12 serves as a guide for a quick overview of physical security. Figure 2.13 provides a complete summary in checklist form.

FIGURE 2.12
Components of a
Physical Security
System

☑ Identification/authorization
mechanisms

☑ Entrance control

☑ Guard patrols

☑ Television surveillance

☑ Library control

☑ Environment alarms

☑ Fire suppression system

☑ Police/fire communication

☑ Public address

☑ Emergency evacuation system

☑ Remote site file backup

☑ Disaster recovery plan

FIGURE 2.13
Physical Security Checklist

✔ Locate computer facilities in low-traffic areas.

✔ Avoid locating computer facility near outside walls and windows.

✔ Install power sources, heating, and air conditioning independent of other areas.

✔ Familiarize building maintenance personnel and facility engineers with computer facilities' special requirements.

✔ If possible, avoid placing computer facility near hazardous sites (e.g., airports, chemical storage facilities, rivers, gas/water/sewage).

✔ Do not mark computer facility conspicuously.

✔ Make ceilings and walls watertight.

✔ Install drainage pipes in floor, as well as water-level detectors.

✔ Place shredders in or near computer facilities.

✔ Operate fire plans independently from the general facility and "practice" them periodically.

✔ Prohibit smoking, eating, and drinking in or near the computer facility.

✔ Install flame-resistant materials throughout facility, as well as non-shattering glass.

✔ Clearly mark master power switches.

✔ Place portable fire extinguishers liberally throughout facility.

✔ Test smoke and fire detection equipment regularly, as well as alarms.

✔ Place breathing apparatus units liberally around facility.

✔ Install flame-retardant plastic throughout facility.

✔ Locate paper supplies outside computer facility.

✔ Conduct frequent, thorough fire drills for *all* personnel.

✔ Limit access to computer facility to approved personnel only.

✔ Place receptionist or security guard at all computer facility access points.

✔ Inspect all packages, deliveries, and parcels entering computer facility.

✔ Back up all critical files and locate off premises. Keep backups current.

In companies with large computer centers, fences and guarded gates around the building are common. Smaller computer centers may not have fences. Entrances to the computer center should be locked at all times to restrict access. A guard at the entrance to log personnel in and out may be used, especially during normal working hours. This also allows for a form of personal surveillance.

Smart Cards. Unlocking the door(s) to get into the computer center may be done in several ways. A more common access device is electronic card keys, or **smart cards**. These cards are the size of a credit card with embedded microchip memories and processors. Volumes of personal data and authorization codes that are hard to fake are programmed onto the card. Knowledge of a secret code can be used with these ID cards. Smart card systems can restrict access to authorized individuals at specific entrances during specified hours of selected days. Card access systems can also be used to log each entrance and exit attempt. This process can double-check or eliminate guard duty at the door. Badges can be read with the same type of information as the smart card and can also have a photograph of the worker.

Biometric Devices. Alternatives to card access lock systems are available. A group of devices that provides greater security consists of **biometric devices**.

Current users of biometric systems include the United States military, nuclear plants, research labs, and banks. Sales of biometric devices to these users are climbing, with a predicted annual growth rate of 40 percent[3]. Sales, however, have not always been so impressive. When first introduced in the mid-1970s, biometric security devices were slow, inaccurate, and outrageously expensive. In the mid-1980s the manufacturers changed gears and targeted the commercial market. However, the devices still were not up to acceptable performance levels. Businesses hesitated to employ the devices because of their difficulty of use and their inability to compete with the cost of traditional technologies, such as passwords and keys. Today, biometric devices are faster, smaller, cheaper, and more accurate. All of these factors contribute to the increasing popularity of biometric security devices.

There are two types of automated biometric systems. Physiological, or active, devices analyze unique personal characteristics; behavioral, or passive, devices analyze unique characteristics related to behaviors. Each type of biometric device employs three methods of user authentication. The physiological devices include retina scans, fingerprint scans, and hand geometry. Signature, voice, and keystroke dynamics all fall under behavioral devices.

Retina Pattern Matching. **Retina pattern matching** is used mainly for high-security applications, such as those involved in government operations. The technology behind the device consists of an infrared light that is aimed through the pupil to the back of the eye. The blood vessels found here reflect the infrared light at a lower intensity than the surrounding tissue does, forming a distinct pattern. To create an ID for an individual, the device picks 320 points at which it measures the intensity of the reflection and assigns to each point a degree of intensity (0–4,095). These measurements are condensed into an 80-byte code and stored in the computer's memory. Future attempts to access a system require the user to enter a personal identification number (PIN) and look into the scanner's eyepiece. The computer then compiles another ID code and checks it against the original code stored memory. A match occurs when at least 70 percent of the pattern is identical. Companies can alter the percentage requirement as desired, but a requirement of 100 percent is not reasonable because of the physiological changes that occur over time.

The entire process takes only a matter of seconds, which is an advantage over other access methods. Seeking such speed and accuracy, many firms have turned to retina scanners to secure their computer centers. American Airlines, for example, uses scanners at its underground computing center in Tulsa, Oklahoma. Currently only one company manufactures the scanners. EyeDentify of Beaverton, Oregon owns the patents to the scanning process.

This process is close to perfect, but use of the scanners suffers from people problems. People have been afraid of getting pink-eye when coming in contact with the eyepiece. (This issue forced the manufacturer to create a noncontact device.) Another misconception occurs because people think the scanners use a laser to scan the eye. Some users may fear that this "laser" could damage their eyesight.

Fingerprint Matching. **Fingerprint scanners** work in the same fashion as the retina scanner, but they do not use an infrared light. The scanner compares specific points along the ridges of the fingerprint with points stored as code in the system. The code for each individual is created by scanning an image of the person's fingerprint, digitizing the image, encrypting it, and then storing it in the computer's memory. Currently two companies manufacture fingerprint scanners. Identix makes IDX machines that record the entire fingerprint image, and Fingermatrix makes the Ridge Reader, which uses the method previously stated. Fingerprint scanners are used widely in the United States and the rest of the world. For example, in Denver, a health spa uses a fingerprint scanner to record how frequently its members use the spa. A Japanese builder installs the fingerprint scanners in luxury homes as a security selling point.

Fingerprint scanners are subject to ingenious attempts to deceive the system. Illegal access attempts include using a photograph of a fingerprint and using the severed finger of a valid user. Advances in technology, however, have taken care of these problems. For instance, many devices now have a three-dimensional mapping process that a two-dimensional photograph cannot reproduce. Cutting off a valid user's finger will not gain an illegal user access because new devices can detect whether a finger is alive by measuring natural color changes in hemoglobin. (In addition to criminal intentions, there are some situations in which a fingerprint scanner will deny access to a valid user. Dirty or cut fingers may cause prints not to match.)

Hand Geometry. **Hand geometry** is based on finger length, palm dimensions, and skin translucency. All are measured to identify each individual. As with fingerprint and retina scanning, the measurements are stored as codes in the computer's memory and used for comparison each time access is requested. Even though hand geometry is a fairly new biometric device, several companies manufacture the item. Stellar Systems makes the Identimat, Recognition Systems produces the ID-3D, and Mitsubishi Electric Sales America manufactures the Palm Recognition System. There is a great deal of competition among these companies because a hand geometry device is seen as being more acceptable to the user than other types of devices.

Voice Pattern Matching. The voice verification device works by creating a digital picture of an individual's vocal tract. Impersonators cannot gain access because the machine recognizes the physiological characteristics that produce speech, not sound or pronunciation. It is possible, however, for a sore throat, alcohol, mood swings, or nervousness to trip up this verification device. Voice verification devices can be used in many ways. They can be used in homes to secure door locks, in offices to secure workstations, and in ATMs to ensure that only valid users make transactions. Only two companies currently manufacture voice verification devices. Ecco Industries and Voxtron Systems serve customers such as Hertz and Martin Marietta. These companies take advantage of voice verification features that include time zone programming, door alarm monitoring, and time-stamped audit trail reports.

Signature Dynamics. A **signature dynamics** system uses a special pad and pen to verify an individual's writing techniques. The elements used to create a user's ID include writing speed, pressure applied, crossing of t's, and dotting of i's. An image of the signature is stored in the computer's memory. Each time a user accesses the system, he or she signs the pad and the images are compared. As with the retina scan, physiological changes occur over time and cause variations in the images. To account for this, the signature system updates the stored image each time a user signs onto the system. Currently these devices are used at retail counters to verify credit card and personal check signatures. Several firms—Confirma, Thomas De La Rue, Inforite, and Signify—manufacture signature devices for commercial use.

Keystroke Dynamics. **Keystroke dynamics** involves the observance of typing patterns and rhythms. It is unlikely that any two people type the same way, so common groups of letters, such as "i-n-g" and "t-i-o-n," are monitored to verify who is typing. Keystroke devices are often used in offices where data entry is an important function. These offices use log-on systems as well as systems that monitor a user continually.

Biometric Failures. Even though they seem foolproof, biometric devices are not perfect. Malfunctions do occur; the resulting errors are classified into two types. Type I errors occur when a valid user is denied access, and Type II errors occur when an invalid user is granted access. Both kinds of errors, but especially Type I errors, tend to annoy users.

Other aspects of the systems, such as increased complexity of log-ons, also bother users. Some people just do not trust certain devices because of prior misconceptions. Others worry about potential invasion of personal privacy. Members of the American Civil Liberties Union (ACLU), for example, are working on a Project on Privacy and Technology to ensure that biometric information is used only for security purposes. It is true that the increasing use of biometric devices signifies the need for standards and regulations regarding privacy.

Most people adapt to biometric devices once they become familiar with daily usage. To some employees, however, the introduction of biometric devices suggests aspects of George Orwell's totalitarian nightmare.

The Future of Biometrics. Just like the price of hand calculators, the price of biometric devices is coming down as the technologies advance. Now biometric devices can be purchased for almost the same price as card readers. A biometric device is far less expensive than a human security guard. Many of the six types of devices can be purchased for less than $1,000. In fact, keystroke monitoring systems can be bought for about half that amount. Now that biometric devices are easier to cost-justify, increasing numbers of firms are considering them.

Future of biometrics success is in the hands of manufacturers and organizations like the International Biometrics Association. These entities must develop standards for error measurement and system interfaces and take responsibility for providing information to users about the industry. Manufacturers are in the process of developing several new devices:

- A device that uses neural network technology to recognize faces
- A device that analyzes a person's genetic pattern, or DNA fingerprint
- An ID system based on wrist vein pattern
- A device to identify the unique lipids and fatty acids found in skin oils extracted from forehead sweat[7]

All of the new recognition technologies can enhance and further the use of biometrics in securing areas. Ultimately, however, the success of biometrics will depend on broad business and consumer acceptance.

People may lose their keys or forget a password, but biometric devices provide an efficient and effective means of securing an area. Through their use of unique characteristics, biometric devices give businesses the extra protection needed to guard their computer systems. The future of the industry looks promising, but as with any security measure the technology is not foolproof. An increase in user awareness, coupled with the development of standards, is needed to ensure biometrics' future success.

SUMMARY

The protection of the physical facility in which the computer center is located is the logical first step in securing the overall operation. Once the physical environment is protected, additional measures can be added to reinforce and augment the security.

Natural disasters, such as flood, fire, earthquakes, and tornados, present a real threat to computer centers because they are highly unpredictable. Various measures can be instituted to avoid the disaster or lessen the impact on the computer center equipment and media. An example: **Halon fire extinguishers** are recommended for computer facilities so that gas is substituted for water to suppress and extinguish the fire.

Backup copies of all vital data should be run and stored off premises to ensure reconstruction of operations once a disaster has passed.

Computer center physical protection involves the building, computer room facilities, computers and related equipment, and media libraries containing tapes, disks, and backups. Considerations include air conditioning equipment, humidity control, and power and electrical supply.

A complete **document destruction program** should be adopted that utilizes **shredders**, **disintegrators**, or **incinerators**.

To thwart unwanted or unauthorized persons from entering the facility, **access controls**, such as locks, automatic door closings, policies regarding visitors, **smart cards**, and **biometric devices**, can be used.

By thoroughly assessing the entire physical environment housing the computer operation and instituting ongoing (and updated) protection, the possibility of disasters can be lessened considerably.

REFERENCES

1. Alexander, Michael, "Biometric System Use Widening," *Computerworld*, January 8, 1990, p. 16.

2. Alexander, Michael, "Smart Cards Still Fighting an Uphill Battle," *Computerworld*, May 28, 1990, p. 22.

3. Backler, Michael A., "Biometrics: Science Fiction or Fact?" *Security Management*, August 1988, p. 126.

4. Baskerville, Richard, *Designing Information Systems Security* (New York: John Wiley and Sons, 1988).

5. Berg, Kenneth E., "A Successful Equation for Protecting Computers," *Risk Management*, September 1987, pp. 52-54.

6. Dawber, Fred, "A Card Access Education," *Security Management*, June 1990, pp. 81-84.

7. Dial-Guard, "Information Security Solutions," Dial-Guard, Inc.

8. Diamond, Sam, "Biometric Security: What You Are, Not What You Know," *High Technology*, February 1987, p. 54.

9. "EDP Security: Data, Facilities, Personnel Control," *FTP Technical Library*, Volumes I, II, III.

10. Farhoom, Al F., "Managing Computer Security," *Datamation*, January 1, 1989, pp. 67-68.

11. Fiderio, Janet, "Voice, Finger and Retina Scans: Can Biometrics Secure Your Shop?" *Computerworld*, February 15, 1988, p. 81.

12. Fort, James, "Access Control: A Case Study," *Security Managment,* May 1990.

13. Gallup, James G., "Ozone-Layer Pact Could Change Halon Use," *National Underwriter Property and Casualty/Employee Benefits*, August 15, 1988, pp. 16-17.

14. Garcia, Cristina, "Putting the Finger on Security," *Time*, April 3, 1989, p. 79.

15. Gassaway, Paul, "Businesses Get Physical to Thwart Thieves," *Access*, 4th Quarter, 1989, pp. 36-40.

16. Gast, Bruce M., "Safely Storing Data," *Systems Security*, January 1990, pp. 59-60.

17. *Guide to Paper Shredding*, Cummins-Allison Corporation, Mt. Prospect, IL, 1987.

18. Hafner, Katherine M., " Is Your Computer Secure?" *Business Week*, August 1, 1988, pp. 64-72.

19. Hein, Steve, "The Computer Room," *Security Management,* Vol. 91, No. 8.

20. Hof, Robert D., "Forget the ID—Let's See Your Eyeball," *Business Week*, November 21, 1988, p. 109.

21. Hsiao, David K., *Computer Security* (New York: Academic Press, 1979).

22. "Identix to Help Produce Smart/Biometric Terminal," *Newsbytes*, June 19, 1990.

23. Johnston, R.E., "How to Select and Implement a Data Security Product," *Info Systems*, January 1984, pp. 54-56.

24. Klopp, Charlotte, "More Options for Physical Access Control," *Computers and Security*, September 1990, pp. 229-232.

25. Lauffer, John B., "Fire Protection: Sprinklers or Halon?" *National Underwriter Property and Casualty/Risk and Benefits Management*, August 7, 1989, p. 11.

26. "Look Me in the Eye," *Discover*, February 1990, p. 8.

27. Marsh, Gerard V., "The Practitioner and the Computer," *The CPA Journal*, August 1987, p. 107.

28. Mayfield, Charles, "Who Goes There?" *Security Management*, March 1989, p. 36A.

29. Moulton, Rolf T., *Computer Security Handbook* (Englewood Cliffs, NJ: Prentice-Hall, 1986).

30. Olson, Ingrid, and Marshall D. Abrams, "Computer Access Control Policy Choices," *Computers and Security*, September 1990, pp. 699-714.

31. "PC Security Bull Team to Secure Personal Computer Via Smart Cards," *Computergram International*, June 19, 1990.

32. Pfleeger, Charles P., *Security in Computing* (Englewood Cliffs, NJ: Prentice-Hall, 1989), pp. 1-538.

33. Sanderson, Andrew, "Security Matters," *Accountancy*, December 1988, pp. 134-138.

34. Shalowitz, Deborah, "Data Equipment Needs Safety Plan," *Business Insurance*, June 3, 1991, pp. 18-20.

35. Smith, Cherie, "A Trio of Unexpected Triumphs," *ISSA Access*, 4th Quarter, 1989, pp. 1-6.

36. Smith, James H., "Baton Waver Wins Converts to 'Wallet-Top Computer'," *Government Computer News,* June 11, 1990, p. 10.

37. Smith, James, M., "Feds Underestimate Smart Cards, Backers Say," *Government Computer News*, June 25, 1990, pp. 79-80.

38. Snyders, Jan, "How Safe Is Safe?" *Infosystems*, June 1984, pp. 64-66.

39. Sowder, Kathleen A., and Joseph A. Barry, "Update on the Biometric Beat," *Security Management*, June 1988, p. 53.

40. Spitz, S. Leonard, "Sincerely Yours, Chris Goggans," *ISP News*, September-October 1991, pp. 14-18.

41. Svigals, Jerome, *Smart Cards* (New York: Macmillan, 1987), pp. 1-193.

42. "Thieves Window Shop, Too," *Modern Office Technology*, December 1988, p. 30.

43. Weber, Ron, *EDP Auditing* (New York: McGraw-Hill, 1988).

44. Wood, Lamont, "Smart Cards: Smarter Than Passwords," *Datamation*, July 15, 1991, pp. 69-70.

45. Wood, Patrick, "Safe and Secure?" *Byte*, May 1989, pp. 254-258.

46. Zviran, Moshe, and William J. Haga, "Cognitive Passwords: The Key to Easy Access Control," *Computers and Security*, September 1990, pp. 723-736.

REVIEW QUESTIONS

1. Briefly identify the dangers presented to computer centers from natural disasters such as fire, flood, and earthquakes.

2. Compare natural disasters to artificial disasters regarding their implications for security.

3. How do Halon fire extinguishers function, and why are they used in computer centers rather than traditional extinguishers?

4. What data/information should be backed up, and where are these copies normally stored?

5. What are some of the criteria used for selecting and preparing the physical location of the computer facility?

6. What are some of the reasons a UPS should be installed in a computer facility?

7. How are shredders, disintegrators, incinerators, and other document destruction devices used?

8. What are some of the access controls that become part of the standard policy of computer center operations?

9. Describe how smart cards function.

10. Describe some of the biometric devices that are currently being used.

DISCUSSION QUESTIONS

1. It has been said that fires present the greatest natural disaster to computer centers and other business locations. Why do fires and their eventual aftermath cause so much harm?

2. Why did earthquakes and hurricanes cause so much damage to areas that have experienced these disasters? Are there methods that could be instituted to lessen the damage and allow for faster resumption of activities?

3. Where would you place Halon fire extinguishers in the computer center facility and adjoining areas, considering some of the pros and cons of Halon? Why?

4. How can computer centers avoid the "Fort Knox Syndrome" by adding so much computer center physical security that the employees and visitors feel as if they are existing in a completely closed environment?

5. If your organization decided to adopt a biometric device for access control, what considerations should be addressed? Would you involve the employees in this decision?

EXERCISES

1. Select a computer center on your campus or in your vicinity and observe some of the physical security items cited in this chapter, such as locks, badges, and guards.

2. Interview someone in your campus computer center or a local business to gain information about what his or her facility is doing to promote good physical security practices.

3. Locate an article in a current periodical in your library relating to biometric devices that are currently being used as access controls.

4. Locate an article in a computer security journal or a computer-related journal that reports the aftermath of either Hurricane Hugo or the San Francisco earthquake of 1989.

5. Locate a local dealer that offers document destruction devices and make an appointment to view the facility and have a demonstration of these devices.

PROBLEM-SOLVING EXERCISES

1. Many people feel that use of biometric devices creates a danger to employees' health and presents a possible on-the-job risk. If your organization was considering using some of these devices, how would you assess the health issue and relay these findings to the employees and to the management of your organization?

2. Develop a brief physical security plan for a small-sized computer facility. Be sure to include equipment security, access controls, and contingency plans for natural disasters.

3. Develop a risk ratio matrix for your local area in terms of natural disasters, such as hurricanes, fires, floods, blizzards, and tornados. Be sure to include information on the number of instances in the past, destruction levels, and local facilities aiding in the recovery.

4. Develop a plan that supports the use of document destruction devices. What are some of the pros and cons of each device (e.g., local regulations, recycling of shredded paper, fire regulations, chemical disposal laws, and environmental issues)?

5. What privacy issues should be considered if your organization instituted a system of guards, monitors, or telephone tracking to be used for access control? Do some of these policies violate the employees' right to privacy?

CASES
Alderman Electronics

Alderman Electronics is a fictitious international electronics firm that develops computer chips. The building that houses Alderman's facility includes its administrative headquarters, which houses all its senior executives, as well as computer information on their products and research. Fred Alderman, president, is becoming increasingly concerned with the security of the facility. Numerous stories have been made public about research facilities being penetrated and expensive, time-consuming research and development ideas compromised. Mr. Alderman would like to keep the competitive edge that his organization currently enjoys, and he is intent on reassessing the entire facility for possible vulnerabilities.

All the assets of Alderman Electronics are similar to those of most high-tech companies. Its personnel, property, and proprietary information are housed in the executive offices, computer rooms, and archives. The security department's daily concerns include competitor intelligence gatherers and their technology, internal and external theft of information and equipment, and other white-collar crime.

The basic layout of the facility is as follows:

Perimeter

- Suburban area
- East, north, and west sides of property bounded by woods
- South-side driveway entrance off a major four-lane highway
- Five-hundred-car parking lot on the east side of the building
- Main driveway to the building on the south side with access to the parking lot on the east side

Building

- Three hundred employees, 20 of whom are proprietary security officers staffing the facility 24 hours a day
- Ten floors, plus basement
- Two elevators in the center of the building
- Freight elevator adjacent to the shipping and receiving entrance on the first floor, north side
- Two stairways inside the shell of the building, one each on west and east sides
- Windows on every floor
- Security control center in first floor lobby area

Entrances

- Employee and visitor access through main lobby, first floor, south side
- Shipping and receiving entrance on first floor, north side, with access by way of parking lot (entrance faces wooded area)
- Two additional entrances, one each on west and east sides, and fire doors where stairways meet on first floor

Special Considerations

- Archival storage in basement
- Fourth floor computer room
- Eighth floor executive offices

Provide solutions for controlling access to specific areas at Alderman. (Many professionals might caution that it is not feasible to formulate independent security plans for the various sections of the facility—all areas must be considered as a whole. Point out some of the vulnerabilities in this design and suggest solutions.)

Thompson Financial Services

Thompson Financial Services provides credit reports, loan servicing information, and other financial data to banks and other institutions whose customers are seeking credit for purchases. Mr. Matthew Thompson, President of TFS, is interested in undertaking a project that would ensure that all vital backups of data and information are secure so that the organization can provide timely and correct information to its clients, while maintaining the confidentiality of information contained in these reports. A complete backup treats the information as a whole and copies it in its entirety to the backup medium; individual files are not identified. A partial backup identifies files to be copied and transfers them to the backup device. In most computer centers, partial backups are used only for those files that need updating. In TFS' situation, the backups must be available on short notice, yet secured against disasters and break-ins.

Develop a plan that can delineate and enforce a corporate policy to safeguard against potential disasters. Identify the programs and data to be stored, the media on which the files are to be stored, the frequency of backup, and who is responsible. The backup procedure should be classified as either complete or partial backup.

How would you institute a backup program and yet protect the privacy of these credit reports so that information is not compromised? Where would you suggest that this backup information be stored?

SYSTEMS SECURITY
AND CONTROL

HARDWARE SECURITY CONTROLS

LEARNING OBJECTIVES

After studying this chapter, you will be able to:

1. Understand the various types of breaches and vulnerabilities that threaten computer systems.
2. Identify the various types of computer losses and where they occur.
3. Explain the four levels of hardware controls that should be instituted to protect hardware.
4. Explain the principles of confidentiality, integrity, and availability that should be applied to information being processed.
5. Apply the rules of criticality and sensitivity to data handling.
6. Describe the parameters of operating systems controls.
7. Explain the function of a security kernel.
8. List the various types of access controls that should be applied to computer systems.
9. Explain the special types of protection that should be applied to remote devices.
10. Detail the various basic rules relating to password protection.
11. Describe the various identification devices that are currently being used.
12. Define the following terms.

TERMS

- Hardware Controls
- Prevention
- Detection
- Limitation
- Recovery
- Confidentiality
- Integrity
- Availability

- Criticality
- Sensitivity
- Operating System
- Security Kernel
- Password
- Default Password
- Audit Trail
- Login Security

TERMS (continued)

- Call-Back Procedure
- Key Card
- Encrypted Codes
- UNIX

- VAX/VMS
- IBM MVS
- VM/370

PROTECTING CONFIDENTIAL IRS DATA

For the past quarter century, the Internal Revenue Service (IRS) has attempted to modernize its systems to keep up with changing technology. Not only have those efforts largely been foiled by the speed of technology improvements as well as the massiveness of the agency and its paper-handling requirements, but a recent report alleges that the efforts to stay current have caused the IRS to ignore or forego security needs.

According to the report released by U.S. Senator John Glenn in August 1993, 368 employees of the IRS Atlanta office exploited "ineffective security controls" to snoop through the computerized tax accounts of friends, relatives, neighbors, and celebrities. Glenn also said that five additional employees had violated the system in order to engineer fraudulent returns; one of these workers triggered at least 200 false tax refunds. The investigation, which expanded to other geographic regions, has uncovered another 100 IRS workers suspected of unauthorized access to taxpayer returns, according to Glenn.

Senator Glenn, who expressed concern that "this is just the tip of a very large iceberg," also said he was troubled that the IRS has focused its investigation on just one region and on only one of 56 ways the database system can be accessed by employees. Ironically, the report said that an enhancement to the existing system "greatly increases the risk of employee browsing, disclosure and fraud" by making more data accessible to more employees.

IRS Commissioner Margaret Milner Richardson said that the 20-year-old system, used by 56,000 employees, has almost all the security features required by federal law, including passwords. Deputy Commissioner Michael P.

Continued

PROTECTING . . .

From page 61
Dolan said the Atlanta office discovered the illegal activity by using software that analyzed a database of audit trial information to detect suspicious use patterns. He added that widespread use of the detection technique was previously inhibited by the difficulty of processing tape archives, which grow by 100 million transactions per month. Dolan said that each regional IRS center will soon be equipped with the pattern detection software and high-capacity optical disk hardware to allow managers to monitor use—and possible misuse—by IRS employees.

As reported in "IRS Uncovers Bogus Access to Tax Records," by Gary H. Anthes, *Computerworld*, August 9, 1993, p. 15.

THE TOTAL SYSTEM NEEDS SECURING

Now that we have looked at the physical security of the computer facility, we need to address the issue of what actually happens inside the computer to cause vulnerability and disclosure. The approach of "going from the outside in" gives several layers of needed protection. (See Figure 3.1.) Only when all levels are secured can an organization feel it is protected. If security and control are weak at any of these levels, all the connecting levels are vulnerable.

Hardware controls involve operating systems, protection of memory and addressing, protection of access to general objects, file protection mechanisms, file authentication routines, and password protection. Hardware security should be incorporated in the basic system design and updated as changes are made to the system. Only by constant review and updating can hardware security be considered truly a "fixed" part of the system, rather than an "add-on" at the end.

The past few years have witnessed the very rapid growth and increased usage of computers in organizations of all sizes and functions. As this prolific growth occurred, however, security was not always considered at the system design stage. Except where security was obviously of major importance (e.g., in banking systems or top secret military or defense projects), designers of computer systems too often ignored it. Company management, most likely, did not understand how their computer systems worked and so did not recognize their vulnerability to fraud and malicious interference. However, it recently has been recognized that computing systems are easily compromised. This is especially true because most security sys-

FIGURE 3.1
Layers of Protection

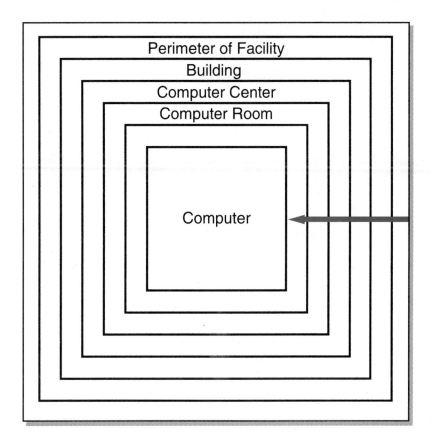

tems have evolved on an ad hoc basis with "patches" made in elements of the system to thwart any perceived weaknesses.

Losses from Computer Fraud

Estimates of losses from computer security violations have reached staggering figures. One source describes the extent of computer crimes as follows:

> The expectations of annual losses from computer fraud range from an estimate of $100 million by the U. S. Chamber of Commerce to an estimate of $3 billion in a recent article published by the *Harvard Business Review*. According to the U. S. Department of Commerce, only one of 100 such crimes is detected so these figures are only gross estimates. It is further estimated that only 20 percent of computer crimes are reported and even fewer are prosecuted.[16]

Another expert states:

> There are no accurate estimates of the total losses to business and to the government each year through computer crimes . . .Many companies and agencies are reluctant to admit that their computer systems are vulnerable to thievery . . . Experts say that only one computer theft in 10 is discovered.[44]

Classification of Breaches

There are three general types of security breaches:

1. **People inside the organization** who have enough technical knowledge to enter false commands or to change programs, usually to steal money or materials

2. **People outside the organization** who learn enough about the system to break the security codes and enter false commands, usually to steal money

3. **People outside the organization**, such as business competitors, who breach the system's security codes to obtain information they can use to their own advantage[44]

Business Installations

Every computer installation is vulnerable to criminal activity. In the computer environment today, the possibilities of fraud, abuse, and theft are limited only by the criminal's imagination.

> White collar thieves have misused computers to embezzle funds, pilfer time-sharing services and programs, eavesdrop on the bids of business competitors, divert inventory, disclose tax and banking records, snatch valuable mailing lists, monitor private medical and pharmaceutical records, print payroll checks and other documents that can be converted into ready cash, reduce and eliminate premiums on insurance and other installment-type payments, and alter transcripts at colleges and universities.[12]

Too often information systems managers concentrate on hardware and software rather than on personnel as a means of checking computer abuse. "But security is, first and last, a people problem," says Donn B. Parker, Senior Management Systems Consultant at SRI International, a research and consulting firm in Menlo Park, California.[41] The computer environment, from an overall point of view, is defined by ". . . corporate policies, operative procedures, and daily practice. From a practical point of view, the security environment has been characterized by and combined with the concept of controls."[23]

Computer crime does not always involve just the loss of money. Loss of vital data—sometimes forever—or such things as invasion of corporate or personal privacy can occur. Embezzlement of funds, believed to be the most common form of computer crime, is not the only way in which organizations can suffer; other computer crimes include:

1. **Theft of services.** As employees use computer time for work other than company business

2. **Selling or changing information** stored in the employer's computer

3. **Invasion of privacy** whose victims might sue the negligent organization[10]

There are three general ways to gain unauthorized access to a computer system:

1. Accidental access
   - Someone, in normal use of a terminal, accidentally transmits a security code that causes another person's data to be sent
   - Often, the person who has done this does not realize what he or she has done or does not want to get into someone else's data.
2. Actively breaking in
   - Impersonation of an authorized user
   - Entry by people who understand the security safeguards and know how to get around them
3. Passively breaking in
   - Wire-tapping a telephone line between a terminal and the computer
   - Unauthorized examination of a printout
   - Use of microwave receivers to intercept data transmissions from satellites or from long distance telephone relay transmitters[44]

The actual abuses of computers and the potential for even more occurrences are well documented; research indicates repeatedly that businesses must be aware of the security issue and ward off possible violations *before* these intrusions occur.

Organizational management should formally recognize the need for a computer security program, thus ensuring the implementation of effective countermeasures.[50]

Above all, management must start taking the break-in problem seriously. Many experts put the blame for poor computer security squarely on top management. To executives at most companies "security is one of the lower priorities"[9], due mainly to cost justification and concentration on the bottom-line profit.

In addition to guarding the central mainframe computers, organizations must pay attention to securing the personal computers spreading throughout their organizations.

Computer Losses

Although it is difficult to accurately access the actual losses from computer infractions (both intentional and accidental), the estimates are mind-boggling. Because a great number of these losses are not reported, the figure is even more staggering. Figure 3.2 gives a realistic portrayal of the severity of the problem. Human errors, accidents, and omissions constitute the majority of computer losses. Careful analysis and planning must be instituted on a corporate-wide basis to ensure compliance with standards of operation.

Plan of Action

To fully assess the situation, a plan of action must be drawn up and studied in detail. Each of its points must be carefully analyzed, stated, and followed to gener-

FIGURE 3.2

Computer Losses

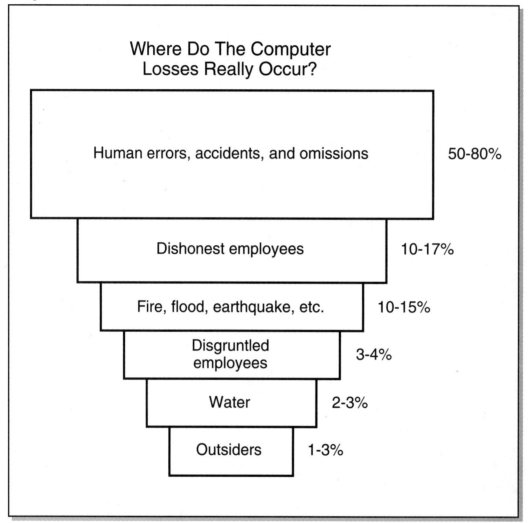

Where Do The Computer Losses Really Occur?

Human errors, accidents, and omissions	50-80%
Dishonest employees	10-17%
Fire, flood, earthquake, etc.	10-15%
Disgruntled employees	3-4%
Water	2-3%
Outsiders	1-3%

ate a fully operational plan of action. The plan of action should include objectives, overview, value, definitions, and standards.

LEVELS OF HARDWARE CONTROLS

The four levels of hardware controls that should be instituted are as follows:

- **Prevention.** These controls restrict access to information and technology to authorized personnel who perform only authorized functions.
- **Detection.** These controls provide for early discovery of crimes and abuses if prevention mechanisms are circumvented.

- **Limitation.** These controls restrict losses if crime occurs despite prevention and detection controls.
- **Recovery.** These controls facilitate efficient information recovery through fully documented and tested contingency plans.

Requirements for Security

Although the requirements for security will vary from one installation to another due to the nature of the information being processed, three principles—*confidentiality*, *integrity*, and *availability*—should be evaluated carefully.

- **Confidentiality** means preventing sensitive information from being disclosed to unauthorized parties. The reasons for confidentiality can involve national security (nuclear weapons or defense data), law enforcement (undercover agents, informants), competitive advantage (marketing and research data), or personal privacy (credit or medical histories).
- **Integrity** means that information and programs are changed *only* in a specified and authorized manner. Data must be kept consistent and changed *only* in an approved manner. The degree of accuracy of the data should be fully specified. Integrity policies should involve prevention of errors and omissions and control of program changes.
- **Availability** ensures that systems work in a timely manner and service is not denied to authorized users. Adequate response time for various applications should be fully described. In the case of life-threatening situations (air traffic control, critical care units of hospitals), timeliness is of the utmost concern.

Once these three parameters have been established, the data should be evaluated in further detail for each application in terms of criticality and sensitivity.[30]

- **Criticality** should be evaluated in terms of what would be affected if the system were to become unavailable. First, divide the system into subelements (i.e., applications) that are related to users or business functions. Then evaluate each application to define the impact on the user if computer support was lost. Include such factors as the effectiveness of the particular function, additional cost of doing business, lost revenue, possible legal problems, and the effect of loss on image and public goodwill of the organization.
- **Sensitivity** measures the impact of a nonauthorized person gaining access to the information. Sensitive areas may also include trade secrets, formulas, financial data, and corporate planning information that may be of significant value to competitors.

Some techniques that will protect the computer (and the data stored within) can be built into the computer. Private personal data should not be disclosed without specific authorization. Other protections can be housed in the software. Only when hardware controls and software controls function in tandem can the security of the system be assured.

OPERATING SYSTEM CONTROLS

An **operating system** (OS) is a set of computer programs that runs or controls the computer hardware as an interface with applications programs.[51] (See Figure 3.3.) UNIX, Vax/VMS, and DOS are all examples of operating systems.

FIGURE 3.3
Function of an Operating System

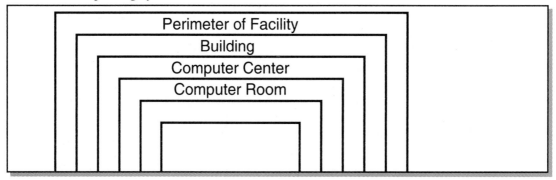

Some of the most common security-related problems for operating systems include the following:

- Design errors
- Implementation errors
- System generation and maintenance problems
- Deliberate penetrations (resulting in modifications)

All of these flaws can produce undesirable effects to the applications system, resulting in disastrous output.

Some of the critical operations that should be carefully (and constantly) monitored include the following:

- User jobs that are read or written outside storage areas
- An OS design or implementation error that allows a user to disable audit controls or to access all system information
- Unauthorized modification to the OS that allows a data entry clerk to enter programs and subvert the system
- Maintenance personnel who bypass security controls while performing maintenance duties. (If controls are bypassed for any period of time, the system is virtually wide open to anyone logged on the system.)
- An OS that fails to maintain an unbroken audit trail
- A user who is able to get into monitor or supervisory mode
- Files that can be read or written without being opened
- Inconsistencies introduced into data because of simultaneous processing of the same file by two jobs

- An OS that does not protect a copy of information as thoroughly as it protects the original

- An OS crash that exposes valuable information such as password lists or authorization tables

- An OS that does not report that multiple copies of output are being run

- An OS that does not report all terminal locations and their previous and present users after a system crash

- An OS that does not erase all scratch space assigned to a job after the normal (or abnormal) termination of a job

Security Kernel

Within the operating system, layering or partitioning can be used to divide the OS into kernels. A **kernel** is the part of the OS that performs the lowest-level function. In a standard OS environment, the kernel implements and monitors standard operations such as:

- Synchronization

- Interprocess communication

- Message switching/passing

- Interrupt handling

A **security kernel** is the hardware and software elements that implement the reference monitoring concept. It must mediate *all* accesses, be protected from modification, and be verifiable as correct.

Security kernels isolate security functions by the following methods:

- **Separations.** By isolating security mechanisms from the rest of the operating system and from the user allotted space, protection from penetration is more easily achieved.

- **Modifiability.** Any changes to the security procedures and protection are easier to install and easier to test.

- **Verifiability.** The kernel, due to its relatively small size, can be easily tested for correctness.

- **Unity.** A single set of codes can be used to test all security functions.

- **Compactness.** The security kernel is likely to be relatively small as it is designed to perform only security functions.

- **Coverage.** Every access to a projected object must pass through the kernel, thus ensuring that each and every access is checked.

On the negative side, because the kernel adds another layer of interface between the user and the operating system, implementation of a security kernel may degrade system performance to a certain degree.

There are two basic design schemes for a security kernel—combining the security kernel with the operating system or employing a separate security kernel.

Operating Systems Penetrations

If a person wants to penetrate an OS, he or she must first find system flaws and then exploit them. Some of the known flaws might be:

- Input/output processing
- Ambiguity in access policy
- Incomplete mediation/authorization
- Lack of uniqueness, leading to trapdoors or standardized procedures

ACCESS CONTROLS

In addition to the various access controls discussed in Chapter 2, additional access controls should be instituted to protect the operating system and other hardware devices.

Vulnerabilities

The following clues should indicate if there are vulnerabilities to the system:

- Security policies and practices do not exist, and no one is assigned responsibility for information security.
- Passwords are posted next to computer terminals, written in obvious places, shared with others, or appear on the computer screen when they are entered.
- Remote terminals, microcomputers, and word processors are left on and unattended during work and nonwork hours.
- There are no restrictions on users of the information, or on application authorization. *All* users can access *all* information and use *all* the system functions.
- There are no audit trails or logs of *who* uses *which* application.
- Programming changes can be made without going through a review and approval process.
- Documentation is either nonexistent or inadequate. In many cases, documentation is extremely outdated and does not reflect the current system or applications.
- Numerous attempts are made to log on with invalid passwords. In dial-up systems (with telephone or modem hookups) computers can be programmed to go through "trial and error" guessing. Some sophisticated hackers have auto-dialers or password checkers that will automatically dial at random until a "hit" is achieved.
- Input data is not verified or checked for accuracy.
- System crashes are excessive.
- Periodic reviews of data and information are not conducted to determine the level of security required.
- Little or lax attention is given to information security.

Information Security Controls

Access to both computer information and applications should be strictly controlled to ensure that only authorized users have access. *Every* user should be verified and *every* possible access should be checked by the following methods.

User Identification. All users should be required to log on to the computer as a means of initial identification.

User Authentication. Nontransferable passwords should be used to authenticate the identity of users. Traceable personal data (names of wife/husband/children, boyfriend/girlfriend's name, pet's name, make of car, etc.) should not be used as passwords. Password management protection controls should be established, and *all* users should be educated regarding common problems and established procedures.

Authorization Procedures. Procedures should be developed that identify which users have access to which information and applications. These procedures should be monitored regularly and kept current as personnel and duties are constantly shifting. Appropriate controls should be instituted that will protect information and applications, yet not burden the user in any obtrusive manner. Management approval must be received before a user can gain access to computer resources, view specific information and applications, and receive a password.

File Protection. In addition to user identification and authorization procedures, steps should be developed to restrict areas to specific data files. File protection procedures should include the following:

- **Labeling** external and internal files to identify the type of information contained and the security level required to access them
- **Restricting** access to related areas containing data files (The restricted areas should include off-site backup facilities, on-site file libraries, and off-line files.) (See Figure 3.4.)
- **Establishing** controls relating specifically to software, hardware, and procedures that will aid in restricting access to on-line files to *only* authorized users

System Protection. To fully protect the system, simple precautions should be taken to limit access:

- **Turning** off idle terminals
- **Locking** rooms where terminals are located
- **Positioning** computer screens away from doorways, windows, and heavily trafficked areas
- **Installing** security equipment devices that limit the number of unsuccessful log-on attempts or that dial back users who attempt to use the system via telephone
- **Programming** the terminal to shut down after a specified period of nonuse
- **Shutting** down the system during nonbusiness hours if the computer operation does not run around the clock.

FIGURE 3.4
Security of Remote Sites

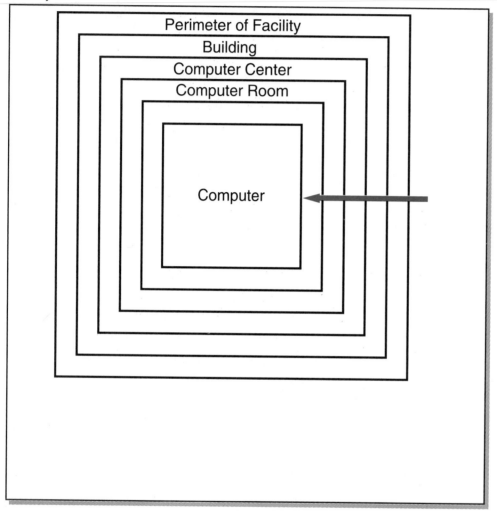

Password Protection. Passwords are the most commonly used type of identification or authorization. Figures 3.5 and 3.6 show some general guidelines that will ensure the protection of passwords.

Other Identification Devices

Passwords are one type of identification—something the user knows. Two other types of identification that are effective are something the user has—such as a magnetic coded card—or a distinguishing user characteristic—such as a voice print.

If the computer has a built-in **default password** (a password that comes built into the computer software and overrides all access controls), it should be changed.

FIGURE 3.5

Guidelines for Protecting Your Password

Protect Your Password
☑ Don't share your password—with anyone.
☑ Choose a password that is hard to guess.
☑ Hint: Mix letters and numbers, or select a famous saying and choose every fourth letter. Better yet, let the computer generate your password.
☑ Do not use a password that is your address, pet's name, nickname, spouse's name, telephone number, or one that is obvious—such as sequential numbers or letters.
☑ Use longer passwords because they are more secure; six to eight characters are realistic.
☑ Be sure that your password is not visible on the computer screen when it's entered.
☑ Be sure that your password does not appear on printouts.
☑ Do not tape passwords to desks, walls, or terminals. Commit yours to memory.

FIGURE 3.6

Guidelines for Managing Passwords

Manage Passwords Carefully
☑ Change passwords periodically and on an irregular schedule.
☑ Encrypt or otherwise protect from unauthorized access the computer-stored password file.
☑ Assign password administration to only the most trusted officials.
☑ Do not use a common password for everyone in an area.
☑ Invalidate their passwords when individuals leave the organization.
☑ Have individuals sign for their passwords.
☑ Establish and enforce password rules—and be sure everyone knows them.

The computer also may be programmed so that when users log on, they are told the last time of its use and the number of invalid log-on attempts since then. This approach makes the user an important part of the **audit trail**, which tracks and records all activity.

Login security should not rely solely on passwords. When users access remote computers, a **call-back procedure** should be used. This procedure provides a high level of security and flexibility. An authorized phone number is recorded in a file and is checked against incoming calls every time access is attempted. If a phone number is not on file, access is automatically denied. This system is flexible enough to accommodate special circumstances. For example, a user who needs access during a temporary business trip can get temporary privilege for a new phone number before he or she departs.

A **key card** can also be used for those who travel but may not know the phone number of their destination. The key card will send in **encrypted codes** that the user does not know. The individual can then input his/her regular password for access.

GENERAL-PURPOSE OPERATING SYSTEMS SECURITY

UNIX

The **UNIX** operating system was not designed for a high level of security. It was designed in 1969 by two programmers, primarily for their own use, to develop, test, and maintain programs; it was intended for "non-hostile" environments. UNIX is widely used in research laboratories and universities where the primary objective is the open sharing of objects and information. UNIX's popularity can be attributed to the variety of UNIX-based products available; however, because of the independent development work that has been done on UNIX, one UNIX product may not be the same as another.

UNIX systems administrators can take several steps to secure their systems, including following simple guidelines and prohibiting the use of certain passwords. A checklist of essential UNIX security activities for system administrators includes the following:

- Follow installation rules.
- Follow configuration rules.
- Enforce password lengths and require frequent password changes.
- Never leave active root or superuser status logged in on an unattended terminal.
- Limit access to privileged passwords.
- Follow vendor recommendations regarding security administration.
- If there are no vendor security recommendations, inquire why these are not available as support.
- Scrutinize and be suspicious of all software programs, especially those requiring privileges.

- Write setup-ID programs with caution and then limit access.
- Do not ignore physical security.[49]

DEC VAX/VMS

The Digital Equipment Corporation **VAX/VMS** operating system is an example of a general-purpose OS that has improved substantially as it matured. The security features of VAX include the following:

- **Access controls** that operate at the single-subject/single-object level
- **Password controls** that can be configured to meet the needs of a particular system
- **Auditing functions** that track security events
- **Monitoring functions** that warn security administrators of suspicious events
- **Encryption**

IBM MVS

The underlying system design for the **IBM MVS** is OS/360, designed in the early 1960s for the 360 family of computers. The 360 architecture then evolved into the 370, and the OS evolved from OS/360-MVT to OS-VS2 release 2, and then to MVS. MVS's security features include the following:

- Access controls, selected by the user, and applied to individual files
- Virtual memory (VM), which provides solid memory protection
- Auditing of accesses

VM/370

Isolation within a multiprogramming environment is the key feature of IBM's **VM/370**. The VM/370 was designed as an OS's OS as it runs other operating systems. Each OS is isolated through logical separation so that it is completely unaware of the existence of the other OS. The VM/370 manages disk space, memory, access to hardware features (i.e., interrupts and processor use) so that one resource can be logically shared among several different OSs.

SOURCES OF ADDITIONAL INFORMATION

The United States federal government publishes a wide variety of information that aids both the public and private sector in establishing security guidelines for both systems and applications.

Government Publications

Most technical publications are available from the Government Printing Office or the National Technical Information Service. Contact:

Superintendent of Documents
U.S. Government Printing Office (GPO)
Washington, D.C. 20402
(202) 783-3238
Hours: 8:00 a.m.–4:00 p.m. EST

You can place an order automatically, any time, if you know the GPO stock number and intend to charge your order to a credit card.

You may also contact:

U.S. Department of Commerce
National Technical Information Service (NTIS)
5285 Port Royal Road
Springfield, VA 22161
(703) 487-4650
Sales Desk Hours: 8:30 a.m.–5:30 p.m. EST
Identification Branch Hours: 9:00–5:00 p.m. EST

You can place an order automatically, any time, if you know the NTIS number and intend to charge your order to a credit card.

The Rainbow Series

The books in the "Rainbow Series" are published by the National Computer Security Center's Technical Guidelines Program. Named for the different colors of their covers, the books in the Rainbow Series include the Orange Book, which defines trusted computer system evaluation criteria used to assess the effectiveness of security controls built into computer system products (described in Chapter 6, *Inside the Orange Book*). Other publications in this series provide detailed interpretations of certain Orange Book requirements and descriptions of evaluation program procedures. To order one complimentary copy of any of the books, contact the Government Printing Office or the following office:

Director, National Security Agency
INFOSEC Awareness
Attention: X71
9800 Savage Road
Fort George G. Meade, MD 20755-6000
(301) 766-8729

SUMMARY

A computer system is secure ONLY if you can depend on it to perform as expected.

Hardware controls involve the protection of operating systems, protection of memory and addressing, protection of access to general objects, file protection mechanisms, file authentication routines, and password protection.

The losses from computer fraud that threaten systems are increasing at an alarming rate each year. Business installations are particularly vulnerable due to the nature of the information maintained and the potential for embezzlement.

The various levels of hardware controls should incorporate **prevention, detection, limitation,** and **recovery**. The principles of **confidentiality, integrity,** and **availability** should be evaluated carefully. Data should be evaluated in terms of **criticality** and **sensitivity**.

The **operating system** should be carefully considered as most computer functions involve the OS to some extent. A **security kernel** monitors all access to the OS.

Access controls add an additional layer of protection to hardware by utilizing user identification, authentication procedures, file protection, system protection, and password protection. Other identification devices can include **login security, key cards, encrypted codes,** and **audit trails**.

REFERENCES

1. Alexander, Michael, "Hackers Promote Better Image," *Computerworld*, June 24, 1991, p. 124.

2. "An Interview with Michael Gorby," *Information Executive,* June 1988, pp. 5-8.

3. Baker, Richard H., *The Computer Security Handbook* (Blue Ridge Summit, PA: Professional and Reference Books,1985).

4. Berman, Alan, "Security in a Dial-up Environment," *Data Security Management*, May 1985, pp. 1-8.

5. BloomBecker, J. J. Buck, "Commitment to Security," National Center for Computer Crime Data, 1989.

6. Branstad, Dennis K., "The Federal Password Usage Standard," *Data Security Management*, May 1984, p. 1-20.

7. Brown, Nander, "Security and Control of Online Systems-Design Objectives," *Data Security Management*, May 1984, pp. 1-12.

8. Clark, D. D., and D. R. Wilson, "A Comparison of Commercial and Military Computer Security Policies," *Proceedings of the 1987 Symposium on Society and Privacy*, Oakland, California, 1987, pp. 184-195.

9. "Computer Crime," *Computer Decisions*, Vol. 13, No. 6, June 1981, pp. 104-124.

10. *Computer: Crimes, Clues, and Controls*, Prepared by the Prevention Committee, President's Council on Integrity and Efficiency, March 1986.

11. *Computer Security Awareness Training*, United States Office of Personnel Management, Washington, D.C., 1988, pp. 1-44.

12. "Computer Security: What Can Be Done," *Business Week*, September 26, 1983, pp. 126-130.

13. *Computer Systems Security*, U.S. Department of Commerce, National Institute of Standards and Technology, September 1989, pp. 1-11.

14. *Computer Systems Security Guidelines*, Page Assured Systems, Inc., Fort Lee, New Jersey, May 1989.

15. Davies, Donald, "Confidentiality, Integrity, Continuity," *Computer Control Quarterly*, Spring 1986, pp. 28-32.

16. Enger, Norman L., and Paul W. Howerton, *Computer Security . . . A Management Audit Approach* (New York: AMACO [A Division of American Management Associations]), 1980.

17. Farr, M. A. L., B. Chadwick, and K. K. Wong, *Security for Computer Systems* (Manchester, England: NCC Publications, 1972).

18. Garfinkle, Simson, and Gene Spafford, *Practical UNIX Security* (Sebastopol, California: O'Reilly & Associates, Inc.), pp. 1-48.

19. Gassaway, Paul, "Businesses Get Physical to Thwart Thieves," *ISSA Access*, 4th Quarter, 1989, pp. 36-39.

20. Graubart, R. D., "On the Need for a Third Form of Access Control," *Proceedings of the 12th National Computer Security Conference*, National Institute of Standards, National Computer Security Center, Baltimore, Maryland, October 10-13, 1989, pp. 296-304.

21. *Guideline for Automatic Data Processing Risk Analysis*, FIPS Pub 65, Department of Commerce, National Bureau of Standards, Springfield, Virginia, August 1979.

22. Henderson, Stuart, "How MVS Security Evolved, and Where It's Going," *ISP News,* March/Spring 1992, pp. 6-9.

23. Hodge, Barton, Robert A. Fleck, Jr., and C. Brian Honess, *Management Information Systems* (Reston, VA: Reston Publishing Company, Inc., 1984).

24. Honan, Patrick, "Data Security," *Personal Computing*, January 1987, pp. 101-107.

25. Hootman, Joseph T., "A Gentle Knock," *Information Strategy: The Executive's Journal*, Spring 1989, pp. 35-38.

26. Hootman, Joseph T., "Identity Authorizes Access," *Information Strategy: The Executive's Journal*, Winter 1990, pp. 48-50.

27. Hootman, Joseph T., "Say the Secret Word," *Information Strategy: The Executive's Journal*, Summer 1989, pp. 35-38.

28. Hootman, Joseph T., "You Are What You Prove You Are," *Information Strategy: The Executive's Journal*, Summer 1990, pp. 50-53.

29. Howe, Charles L., "Coping with Computer Criminals," *Datamation*, Vol. 28, No. 1, January 1982, pp. 118-128.

30. *Information Security Modules*, National Institute of Standards and Technology (NISI), Gaithersburg, Maryland, 1991.

31. "Invasion of the Credit Snatchers," *Information Week*, August 19, 1991, pp. 40-42.

32. Kaplan, Ray, "Security Assessment Tools for VAX/VMS," *ISP News,* March/April 1992, pp. 10-14.

33. Khosrowpour, Mehdi, *1990 Survey Results of Information Security Issues* (Harrisburg, PA: Idea Publishing Group, 1990), pp. 1-12.

34. Kurzban, Stanley, "Making MVS/XA Secure," *Data Security Management*, 1985, pp. 1-8.

35. Margulies, Benson I., "Security in a MULTICS Environment," *Data Security Management*, 1985, pp. 1-16.

36. Millen, J. K., *Models of Multilevel Computer Security*, MTR-10537, The MITRE Corporation, January 1989.

37. Mullen, Jack B., "A Checklist for Online Terminal Security," *Data Security Management*, May 1986, pp. 1-16.

38. Murray, William H., "Security in a VM Environment," *Data Security Management*, August 1985, pp. 1-12.

39. "Not Giving Credit Where It's Due," *Information Week*, August 19, 1991, pp. 36-37.

40. Olson, Ingrid M., and Marshall D. Abrams, *Computers and Security*, Vol. 9, 1990, pp. 699-714.

41. Parker, Donn B., *Computer Security Management* (Reston, VA: Reston Publishing Company, Inc., 1981), pp. 1-380.

42. Parker, Donn B., "20 Principles for Selecting Information Safeguards," *Data Security Management*, May 1985, pp. 1-12.

43. Pfleeger, Charles P., *Security in Computing* (Englewood Cliffs, NJ: Prentice-Hall, 1989), pp. 1-538.

44. Potter, George B., *Data Processing: An Introduction* (Plains, Texas: Business Publications, Inc., 1984).

45. Rogers, Stephen M., "Access for Success," *Security Management*, May 1990, pp. 16A-18A.

46. Schell, Roger R., "Security Kernel Design Principles," *Data Security Management*, May 1984, pp. 1-16.

47. Schweitzer, James A., "Automated Logical Access Control," *Data Security Management*, August 1986, pp. 1-8.

48. Sciullo, Francis D., "Defining Access Control Requirements," *Data Security Management*, May 1984, pp. 1-8.

49. Sherizan, Sanford, "Striving for UNIX Security," *ISSA Access*, 4th Quarter, 1989, pp. 8, 9, 43.

50. Srinvasan, Cadambi A., and Paul E. Dascher, *Infosystems*, Vol. 28, No. 5, Part I, May 1981, pp. 116-123.

51. Stair, Ralph M., *Principles of Information Systems: A Managerial Approach* (Boston, MA: boyd & fraser publishing company, 1992), pp. 1-701.

52. Teodoro, Reynaldo S., "DP Technology Security Impact Statement," *Data Security Management*, May 1984, pp. 1-16.

53. *The Handbook of Information Security*, Arca Systems, Inc., San Jose, California, 1991.

54. Warfel, George H., "Automatic Signature Verification," *Data Security Management*, May 1984, pp. 1-16.

55. Warfel, George H., Jr., "Identification Technology," *Data Security Management*, August 1986, pp. 1-12.

56. Wilkinson, Bryan, "Security Standards Guidelines," *Data Security Management*, December 1985, pp. 1-16.

57. Wilson, David R., "Trends in Information Security," *Computer Security Journal,* Vol. IV, No. 2, pp. 29-38.

58. Wilson, Geoffrey A., "Security in a UNIX System Environment," *Data Security Management*, May 1985, pp. 1-16.

59. Zviran, Moshe, and William J. Haga, "Cognitive Passwords: The Key to Easy Access Control," *Computers and Security*, Vol. 9, 1990, pp. 723-736.

REVIEW QUESTIONS

1. What are the various layers of protection of computer systems that should be instituted to achieve safe systems?

2. What are some of the ways organizations can suffer from computer losses?

3. Where do computer losses generally occur? What/who are some of the possible threats?

4. What items should generally be included in a security plan of action?

5. Describe the four levels of hardware controls.

6. Explain the three requirements for information security: confidentiality, integrity, and availability.

7. Contrast criticality and sensitivity of data.

8. Explain the critical operations of an operating system that should be closely monitored.

9. What is a security kernel? Describe a kernel's basic functions.

10. Explain the various access controls that utilize password procedures.

DISCUSSION QUESTIONS

1. Comment on the statement: "Too much security will protect systems but stifle creativity and productivity."

2. Many organizations adhere to the policy that passwords must be changed on a regular (and frequent) basis, such as every week or every month. If employees have different passwords for multiple files, this presents a dilemma in remembering the passwords. How could this policy be more workable?

3. Hacking can be both an intellectual pursuit and a harmful activity. How could this "two-edged sword" be redefined to appear a little more grey, rather than a black or white issue?

4. Figure 3.1 presents the various layers of protection for computer systems. If the dollars allotted to security are limited, how could these funds be best appropriated and balanced to ensure protection at all layers?

5. Certain employees (operators, systems analysts, programmers, systems operators) in a computer facility have greater access to operating systems. Should these employees be more closely scrutinized than others, or would this "watch dog" approach be counterproductive?

EXERCISES

1. Develop a corporate policy on development of passwords. Include statements on monitoring passwords to ensure that policies are followed.

2. You are employed in a bank as a teller trainee. What are some of the data integrity guidelines that should be followed (i.e., availability, confidentiality, integrity)?

3. Select an organization in your area and interview a key employee in the computer facility to gain information about the operating systems security.

4. Develop a list of 10 possible passwords that you might want to use. Apply the rules of password development to each one to point out strengths and weaknesses.

5. Locate an article in a computer-related periodical relating to operating systems security.

PROBLEM-SOLVING EXERCISES

1. A security kernel should be designed to achieve completeness, isolation, and verifiability. What would be the outcome if these three criteria were not achieved in each of these operations:

   a. An accounting firm

   b. A manufacturing facility

   c. A small business

   d. A branch bank facility

2. Identify the parameters of confidentiality, integrity, and availability of data for each of the following operations:

   a. A hospital pharmacy

   b. A doctor's office

   c. An accountant

   d. A retail store

   e. A bank

   f. A university registrar

3. To ensure 100% accuracy of data, all entries should be verified (preferably by more than one employee). How could a system of random checks or checking every Nth entry be utilized so that accuracy is heightened, yet productivity remains relatively high?

4. One of the current trends is to have the computer generate passwords rather than allowing the users to produce one of their choice. What are the drawbacks and advantages of this procedure?

5. Compare and contrast each of the following operating systems:

   a. UNIX

   b. VAX

   c. MVS

   d. VM/370

CASES

Stillwater Medical Center

Stillwater Medical Center is a regional hospital, serving a city of 30,000 residents and patients from areas outside of the city. All patients are treated in the hospital facility except in special needs cases, where patients are transferred via ambulance or helicopter to a larger facility.

The Computer Center at SMC processes all records dealing with patients, from the time they are admitted to the hospital until they are discharged. All records are stored in a central mainframe computer, housed in the basement of the hospital. Entries are made by remote terminals located throughout the building in admissions, nursing stations, food services, and other departments. The computer center is off-limits to everyone except authorized employees. When patients are referred to the hospital facility by a physician, their records are created once the referral takes place. All records are then sent to the referring physician, as well as to the individual patient.

The records processed daily are:

1. Laboratory tests
2. Financial/accounting data
3. Nutrition/food service data
4. Pharmacy data
5. Surgical data
6. On-floor nursing staff data
7. Personal information relating to patient
8. Physician's records/follow-up
9. Special needs of individual patients
10. Insurance billing information
11. Any other pertinent information

Analyze the various information kept for each patient and assess the level of security that should be attached to each file/record. Bear in mind that some of this information is extremely confidential and some information is somewhat public. What are the legal, moral, and ethical considerations if information is violated or disclosed?

Access for Success

The access control industry has a big future. Its growth is fueled by organizations looking for ways to enhance security. Those organizations have invested time and money in a variety of electronic technologies designed to protect people and property from unauthorized access to buildings and restricted areas. But as the access control industry evolves to meet future security needs, what will be the fate of such investment decisions? For a variety of reasons, an organization's changing need for security can outlive the access control system installed to meet that need. Some of the factors that lead to this aging process are:

- The growing number of users
- Transient populations (e.g., college and university students)
- Site expansions
- Growth of technology
- Expansion of businesses that lead to multiple access points

What are some other factors that could add to this aging process in any organization? Is some aging industry-specific? Are some organizations more vulnerable to increased aging than others?

SOFTWARE CONTROLS

LEARNING OBJECTIVES

After studying this chapter, you will be able to:

1. State the main needs for protection of software used in computer systems.
2. Identify some of the primary sources of software problems that plague systems.
3. List the basic elements composing a secure software design.
4. Describe the most common types of software intrusions.
5. Explain configuration management and its relationship to software security.
6. Explain the concept of trusted systems and show how they are used to control computer systems.
7. Compare modularity and encapsulation design principles.
8. Describe some of the basic information protection measures currently in use.
9. Describe the guidelines being used to evaluate security software.
10. Define the following terms.

TERMS

- Software Security
- Configuration
- Configuration Management
- Trusted Computer System Evaluation Criteria
- Trusted Computer Base
- Modulation/Modularity
- Encapsulation
- Accountability Control Objective
- Accountability

- Auditability
- Integrity
- Usability
- RACF
- ACF2
- Top Secret
- Omniguard
- MULTICS

OOPS! JUST ERASE A FEW ZEROS BEFORE YOU CASH THAT CHECK

A North Dakota farmer got more than he bargained for when he opened his mail one day in the spring of 1992. He was expecting a farm subsidy payment of $31 from the federal government; instead, he received a check for more than $4 million.

"I had a good laugh over it. And then I took the check back," said Harlen Johnson, a farmer in the northwestern Dakota town of Crosby.

Meanwhile, back at the U.S. Agricultural Stabilization and Conservation Service, which issued the generous check, agency head Dale Ihry commented, "It's just one of those freaky deals that sometimes happens. The $4,038,277.04 amount [of the check] is a number that occasionally pops up on agency records without explanation. Our computer program, for whatever reason, sometimes picks up [this] amount and prints it out on something. This is the first time it made it on a check, though."

Johnson, who was in the middle of seeding his crops when the check arrived, photocopied the check, laminated the copy, and hung it on the wall as a memento. But despite the fact the agency caught the error before Johnson even received the incorrect check, Johnson said he still had not received his $31.

As reported in *Daily News Record*, Harrisonburg, Virginia, May 9, 1992.

SOFTWARE SECURITY AND CONTROLS

Security of data and information is one of the most important aspects of computer security. The old adage, "Don't put all your eggs in one basket," applies to our present systems. People can break into computers to retrieve, change, or destroy data. Every attempt should be made to prevent unauthorized access and to protect the computer from harm. Access to software has to be restricted on a need-to-know basis. With a database, access can be controlled by a matrix that grants authorization of specific data and information. Confidential data should be highly protected so that a user does not "fall into" a file accidentally. All essential programs, software systems, data, and associated documentation should be kept under lock and key. This vital information, the life blood of the organization, should also be backed up with extra copies kept in a separate location so that operations can quickly resume if necessary.

In Chapter 3, access control and other hardware restrictions were introduced. This chapter will concentrate on software, data, and program controls. Hardware and software must function in tandem to achieve the desired objectives of secure systems.

This chapter will discuss information access related to software; types of software intrusions; configuration management; modularity and encapsulation; independent software testing; program accuracy, integrity, and reliability; trusted software; programming standards; testing and acceptance of software; and software designed specifically to protect the security of systems.

Need for Software Protection

The increased use of computers and the computer literacy level of users have heightened public awareness of the sensitive nature of information that is processed and stored in both mainframe and personal computers. The constant exposure by the media of computer violations, breaches, and accidents has

> shattered the illusion that computers are inaccessible to all but the technical experts in charge of machines. At the same time they have created another illusion, that anyone with even a minimum amount of electronic know-how and a telephone can gain access to information stored in the most secure computer system. Both illusions are incorrect, and yet, as with other possessions, the more data that are accumulated, the more there are to protect.[25]

In a collaborative effort in 1991, the Systems Security Study Committee, the Computer Science and Telecommunications Board, the Commission on Physical Sciences, Mathematics and Applications, and the National Research Council offered definitive guidelines on system and software security measures. The Generally Accepted System Security Principles (GSSP) for computer systems presented the idea that" . . .it is possible to enunciate a basic set of security related principles that are so broadly applicable and effective for the design and use of systems that they ought to be a part of any operational requirements."[14] (See Figure 4.1.)

In *Computers at Risk,* published by the National Research Council (1991), this collective group stated:

> Successful GSSP would establish a set of expectations about the requirements for good practice that would be well understood by system developers and security professionals, accepted by government, and recognized by managers and the public as protecting organizational and individual interests against security breaches and lapses in the protection of privacy.[14]

Sources of Software Problems

Threats to software can originate from various sources. (See Figure 4.2.) Hardware violations do exist, but they represent the smallest threat to the overall functioning of the software. Programming bugs and errors introduce a larger threat due to their ability to spread throughout the system. Data entry errors, which represent the biggest threat, are often the easiest to correct by following very simple measures. Only when all functions are running smoothly can we be assured that the software is doing its job.

FIGURE 4.1
Potential Elements of Generally Accepted System Security Principles

☑ Quality Control

A system is safe and secure only to the extent that it can be trusted to provide the functionality it is intended to supply.

☑ Access Control on Code as well as Data

Every system must have the means to control which users can perform operations on which pieces of data and which particular operations are possible.

☑ User Identification and Authentication

Every system must assign an unambiguous identifier to each separate user and must have the means to assure that any user is properly associated with the correct identifier.

☑ Protection of Executable Code

Every system must have the means to ensure that programs cannot be modified or replaced improperly.

☑ Security Logging

Every system must have the means to log for later audit all security-relevant operations on the system.

☑ Security Administrator

All systems must support the concept of a special class of users who are permitted to perform actions that change the security state of the system, such as adding users or installing trusted programs.

☑ Data Encryption

While data encryption is not, in itself, an application-level security requirement, it is currently recognized as the method of choice for protecting communication in distributed systems.

☑ Operational Support Tools

Every system must provide tools to assist the user and the security adminstrator in verifying the security state of the system.

☑ Independent Audit

At some reasonable or regular interval, an independent, unannounced audit of the on-line system, operation, administration, configuration control, and audit records should be invoked by an agency unrelated to that responsible for the system design and/or operations.

☑ Hazard Analysis

A hazard analysis must be done for every safety-critical system.[20]

FIGURE 4.2
Sources of Software Threats and Errors

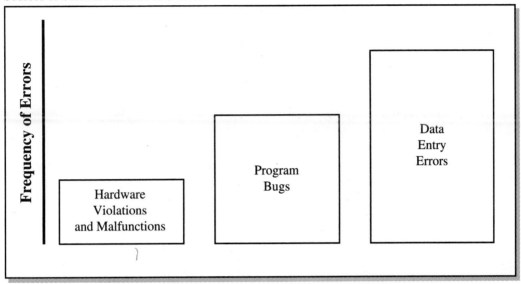

Definition of Software Security

Software security can be defined as the protection of data and programs in a computer system. There are three major elements to be considered:

1. Protection of programs by software
2. Protection of data in on-line systems by software
3. Protection of data in conventional batch processing systems by software

Security by means of software is provided by computer programs designed to protect data and to protect themselves.[25]

TYPES OF SOFTWARE INTRUSIONS

To illustrate the vulnerability of both systems and the accompanying software, consider the impact of some of these recent events relating to software.

Software Publisher's Association Award Denied

The board of the 140-member organization of microcomputer industry developers denied Central Point Software of Portland, Oregon, certification under its gold award honoring best-selling software. Central Point's popular Copy II series was designed to enable users to copy most copy-protected disks. Uneasy board members voted "no" in order to protect the integrity of the certification program. A copy-protection breaking program, according to SPA Executive Director Kenneth A. Wasch, was not the type of program the board considered eligible for the award.[54]

Leaking Private Health Data

The Los Angeles Times reported on November 4, 1989, that during that month, the Santa Barbara, California sheriff's detectives searched the home of a University of California-Santa Barbara student suspected of supplying confidential health care information to a campus newspaper. Detectives removed more than 30 computer disks from the home of the student, who was a former employee of the County Health Care Services.

Health officials sought the search warrant after a reporter from the university's student newspaper, *The Daily Nexus,* began questioning them about the alleged dumping of patients into neighboring counties. Authorities said they suspected the former employee of leaking county documents to the newspaper because he was friendly with a reporter and had placed calls from the county health office to the *Nexus.* The former employee, however, said he was never in possession of the documents in question. He said the seized computer disks contained personal material, including homework, and that he might seek court action for their return.[52]

The Cuckoo's Egg

A highly acclaimed and widely read book by Cliff Stoll brings out all too clearly the potential for harm when software and systems are breached by malicious outsiders. *The Cuckoo's Egg* is a story of computer espionage on an international level. For months, Cliff Stoll, an astrophysicist at Lawrence Berkeley Labs (LBL), stalked an elusive hacker who had invaded computer networks across the nation. A 75-cent discrepancy in an account report piqued Stoll's curiosity as he set out to find the perpetrator. His elaborate and often tedious recording of evidence, which eventually led to the hacker's conviction, is detailed in the book. Each time the hacker broke into another machine, copied another password file, or tried another back door, Stoll was called upon to create yet another unique technique to track the progress.

The espionage detailed in *The Cuckoo's Egg* began with a trivial accounting error and led to a worldwide search to track the invader through exhaustive manual analysis of account records. During the search, Stoll discovered the presence of an unauthorized user called Hunter. Lawrence Berkeley Labs then received a complaint that one of its users was attempting to break into a remote facility. More manual analysis uncovered another discrepancy in accounting records and activity by a user known to be travelling overseas and not likely to be accessing the system. The intrusion was apparently made through an inactive account: because no one was using regularly the account, no one was monitoring its activity. The hacker was able to make repeated attempts to enter through this inactive account without detection. The surprising use of inactive accounts and the sudden use of recent established accounts in the system provided the first clues that a potential unauthorized access was taking place.

In a manual review of records, Stoll discovered that LBL had several guest accounts set up to facilitate access by visitors. The password was the same as the login (guest/guest). The use of accounts with obvious or easily guessed passwords was common throughout the organization. One of the methods the hacker used frequently was simply trying to log in using obvious combinations, such as

ACCOUNT: Guest PASSWORD: Guest, or ACCOUNT: System PASSWORD: Manager. This weak password protection let the hacker repeatedly access the system and move around various networks.

Stoll concluded that the hacker initially entered the system through a guest account and then placed his own programs and files in the system, including a program to grant privileges. Once the privileges were established through a hole in the editor program, the hacker added a new, privileged account to the system and then used it to read files throughout the system, including unsecured files and unprotected scripts. The hacker eventually downloaded password files and was then able to read more accounts and files.

When Stoll realized that the suspicious activities resulted from a hacker who was obtaining superuser privileges (rather than a curious student or prankster), the decision was made that instead of merely blocking the path of the hacker, LBL would track the hacker and accumulate evidence. Stoll developed custom programs and spent many long hours searching for more illicit, real-time use of "unused" accounts. His constant monitoring revealed that a modem port was being used to access LBL, via a telephone line at 1200 baud. Stoll scavenged terminals, printers, teletypes, PCs, and other equipment to set up a way to capture user sessions coming over LBL's lines. Once the hacker had obtained a privileged account, he was able to move throughout the network undetected, relying heavily on his ability to erase his tracks and leave no evidence.

Stoll tracked the hacker as he downloaded a list of additional LBL computers. Once the hacker accessed Internet and the Network Information Center (NIC), he obtained the addresses of five computers at White Sands Missile Range in 20 seconds. He eventually acquired the network addresses of other systems.

After months of painstaking effort by Stoll, the hacker was eventually caught and prosecuted.[46]

The Search Goes On

These three actual intrusions are only a small sampling of the violations occurring worldwide with increased frequency. Systems, software, and information are extremely valuable assets that should be protected from any potential intrusion by applying sound security measures.

CONFIGURATION MANAGEMENT

The computer and all the connected peripheral devices are commonly referred to as a **configuration**. To take the next step to fully utilize this valuable asset, configuration management principles should be applied to systems, software, and information so that security can be achieved. **Configuration management** is the management system of controlling network topology, physical connectivity, and network equipment, and for maintaining supporting data.[20] Included in configuration management is the allocation of transmission lines to various applications and consolidation of low-speed traffic on higher speed circuits for more economical and secure transmission. (See Figure 4.3.)

FIGURE 4.3
Configuration
Management
Concerns

> ☑ Physical connectivity
>   (terminal, modems)
>
> ☑ Network topology
>   (inventory of equipment, physical location, technical capability, intended application)
>
> ☑ Bandwidth allocation
>   (all applications should be checked)
>
> ☑ Equipment inventory
>   (upgrades and additions should be updated)
>
> ☑ Information on users
>   (must be constantly updated and monitored)
>
> ☑ Vendor data
>   (record service calls, upgrades)[20]

Need for Control

"Configuration management is another aspect of software engineering that offers advantages in security."[38] All changes to a program or documentation are carefully monitored and approved by a group of professionals. Configuration management's primary goal is to guarantee availability and usage of the correct version of all system components—software, design documents, documentation, and control files.

The three general purposes of configuration management, according to Charles P. Pfleeger in *Security in Computing*, are as follows:

1. To guard against inadvertent loss (deletion) of a version of a program
2. To manage the parallel development of several similar versions of one program
3. To provide facilities for controlled sharing of modules that combine to form one system[38]

The primary security advantages of using configuration management, according to Pfleeger, are:

- Protection against unintentional threats
- Protection against malicious intrusions

Trusted Systems

The **Trusted Computer System Evaluation Criteria (TCSEC)** is the standard used for evaluating the effectiveness of security controls in federal government computer systems. The TCSEC is divided into four categories—D, C, B, and A—

ordered in a hierarchical manner with the highest division, A, reserved for systems providing the best available security.

TCSEC requires that all changes to the **Trusted Computing Base (TCB)** be controlled by configuration management. This requirement consists of identifying, controlling, accounting for, and auditing all changes made to the TCB during its development, maintenance, and design. "The TCB is defined as the totality of protection mechanisms within a computer system—including hardware, firmware, and software—the combination of which is responsible for enforcing a security policy."[2]

The TCSEC, under the guidelines offered by the National Computer Security Center at Fort Meade, Maryland, gives the following as the assurance control objective:

> Systems that are used to process or handle classified or other sensitive information must be designed to guarantee correct and accurate interpretation of the security policy and must not distort the intent of that policy. Reassurance must be provided that correct implementation and operation of the policy exists throughout the system's life cycle.[2]

Configuration management maintains control of a system throughout its life cycle, ensuring that the system in operation is the correct system, implementing the correct security policy. The assurance control objective, as it relates to configuration management, leads to the following control objective that may be applied to configuration management:

> Computer systems that process and store sensitive or classified information depend on the hardware and software to protect that information. It follows that the hardware and software themselves must be protected against unauthorized changes that could cause protection mechanisms to malfunction or be bypassed completely. (For this reason, changes to trusted computer systems, during their entire life cycle, must be carefully considered and controlled to ensure that the integrity of the protection mechanism is maintained.) Only in this way can confidence be provided that the hardware and software interpretation of the security policy is maintained accurately and without distortion.[2]

Principles to Be Applied

Configuration management consists of four separate tasks:

1. Identification
2. Control
3. Status accounting
4. Auditing

For every change made to a system, the design and requirements of the changed version of the system should be identified. The control task of configuration management is performed by subjecting *every* change to documentation, hardware, and software/firmware to review and approval by an authorized person or committee. Configuration status accounting is responsible for recording and reporting on the configuration of the product throughout the change. Finally, through the process of a configuration audit, the completed change can be verified to be functionally correct and, for trusted systems, consistent with the security policy of the system.

Procedures should be established and thoroughly documented by a configuration management plan to ensure that configuration management is performed in a specified manner. *Any* deviation from the configuration management plan could contribute to the failure of the configuration management of a system entirely, as well as the trust placed in a trusted system. (See Figure 4.4.)

FIGURE 4.4
Configuration
Management
Guidelines

> The following is a sample list of what shall be identified and maintained under configuration management:
>
> ☑ Hardware, software, and firmware
> ☑ *Any* changes to the hardware, software, and firmware since the previous baseline
> ☑ Design and user documentation
> ☑ Software tests, including functional and system integrity tests
> ☑ Tools used for generating current configuration items

Configuration management is indeed a lofty goal for a system administrator, but it is necessary if systems security is to be achieved. In the following sections, various methods and operations will be detailed to illustrate how configuration management is fully realized.

MODULARITY AND ENCAPSULATION

Several methods can be used to separate software (and programs) into segments. Programmers and users do not have access to all segments, so they are unable to make significant changes to the program. Typically, only the systems administrator or chief systems analyst can see "all the pieces of the puzzle." Separation of duties ensures that control of the total program is available only to a small, select group of individuals.

The two primary methods used are:

1. Modularity
2. Encapsulation

Modules

The principle of writing programs in small, self-contained units (or modules) is called **modulation** or **modularity**.

By dividing the task (or program) into small units (subtasks), each module performs a distinctively independent part of the overall task. Modularity offers many advantages for logical program development, following structured guidelines, as well as definite security advantages. "A module can be isolated from the negative effects of other programs with which it interacts."[38] Information can be hidden, using modules, where ". . . other modules know that a module performs a certain task, but not *how* it performs that task."[38] (See Figure 4.5.)

FIGURE 4.5
Principle of
Modularity

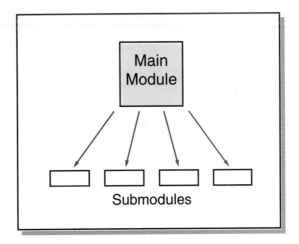

Program units (segments) should be designed in such a way that each unit is only as large as needed to perform its separate function. The advantages of writing a series of small modules are as follows:

- **Maintainability.** If revisions are necessary, only one unit needs to be changed. Because the connections to other units are negligible, the "spillover" effect should be nonexistent.

- **Simplicity.** A series of small units will be easier to understand than one, large, unstructured program.

- **Reusability.** Modules (units) developed for one purpose can often be reused in other programs without any threat to the security of the overall program. Additionally, reuse of correct, existing program modules can significantly reduce the difficulty of programming and testing.

- **Correctness.** Errors can be quickly traced to the cause or source because each module performs only one task.

- **Testworthiness.** Each module should contain well-defined inputs, outputs, and functions. Each operation can be thoroughly tested in isolation, without concern for its effect on other modules.

Integration between modules is nonexistent (or minimal), so a high level of independence can be achieved and maintained. Even though this modulation function may appear to stifle a programmer's creativity, the benefits of independence far outweigh its disadvantages.[38]

Capsules

A slight deviation from the principles guiding modules is to develop programs in small capsules. **Encapsulation** is the process of each module operating as if it were surrounded by a shield that prevents unwanted access from the outside.[38] With encapsulation, there is very little coupling with other routines in the same program. Modules interact only through well-defined interfaces, and each module is entered only at specified entry points, Ideally, a module interacts with as few other modules as possible. (See Figure 4.6.)

FIGURE 4.6
Encapsulation
Design

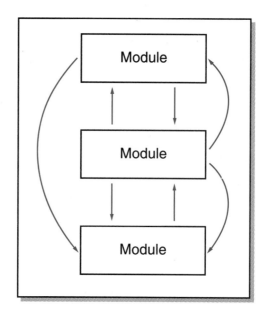

Cautions of Modules and Capsules

"It is not uncommon for a hundred program modules to be invoked from the time a single transaction is entered into a terminal device until a response is returned to the device."[25] The major trade-off in planning security measures is that caution must be exercised so that, in making data and software secure, they are not made inaccessible.

> Technically-oriented individuals are inclined, because of their knowledge and capabilities, to develop complex security procedures that are difficult to breach. This can be counter-productive, especially in emergency situations where it may be necessary to bypass security. The establishment of security procedures involves trade-offs; they should be able to keep sensitive data reasonably secure at reasonable cost.[25]

PROTECTING INFORMATION

All too often, stringent measures are instituted to protect hardware and software, yet, the data that feeds into the system is not closely controlled. Any erroneous or falsified data that enters the system will result in output that is misleading or unusable. The classic responses of "the computer made a mistake" or "it must have been a programming error or machine malfunction" are all too often used to mask the real problem: bad data produces bad output—garbage in, garbage out.

Some very useful measures should be instituted to protect data as it feeds into the system, is processed, and becomes information when output.

To protect the reliability, accuracy, and integrity of data, stringent, ongoing measures should be taken that include automatic checks built in to the system for constant vigilance. (See Figure 4.7.)

FIGURE 4.7
Information
Protection Measures

Is the data and information:

☑ Authorized to be processed?

☑ Accessible by the user?

☑ Incomplete?

☑ Inaccurate or unverifiable?

☑ Not subject to any error checks?

Information Integrity

Procedures should be applied that will compare what was processed against what was supposed to be processed. For example, controls can compare totals or check sequence numbers. In other words, was the *right* operation performed on the *right* data?

Information Accuracy

To check input accuracy, data validation and verification checks should be able to perform the following tasks:

- **Character checks** compare input characters against the expected type of character (e.g., numeric or alpha).
- **Range checks** check input data against predetermined upper and lower limits.
- **Relationship checks** compare input data to data on a master record file.
- **Reasonableness checks** compare input data to an expected standard.
- **Transaction limits** check input data against administratively set ceilings on specified transactions.

In addition, *all* transactions should be traced through the system using transaction lists. Cross-checks of the contents of all files should be performed by record counts or control totals.

Software Protection

If software is shared, protect it from undetected modification by ensuring that policies, developmental controls, and life cycle controls are in place, and that users are educated in security policies and procedures.

Software developmental controls and policies should include procedures for changing, accepting, and testing software *prior to* implementation. Policies should require management approval for any software changes, limit who can make software changes, and address maintaining documentation.

An up-to-date inventory of software applications should be developed, maintained, and periodically reviewed. Stringent controls should be installed that will prevent unauthorized persons from obtaining, altering, or adding programs by using remote terminals.

Bad Data Means Bad Output

Erroneous or falsified input data is the simplest and most common cause of undesirable performance in an applications system. Vulnerabilities occur whenever data is collected, processed (via manual or automated steps), or prepared for entry into the computer.[22] (See Figure 4.8.)

FIGURE 4.8
Examples of Erroneous or Falsified Data

☑ Unreasonable or inconsistent source data values.

☑ Keying errors.

☑ Incomplete or poorly formatted data records.

☑ Records in one format interpreted according to a different format.

☑ An employee fraudulently adds, deletes, or modifies data (e.g., payment vouchers, claims) to obtain benefits.

☑ Lack of document counts and other controls over source data or input transactions allow some of the data or transactions to be lost without detection (or allow extra records to be added).

☑ Records about personnel modified during data entry.

☑ Data which arrives at the last minute (or under some other special or emergency condition) not verified prior to processing.

☑ Records in which errors have been detected are corrected without verification of the full record.[22]

Misuse by Unauthorized Users

End users are people who are served by the computer system. Even though the system is designed for their use, people can misuse the system for undesirable purposes. "It is often difficult to determine whether their use of the system is in accordance with the legitimate performance of their job."[22] (See Figure 4.9.)

Cases in Point

"Most computer crimes are not committed by hackers but by trusted employees—programmers, managers, clerks, and consultants—who turn against their employers, using company computers for extortion, theft, and sabotage."[4]

FIGURE 4.9

Examples of Misuse of Data by Unauthorized End Users

☑ An employee may convert information to an unauthorized use (e.g., selling privileged information to credit agency, insurance company, or competitor).

☑ An individual could use statistics for stock market transactions before public release of the information.

☑ A user whose job requires access to individual records in a file may manage to compile a complete listing of the file for unauthorized use (e.g., selling a listing of employees' home addresses and telephone numbers).

☑ An unauthorized user may use the system for personal benefit (e.g., running a small business).

☑ A supervisor may manage to approve and enter a fraudulent transaction.

☑ A disgruntled or terminated employee may destroy or modify records—possibly destroying backup records in the process.

☑ An authorized user may accept a bribe to modify or obtain information.[22]

Allen Green, a clerk at Girard Bank in Philadelphia, was hired to scan computer printouts for signs of suspicious automatic teller machine (ATM) transactions. His department also received automatic teller machine cards that the Post Office couldn't deliver. According to police, Green made a fake deposit to the account of a man whose card was returned. He then raised the withdrawal limit, made an actual withdrawal of $4,500, and later returned the withdrawal limit to its original $200 ceiling. He repeated these maneuvers three times before an audit program—in a sense, a backup to his own job—finally caught him.

In Washington, Stanley Slyngstad was a programmer who developed software through which the state authorized payments to injured loggers. Slyngstad lost an arm in a childhood accident, and, according to his supervisor, was all the more trusted because of his handicap. However, he used his program to authorize $17,000 in payments to himself and two friends, then erased all record of the fraud. The daughter of one of his friends tipped police to the scheme, and Slyngstad was arrested.

Dennis Williams and Michael Lampert, unhappy with the management at Collins Foods, were accused by police of placing "logic bombs" in the company's two computers. The bombs were set to activate in the future and destroy the operating systems. The company was tipped off by a worker who overheard Williams discussing the scheme. Fortunately, both bombs were deactivated before causing any damage.

An employee of Micro Porcelain Dental Laboratories tampered with the company computer so that it couldn't be started without his help. That condition would cost the company $573 in vacation pay he claimed he was owed. Police filmed the request for money and, with that as evidence, arrested the employee.[4]

Auditing of Systems

In all too many instances, employees are able to circumvent normal operations through a flaw in the software controlling the system. A fully functioning internal audit may lessen the possibility of penetration and misuse in these cases.

The U.S. federal government, describing its audit objective as the Trusted Computer System Evaluation Criteria, gives the following as the **Accountability Control Objective**:

> Systems that are used to process or handle classified or other sensitive information must assure individual accountability whenever either a mandatory or discretionary security policy is invoked. Furthermore, to assure accountability, the capability must exist for an authorized and competent agent to access and evaluate accountability information in a secure means, within a reasonable amount of time and without undue difficulty.[1]

The Accountability Control Objective, as it relates to auditing, leads to the following control objective for auditing:

> A trusted computer system must provide authorized personnel with the ability to audit any action that can potentially cause access to, generation of, or effect the release of classified or sensitive information. The audit data will be selectively acquired based on the auditing needs of a particular installation and/or application. However, there must be sufficient granularity in the audit data to support tracing the auditable events to a specific individual who has taken the actions or on whose behalf the actions are taken.[1]

THE ORANGE BOOK

If you are at all interested in computer security, you will need to know something about the Orange Book, which is the Department of Defense Standard as outlined in *Department of Defense Trusted Computer System Evaluation Criteria,* DOD 5200.28-STD; Library of Congress Number 5225.711, December 1985. This book is part of the Rainbow Series discussed in Chapter 3.[51]

The Orange Book defines four broad hierarchical divisions of security protection. In increasing order of trust they are:

- D—Minimal security
- C—Discretionary protection
- B—Mandatory protection
- A—Verified protection

The purpose of the Orange Book is to achieve the following objectives:

1. **Measurement.** To provide users with a metric with which to assess the degree of trust that can be placed in computer systems for the secure processing of classified or other sensitive information.

2. **Guidance.** To provide guidance to manufacturers as to what to build in to their trusted commercial products to satisfy trust requirements for sensitive applications.

3. **Acquisition.** To provide a basis for specifying security requirements in acquisition specifications. . . . A customer can be confident that the system acquired has already been checked out for the needed degree of security.

System Requirements

The fundamental computer security requirements detailed in the Orange Book outline six basic requirements:

1. **Security Policy.** There must be an explicit and well-defined security policy enforced by the system.

2. **Marking.** Access control labels must be associated with objects.

3. **Identification.** Individual subjects must be identified.

4. **Accountability.** Audit information must be selectively kept and protected so that actions affecting security can be traced to the responsible party.

5. **Assurance.** The computer system must contain hardware/software mechanisms that can be independently evaluated to provide sufficient assurance that the system enforces requirements 1–4.

6. **Continuous Protection.** The trusted mechanisms that enforce these basic requirements must be continuously protected against tampering and/or unauthorized changes.[51]

Disagreements and Comments

Some respected security practitioners disagree with the Orange Book specifics and with the government's reliance on this book alone as a way of measuring trust. Some of the complaints include these comments:

- Model does not work in industry environments
- Secrecy is the only aspect of security considered
- Too much emphasis on unauthorized access from outsiders
- Networking issues are not addressed
- The number of security ratings is limited

SELECTING SECURITY SOFTWARE

In the past few years, the importance of implementing data security software in major installations has become increasingly apparent. Today, the objective is to decide *which* data security package will best meet our needs, not *whether* or not we should use one.

The following precautions should be followed as guidelines to security implementation:

- A company's data is not secure merely because a security system is installed.
- The process of security implementation is a long, complicated process and requires a great deal of support and effort.

- Security should be the responsibility of all people creating and using information—not just the security administration staff.
- Maintenance of the security system is an ongoing process.

Criteria Used in Selection of Security Software

When selecting and installing a security software package, the following general guidelines should be considered:

- The computer environment should always be protected when it is available for use.
- When security software is disabled in any way, notification should be given to the appropriate parties immediately.
- Bypassing security procedures should be restricted.
- Systems resources should be properly protected by the security administration personnel.
- Vendor-supplied system software should be implemented only by security administration personnel.

Meeting Objectives

After the functional and operational requirements that the software must meet are established, the user must establish the technical area for selection.[18] Regardless of the system or information to be protected, the selection of a security software package must be based on its ability to support certain administrative objectives, such as:

- **Accountability.** Users are responsible for their actions.
- **Auditability.** Audit trails should be a routine event to show access and use.
- **Integrity.** Circumvention of the system should be closely monitored.
- **Usability.** Human factors and cost should be practical and non-restrictive.

The basic requirements of a security software package are a function of the specific hardware, software, and data communications environment. Critical to this process and desired outcome is the proper definition of the environment. A thorough definition should include key factors, such as:

- Both hardware and software *compatibility* should be evaluated
- The selected software should use as little overhead expense as possible
- The system and user documentation supplied by the vendor must be thorough and easily understood

ANALYSIS OF SOFTWARE PRODUCTS

The dominant access control software products for the IBM and Honeywell operating systems environment are:

- RACF
- ACF2

Dr. G. Robert Redinbo
Dept. of Electrical and
Computer Engineering
University of California
Davis, California 95616

- Top Secret
- Omniguard and Omniguard/CICS
- MULTICS

RACF

IBM's Resource Access Control Facility (RACF) is designed for use in large-scale IBM systems to control access to system resources through the employment of user identification codes and passwords. **RACF** identifies, verifies, and limits the user to the authorized resources. User resources interactions are recorded and reported, and RACF permits both protected and unprotected resources to coexist.

ACF2

Offered by Cambridge Systems Group, Inc., Access Control Facility (ACF) is designed to provide data security to computers using IBM OS/MVS or OS/VS1 operating systems. A VM version is also available. Resources are protected by default as all access to data or resources is denied unless specifically permitted by a given user or user group.

ACF2 is a "resource-oriented" product that looks at security from the perspective of which resource needs protection and then identifies the users who have access to it. For example, if a user defines the security to be placed on personnel system files, ACF2 would start an identification of all personnel files. Once the files have been identified, ACF2 needs to be able to identify the users who should have access to the personnel files.

Top Secret

Developed and marketed by Software Products Group, **Top Secret** provides logical protection of resources and facilities under control of IBM's OS/VMS by validating resource access.

Top Secret looks at access security by starting with controls over a user or group of users and then maintains information about a user or group of users and calls that collection of information an "accessory id," or "acid."

Omniguard

Designed for use in large IBM systems and IBM CICS subsystems, **Omniguard** is a complete host access system that is offered for different versions of IBM operating systems. Control of access to resources is handled by default.

Omniguard/CICS is a special version of the software package designed to protect only the IBM CICS environment as CICS lacks adequate security, good audit facilities, and limitations on user access to transactions and programs. Omniguard/CICS (formerly known as Guardian) can function as an independent CICS security system, or it can run under global packages such as ACF2 or Top Secret.

MULTICS

Multiplexed Information and Computing Service (MULTICS) was designed for use on Honeywell systems. **MULTICS** is an integrated security policy and is enforced by software and hardware. MULTICS protects against unauthorized use by these mechanisms:

1. Access Control List for discretionary access
2. Access Isolation Mechanism for nondiscretionary access, activated for highly sensitive information
3. Ring Protection Mechanism to control intraprocess access

MULTICS is a total system with high-level security incorporated into the design of hardware and software.

SUMMARY

Ensuring the integrity of complex computer operations and the software used is a difficult but essential task in today's increasingly challenging (and threatening) environment. Sophisticated operating systems and support software often involve a complex array of program products, provided from a wide range of sources. These products must be integrated into a computer environment without impairing security or integrity.

Increasingly, technically knowledgeable thieves or system abusers are turning to sophisticated attacks against computers and their support software. Individuals' demonstrated ability to gain unauthorized access to privileged systems and thus seize control of the systems can have a catastrophic effect on any organization.

Some of the principles of generally accepted system security applied today to protect software include quality control, access control, user identification and authentication, protection of executable code, security logging, security administrators, data encryption, operation support tools, independent audits, and hazard analysis. The sources of software threats and errors originate from hardware violations and malfunctions, program bugs, and data entry errors.

Configuration management includes the computer and all the connected peripheral devices and is designed to control all systems and supporting data.

Trusted Computer System Evaluation Criteria (TCSEC) is the standard used to evaluate the effectiveness of security controls built in federal government computer systems.

Modularity and **encapsulation** are ways of writing programs in smaller units for greater protection.

The data and information that feeds into the system must be protected as thoroughly as the software and hardware. The integrity and accuracy of the data and information should be verified constantly to ensure a reliable outcome from the computer operation.

The Orange Book provides stringent guidelines for federal computers and thoroughly outlines operations of trusted systems.

Many security software programs currently available will provide protection of computer systems. Included in this growing offering of software are **RACF**, **ACF2**, **Top Secret**, **Omniguard**, and **MULTICS**.

Only when hardware, software, data, and information are secure and working in sync can administrators of computer systems feel reasonably safe from misuse and error.

REFERENCES

1. *A Guide to Understanding Audit in Trusted Systems*, National Computer Security Center, Fort George G. Meade, Maryland, NSCC-TG-001, Version-2, June 1, 1988, pp. 1-25.

2. *A Guide to Understanding Configuration Management in Trusted Systems*, National Computer Security Center, Fort George G. Meade, Maryland, NCSC-TG-006, Version-1, March 28, 1988, pp. 1-31.

3. *A Guide to Understanding Discretionary Access Control in Trusted Systems*, National Computer Security Center, Fort George G. Meade, Maryland, NCSC-TG-003, Version-1, September 30, 1987.

4. Ahl, David H., "Employee Computer Crime on the Rise," *Creative Computing*, June 1985, p. 6.

5. Baker, Richard H., *The Computer Security Handbook* (Blue Ridge Summit, PA: TAB Professional and Reference Books, 1985).

6. Bell, Jack, "Beware of 'Snuffware'," *Personal Computing*, February 1986, p. 85.

7. Biba, K.J., *Integrity Considerations for Secure Computer Systems*, Electronic Systems Division, U.S. Air Force, Hanscom Field, Bedford, Massachusetts, ESD-TR-76-372, 1977. (Available from NTIS: AD-771543. National Technical Information Service, U.S. Department of Commerce, 5285 Port Royal Road, Springfield, Virginia 22161).

8. BloomBecker, J.J., Buck, J.D., *Commitment to Security*, National Center for Computer Crime Data, Santa Cruz, California, 1989.

9. Branstad, Dennis K., "The Federal Password Usage Standard," *Data Security Management*, May 1985, pp. 1-20.

10. Brock, John L., "Hackers Penetrate DOD Computer Systems," Testimony Before the Subcommittee on Government Information and Regulation, Committee on Governmental Affairs, United States Senate, Wednesday, November 20, 1991. Published by United States General Accounting Office, GAO/T-IMTEC-92-5.

11. Bucken, Mike, "Bank Finds Security Offerings Hacking," *Software Magazine*, February 1989, p. 46.

12. Chalmers, Leslie S., "Managing the Implementation of a Security Software Package," *Data Security Management*, June 1984, pp. 1-8.

13. Clark, D.D., and D.R. Wilson, "A Comparison of Commercial and Military Computer Security Policies," *Proceedings of the 1987 Symposium on Security and Privacy*, IEEE Computer Society, Washington, D.C., 1987, pp. 184-195.

14. *Computers at Risk . . . Safe Computing in the Information Age* (Washington, DC: National Academy Press, 1991).

15. *Considerations for Success . . . Security Software*, CGA Computers, Inc., Software Products Group, Holmdel, New Jersey, 1984, pp. 1-85.

16. Deans, P. Candace, and Michael J. Kane, *Information Systems and Technology* (Boston, MA: PWS-Kent Publishing Company, 1992).

17. Dietz, Lawrence D., "ACF2—Concepts and Implementation," *Data Security Management*, No. 84-02-09, 1984, pp. 1-12.

18. Edwards, Robert W., and Robert E. Johnston, "Selecting Data Security Software," *Data Security Management*, No. 84-02-01, 1984, pp. 1-12.

19. Enger, Norman L., and Paul W. Howerton, *Computer Security . . . A Management Audit Approach* (New York: AMACOM, A Division of American Management Associations, 1980).

20. Frenzel, Carrol W., *Management of Information Technology* (Boston, MA: boyd & fraser publishing company, 1992).

21. Garfinkel, Simson, and Gene Spafford, *Practical UNIX Security*, O'Reilly and Associates, Inc., Sebastopol, Califorinia, 1991.

22. *Guideline for Automatic Data Processing Risk Analysis*, FIPS Pub. 65, Federal Information Processing Standards Publication, U.S. Department of Commerce, National Technical Information Service, Springfield, Virginia, August 1979.

23. Hoheb, Albert C., "Integrating Computer Security and Software Safety in the Life Cycle of Air Force Systems," *Proceedings of the 13th National Computer Security Conference*, Washington, D.C., October 1-4, 1990, pp. 515-525.

24. "How to Detect Software Theft," *Computer Control Quarterly*, Vol. 9, No. 4, 1991, pp. 6-7.

25. Hutt, Arthur, Seymour Bosworth, and Douglas B. Hoyt, *Computer Security Handbook* (New York: Macmillan, 1988).

26. Kurzban, Stanley A., "Making MVS/XA Secure," *Data Security Management*, August 1985, pp. 1-8.

27. Lewis, Barry, "Things to Consider for an Ideal Security Match," *Software Magazine*, February 1989, pp. 43-45.

28. Marcus, Alan I., and Howard P. Segal, *Technology in America . . . A Brief History* (New York: Harcourt, Brace, Jovanovich Publishers, 1989).

29. Margulies, Benson I., "Security in a MULTICS Environment," *Data Security Management*, September 1985, pp. 1-16.

30. *Microcomputer Applications,* 3167 E. Otero Circle, Littleton, Colorado, 80122.

31. Murray, William H., "Security in a VM Environment," *Data Security Management*, No. 84-04-03, 1985, pp. 1-12.

32. Oelrich, Patricia A., "Control Aspects of Program Library Management Software," *Data Security Management*, June 1985, pp. 1-8.

33. Parker, Donn B., *Computer Security Management* (Reston, VA: Reston Publishing Company, Inc., 1981).

34. Parker, Donn B., "20 Principles for Selecting Information Safeguards," *Data Security Management*, No. 81-01-03, 1985, pp. 1-12.

35. *Password Management Guideline*, Department of Defense, CSC STD-002-85, April 12, 1985.

36. Penzias, Arno, *Ideas and Information: Managing in a High-Tech World* (New York: W.W. Norton and Company, 1989).

37. Peterson, A. Radgett, "Disk Compression," *Virus-L Bulletin Board,* March 10, 1992, and March 16, 1992.

38. Pfleeger, Charles P., *Security in Computing* (Englewood Cliffs, NJ: Prentice Hall, 1989).

39. Russell, Deborah, and G.T. Gangemi, Sr., *Computer Security Basics*, (Sebastopol, CA: O'Reilly and Associates, Inc., Sebastopol, 1991).

40. Saydjari, O. Sami, Joseph M. Beckman, and Jeffrey Leaman, "Locking Computers Securely," *Proceedings of the 10th National Computer Security Conference*, September 21-24, 1987, pp. 129-141.

41. Scannell, Ed, "Apprentice 2.0 Adds Security, Archiving, Compression Features," *Infoworld*, October 21, 1991, p. 19.

42. Schweitzer, James A., "Automated Logical Access Control," *Data Security Management*, No. 84-02-20,1986, pp. 1-8.

43. Sciullo, Francis D., "Defining Access Control Requirements," *Data Security Management*, May 1984, pp. 1-8.

44. Sibert, W. Olin, Deborah D. Downs, Holly Traxler, and Jeffrey J. Glass, "UNIX and B2: Are They Compatible?" *Proceedings of the 10th National Computer Security Conference,* September 21-24, 1987, pp. 142-149.

45. Sieber, Ulrich, *The International Handbook on Computer Crime* (New York: John Wiley and Sons, Inc., 1986).

46. Stoll, Cliff, *The Cuckoo's Egg* (New York: Pocket Books, 1990).

47. "System Security," Advanced Information Management, Inc., Woodbridge, Virginia.

48. *The Handbook of Information Security*, Arca Systems, Inc., San Jose, California, 1991.

49. Theofanos, Mary Frances, "A Systematic Approach to Software Security Evaluations," *Proceedings of the 12th National Computer Security Conference*, Baltimore, Maryland, October 10-13, 1989, pp. 423-432.

50. Tompkins, Frederick G., and Russell Rice, "Integrating Security Activities into the Software Development Life Cycle and the Software Quality Assurance Process," *Advances in Computer System Security*, edited by Rein Turn (Norwood, MA: Artech House, Inc., 1988), pp. 26-50.

51. *Trusted Computer System Evaluation Criteria*, Department of Defense, DOD 5200 (28-STD), Library Number 5225, 711, December 1985.

52. "UCSB Student Suspected of Leaking Private Health Data," *The Los Angeles Times*, November 4, 1989, p. A27.

53. Verducci, Frank M., "Implementing RACF—A Case Study Revisited," *Data Security Management*, No. 84-02-13, 1984, pp. 1-12.

54. Wilson, Geoffrey A., "Security in a UNIX System Environment," *Data Security Management*, October 1985, pp. 1-16.

REVIEW QUESTIONS

1. What are some of the basic elements constituting system security?

2. What are some of the threats to software?

3. Briefly describe the purpose of configuration management.

4. What is a Trusted Computer System?

5. Briefly describe the principle of writing programs using modularity.

6. Briefly describe the principle of writing programs using encapsulation.

7. Contrast information integrity and accuracy.

8. How is the Orange Book used to ensure Trusted Systems evaluation?

9. What are some of the guidelines that should be used to select security software?

10. What are some of the security software products currently being used?

DISCUSSION QUESTIONS

1. Software access control packages such as ACF2, Top Secret, and RACF can provide MIS departments with an effective tool for controlling the use of sys-

tems and data. To be successful, however, these packages must be tailored to each environment. What would some of the tailoring factors be?

2. To keep security software current, the personnel department must follow a stringent policy of always notifying the computer operation area when personnel changes take place. How could this notification process be automated (built into the system) to ensure 100% accuracy?

3. Many times, user IDs are tied to functions or projects, rather than to individuals. When would this policy be useful? What are some of the pros and cons of the approach?

4. Computer pirates are scorned and glamorized—at the same time. Could there be a "good pirate"? Defend your approach.

5. Is the federal government's involvement in setting standards a type of "Big Brother is watching you" intrusion?

EXERCISES

1. Develop a checklist of 10 items that should be considered when selecting security software.

2. Select an organization/company and interview an employee who works closely with security software packages to gain insight into how these packages function.

3. Locate an article about a security software package (e.g., RACF, ACF2) and write a short review of the article.

4. Locate articles about ACF2 and RACF, and then develop a comparative profile of these two packages. Your profile should include:

   ▪ Strengths and weaknesses

   ▪ Cost

   ▪ Ease of use

   ▪ Training of users

   ▪ Documentation

   ▪ Dependability

5. Discuss the following statement:

   After establishing the functional and operational requirements that the software must meet, the users must establish the technical criteria for selection. These requirements are a function of the specific hardware, software, and data communications environment. The proper definition of this environment is critical to the selection of an effective security software package.

PROBLEM-SOLVING EXERCISES

1. If employees find it rather simple to disable a security software package, some may do so to improve processing times or eliminate the "bother," or

for malicious reasons. Develop a statement outlining to employees your company's policy on this matter.

2. In a functional comparison of RACF and ACF2, data set protection is defined as:

   RACF: Protects by exception. Each data set is defined by a manual transaction or by using the Automatic Data Set Protection feature.

   ACF2: Protects by default. All data set accesses are monitored. A user must take action to "unprotect" his or her data.

   Analyze each of these packages by comparing these two methods.

3. To ensure that security software packages are fully functional, the proposed security software should also be compatible with the installation's applications systems. The following areas should be considered:

   a. Data structures

   b. Processing routines

   c. User access

   Fully outline a policy on each of these three areas that could be used for compatibility.

4. Any new installation or application must be evaluated for its cost/benefit ratio. What would be unique about this ratio when applying it to security software packages?

5. Security software packages are used to guarantee the following three functions:

   a. Completeness

   b. Isolation

   c. Correctness

   Fully describe each function.

CASES

Slayer Enterprises

Slayer Enterprises, located in Seattle, Washington, is a medium-sized business specializing in building concrete structures (storage buildings, bridges, utility buildings, and various defense-related installations).

Peter Slayer, president of the company, states that security of its computer systems is critical due to the competitiveness of the commercial concrete industry and the need to protect defense-related information. SE's computer system, as is typical of many, grew on an ad hoc basis as the company grew. Security was not an issue in the early days when there were only a few key employees and the day-to-day, hands-on management style assured constant vigilance. Mr. Slayer now feels that a solid security package should be added to ensure the ongoing protection of the computer operation.

Mr. Slayer has assigned the task of evaluating several security packages to the director of computer operations, James Shifflett. The four objectives that Mr. Shifflett has defined for the package selected are:

1. Accountability
2. Auditability
3. Integrity
4. Usability

What factors should be included in an official policy statement sent out with the request for proposal to the vendors? Write a clear and concise statement so that the vendor is able to respond correctly.

Pierson Advertising Agency

Pierson Advertising Agency's chief of computer operations, Lynn Pierson, has recently purchased RACF to protect the system. One of his primary considerations is controlling the assignment and use of user IDs.

Before RACF's purchase, user IDs consisted of the first six characters of the user's last name plus a number to ensure uniqueness. RACF, however, controls access by maintaining an access list of authorized user IDs. This access list, together with additional statistics, is called a profile. A RACF profile exists for every protected user.

To avoid a potential administrative burden resulting from changes in user responsibilities, a new system of assigning user IDs had to be developed. User IDs should be tied to a function as well as to a person. When a user is transferred and changes responsibilities, the user ID should not have to be removed from all former user profiles and added to new ones.

The user IDs were converted to RACF user IDs. The first character is alphabetic and represents the type of employee (e.g., programmer, systems programmer, end user, manager). The next two characters are the department number, followed by a two-character section number. The last two characters are a unique identification number.

What are some of the advantages and disadvantages of this new password scheme? If the company continues to grow and new employees are added, can this present system keep pace?

ENCRYPTION TECHNIQUES

LEARNING OBJECTIVES

After studying this chapter, you will be able to:

1. Define the various terminologies used for encryption techniques.
2. Relate the historical perspective of cryptology to its present use.
3. Compare the different types of ciphers—substitution and transposition.
4. Describe how algorithms aid in the encryption process.
5. Describe several different, simple encryption techniques.
6. Explain how keys are used in the ciphering process.
7. Compare public keys, private keys, and one-time pads.
8. Consider the historical and present status of the Data Encryption Standard.
9. List several points as guidelines when using encryption.
10. Consider the cost vs. benefit ratio of encryption.
11. Define the following terms.

TERMS

- Cryptography
- Cryptology
- Encryption
- Cipher
- Code
- Encipherment
- Encoding
- Cryptoanalysis
- Substitution
- Transposition
- Plaintext

- Ciphertext
- Encryption Key
- Decryption Key
- Encryption Algorithm
- Decryption Algorithm
- Vernam Technique
- Vigenere Cipher
- Private Key
- Public Key
- One-Time Pad
- Data Encryption Standard (DES)

NEW XEROX ENCRYPTION DEVICE FOR NETWORKS

The Xerox Encryption Unit (XEU), developed by Xerox Corporation, represents a new step in assuring security on local-area networks (LANs). The XEU is installed between a workstation or PC and the network via cabling plugged into a protected panel on the back of the XEU. Transparent to the workstation and the user, the XEU is activated by inserting an electronic key into a slot on the front of the unit. Once activated, the XEU encodes data as it passes from the workstation to the network and decodes data it receives.

According to Xerox Integrated Systems president Jay Nussbaum, the XEU allows network users to send both classified and unclassified information on the same LAN, thus eliminating the need for and expense of two separate networks.

The XEU device is relatively small—3 x 10 x 12 inches—and encased in a plastic housing. If an intruder tampers with the unit, built-in circuitry detects the attack and immediately clears all keying information, then sounds an alarm and causes a red light on the unit to blink.

Xerox says the XEU is the first LAN product developed under the U.S. government's commercial Comsec endorsement program that meets federal standards for data protection. This is especially important to the National Security Agency (NSA), which deals with highly classified and sensitive information.

As reported in *Federal Computer Week,* February 26, 1990, p. 19.

ENCRYPTION OVERVIEW

Now that we have studied hardware and software controls, now is a logical time to consider one of the most time-tested means of protecting transmittal of information—encryption. This chapter will cover the basics of encryption; we will follow up with a thorough discussion of data transmission encryption in Chapter 7, which covers network and telecommunications security.

Encryption Defined

Different experts define encryption differently, but they all agree that encryption is a method of scrambling data in some manner during transmission.

Some definitions are:

Cryptography is the art and science of secret communications, embracing all kinds, including codes and signaling. **Cryptology** deals almost exclusively with letters and words, and can include such secret communications as those involving the telephone, facsimile by wire and radio, and certain aspects of television.[2]

Cryptography is the art of keeping data secret, primarily through the use of mathematical or logical functions that transform intelligible data into seemingly unintelligible data and back again. Cryptography is probably the most important aspect of communications security and is becoming increasingly important as a basic building block for computer security.[11]

Encryption transforms original information into transformed information."[43]

Encryption is a process of encoding a message so that the meaning of the message is not obvious.[40]

Russell Kay, editor of *ISP News*, states that "perhaps no aspect of information security is more technical, more obscure, or more arcane, than encryption."[33] Part of the mystery, and resulting confusion, stems from the interchangeable terms that are commonly used. Kay offers the following definitions:

Cipher is a cryptographic system in which the cryptographic process is applied to plaintext elements of equal length. For example, a system which encrypts one letter at a time, or which deals only with two-letter pairs.

Code is a communication system in which message elements of varying length (such as individual letters, words, or sentences) have been transformed into arbitrary, meaningless combinations of letters and numbers.

Cryptography is the discipline (whether considered an art, service, or field of study) involving principles, means, and methods for transforming information in order to hide its information content and context, prevent its alteration, disguise its presence, and/or prevent its unauthorized use. Cryptography is the general term for the study of both the encryption and decryption process.

Encipherment, encoding, encryption is the protection of information by converting it into a form that is unintelligible until it is converted back to its original form. The conversions employ cryptographic procedures. It is common, if technically incorrect, to use "encryption" to mean either encryption or decryption. (Note: The International Standards Organization uses the term encipherment, because encryption does not translate adequately into French.)[33]

Cipher A. Deavours, a well-known expert on encryption, offers these distinctions to further clarify the interchangeable terms applied to encryption: "Data encryption is an integral part of the service of cryptology, the study of concealed and secret writing. Cryptology is divided into two major areas: **cryptography** and **cryptoanalysis**." Cryptography is the study of systems for putting "plaintext" into a coded ciphertext form; cryptoanalysis is the study of methods for recovering the plaintext without knowledge of the specific "key" involved.[30]

Historical Perspective

The confusion that results from various terms being applied to somewhat the same function stems, in part, from encryption's long history of development.

"The earliest ciphers date back to the early Egyptian days—around 2000 B.C., when funeral messages consisting of hieroglyphs were carved into stone—not to keep the messages a secret, but to increase their mystery."[43] Karl Andreassen, in his book *Computer Cryptology*, states that "the origins of secret writing can be traced back 4,000 years in history, since the Egyptian hieroglyphic writings are known to be cryptic in nature."[2] In the Bible, Jeremiah 25:26, the word "sheshach" is ciphertext, but for what reason is anyone's guess. "Deciphered, 'sheshach' becomes 'Babel,' a locale and tower that has been the subject of numerous Sunday-school lessons and pastors' sermons."[2]

Julius Caesar was reported to use a substitution algorithm in which every letter in the original message (known now as plaintext) was replaced by the letter that occurred three places later in the alphabet (i.e., A was replaced by D, B was replaced by E). "The plaintext VENI VIDI VICI would yield YHQL YLGL YLFL. The resulting message, now known as ciphertext, was then couriered to an awaiting centurion, who decrypted it by replacing each letter with the letter that occurred three places before it in the alphabet."[11]

David Kahn, in his book *The Codebreakers*, traces cryptography from ancient Egypt to India, Mesopotamia, Babylon, Greece, and into Western civilization and eventually into the computer age.[31] "From the Spartans to Julius Caesar, from the Old Testament ciphers to the Papal plotters of the fourteenth century, from Mary, Queen of Scots to Abraham Lincoln's Civil War ciphers, cryptology has been a part of war, diplomacy, and politics."[43] Many historians believe that Mary, Queen of Scots, was killed in the sixteenth century because she sent an encrypted message from prison that was intercepted and deciphered. Benedict Arnold used a code book cipher to communicate with the British during the Revolutionary War.

In periods of war, the use of encryption becomes paramount so that messages are not intercepted by the opposing forces. The Enigma machine was first introduced during World War I by a German electrical engineer, Arthur Scherbius. The German Navy began using the Enigma machines early in 1926. The Enigma machine worked as follows: An operator typed the original text of the first letter of the message to be encrypted on the machine's keyboard-like set of buttons. The battery-powered machine encrypted the letter and, using a flashlight-type bulb, illuminated a substitute letter on a glass screen. Rotors that contained electrical contacts were wired in such a way that turning the rotors would change the correspondence between letters. Once the message was encrypted, it would be transmitted by radio to its destination, usually a U-boat in the Atlantic. Enigma machines continued to be used at the start of World War II. Marian Rejewslei of Poland, in the late 1920s, formed a cryptoanalysis unit that began to work on breaking the German codes. In the early 1930s in France, a German named Hans-Thilo Schmidt offered French intelligence some information about setting the Enigma keys.[8]

The French cryptoanalysts didn't have the resources to take advantage of this information; the British also rejected the information as being insufficient.

Eventually the French offered the information to Poland, and Rejewslei used it to make other advances in breaking the Enigma codes. After the fall of Poland in 1939, the Poles passed their information on to the French and the British. The Germans continued to change keys and to modify the design of the Enigma machine. "Under the direction of mathematician Alan Turing, and with the help of Enigma documents captured from U-boats sunk during the remainder of the war, the highly secret 'Ultra' project began to decrypt German naval messages on a regular and timely basis."[43] Americans made their contributions in the 1940s when some people from IBM reconstructed Japanese diplomatic cipher machines through the "Purple" project in the United States.[43]

The use of computers to break codes and ciphers has contributed to military intelligence at a growing pace. The U.S. National Security Agency employs a number of personnel to work with crypto techniques. "According to Kahn, in his book *Kahn on Codes*, the NSA has assembled 'more computers under a single roof, than probably any other institution in the world.'"[2]

Cryptology, whose history contains many tales of international intrigue, spies, and counterspies, is moving into the arena of higher mathematics and statistics.

TYPES OF CIPHERS

"Encryption is a means of maintaining secure data in an insecure environment."[40] The two basic methods used in encryption are:

1. Substitution

2. Transposition

Substitution is the ciphering process in which one unit (normally a single letter/character, but possibly a group of N characters) of **ciphertext** (unintelligible text or signals produced through the use of an encryption system) is substituted for a corresponding unit of **plaintext** (the intelligible text or signals that can be read without using any decryption), according to the algorithm in use and the specific key.

Transposition is the encryption process in which units of the original plaintext (usually individual characters) are simply moved around; they appear unchanged in the ciphertext except for their relative location.

Basically, one method replaces a character for a character (substitution), whereas the other merely rearranges the order of the characters (transposition). Figure 5.1 illustrates the encryption process.

The **encryption key** is the cryptographic key used for encrypting data and for decrypting data. The **decryption key** reverses the process on the receiving end.

The **encryption algorithm** is a sequence of rules or steps, generally expressed in mathematical terms, used to encrypt a message; the **decryption algorithm** reverses the process on the receiving end.

If any of these components is compromised during transmission, the security of the information being protected is severely lessened. "If a weak encryption algorithm is chosen, an opponent may be able to guess the plaintext once a copy of the ciphertext is obtained."[11]

FIGURE 5.1

The Encryption Process

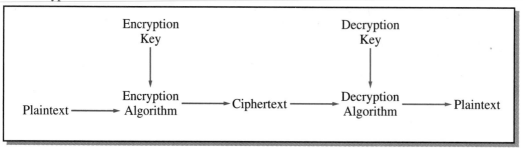

Figure 5.2 depicts the entire process clearly and simply. Typically, the words *encode* and *encipher* are used to mean the same as *encrypt*. There are very slight differences, however, in the meaning of these three words: *Encoding* is the process of translating *entire* words or phrases, while *enciphering* is the process of translating letters or symbols individually. *Encryption* is the generally accepted group term that covers both encoding and enciphering.

FIGURE 5.2

Encryption
Terminology

✔ **Encryption**—process of encoding a message so that meaning is not obvious.

✔ **Decryption**—reverses the process so that message is received in its original form.

✔ **Encode/decode** and **encipher/decipher** are alternate words for **encrypt/decrypt.**

Figure 5.3 points out the encryption process for both original sending/receiving and responses to the original sender. The original form of a message is known as **plaintext,** and the encrypted form is call *ciphertext*.

Cryptographic protection is needed to transmit sensitive data. When a message is encoded, words, numbers, and symbols that have been assigned a special meaning are used to conceal the meaning of the message. Both the sender and recipient of the message should have the same code book that defines the specific meaning of each word, character, or symbol.[20]

Examples of Cryptographics

Figures 5.4, 5.5, and 5.6 illustrate some of the basic techniques of encryption and decryption.

FIGURE 5.3
Sending and Receiving Messages

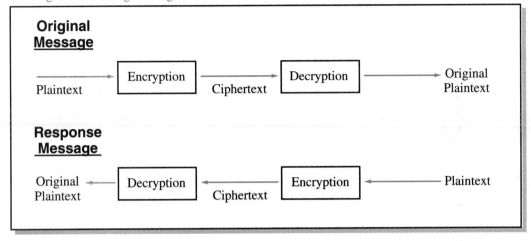

Vernam Technique

The **Vernam Technique** uses an appropriate algebraic equation to combine characters in the message with encoding sequences of characters. The plaintext binary characters are added to the cipher key, producing the cipher text. (See Figure 5.7.)

The Vernam Technique was devised by Gilbert Vernam for AT & T. Another variation of the Vernam Technique uses an arbitrary long, nonrepeating sequence of numbers that are combined with the plaintext. (See Figure 5.8.)

Vigenere Cipher

The **Vigenere Cipher** (or Tableau) is a collection of 26 permutations, represented in a 26 × 26 matrix. All 26 letters are shown in each row and each column. (See Figure 5.9.) To use the Vigenere Cipher, the plaintext letter is located on the top horizontal index alphabet, the key letter directly below noted and then found in the left margin vertical key alphabet.

Other Types of Ciphers

In *The Codebreakers*, David Kahn outlines the development of a number of early transposition and substitution ciphers in ancient civilizations (Spartans, Thucydides, Plutarch, and Xenophon).[31] Figure 5.10 shows another type of transposition cipher.

Julius Caesar's military use of a substitution (Figure 5.11) is one of the earliest documented. Caesar's cipher was a simple form of encryption in which each letter of an original message was replaced with a letter three places further in the alphabet sequence.

FIGURE 5.4

Simple Encryption and Decryption

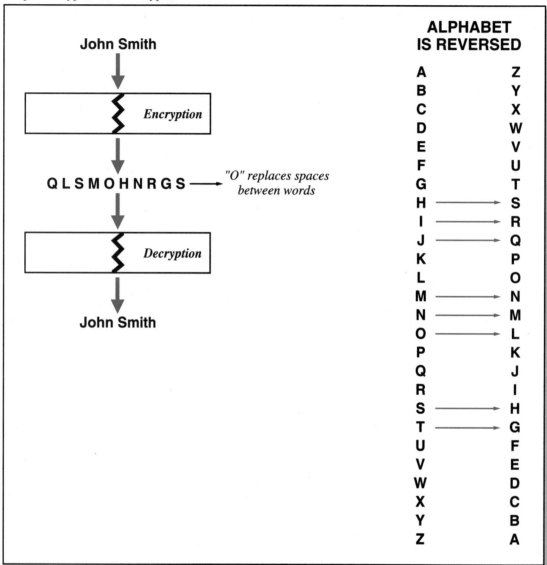

TYPES OF KEYS

More complex ciphers do not use simple substitutions or transpositions, but rather a secret algorithm plus a unique key numbering scheme to control a long sequence of complicated substitutions and transpositions. "The operation of the algorithm upon the original information and the key produces the cipher 'alphabet' that encrypts

FIGURE 5.5
A Simple Transposition Cipher

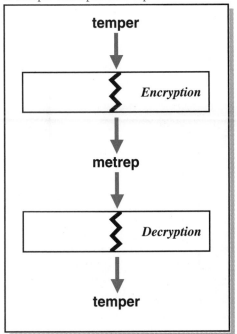

FIGURE 5.6
A Simple Substitution Cipher

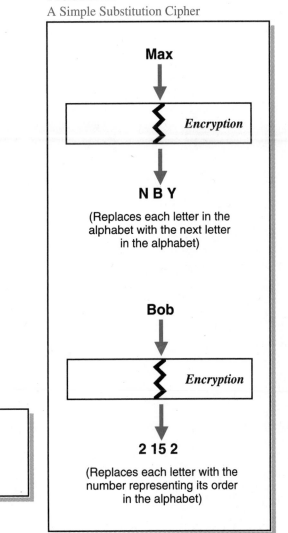

FIGURE 5.7
Vernam Technique Using Binary

☑	Plain Text:	10111
☑	Cipher Key:	01011
☑	Cipher Text:	11100

FIGURE 5.8
Vernam Technique Using Alphabetic Letters

Vernam (Alpha characters represent placement in alphabet)

21	04	17	13	0	12	
76	48	16	82	44	03	(Random number is generated)
97	52	33	95	44	15	
19	0	07	17	18	15	
T	A	H	R	S	P	(Placement of alphabet)

Thus: Vernam is transmitted as TAHRSP

FIGURE 5.9
Vigenere Cipher

```
         A B C D E F G H I J K L M N O P Q R S T U V W X Y Z
         K E Y W O R D K E Y W O R D K E Y W O R D K E Y W O

A        A B C D E F G H I J K L M N O P Q R S T U V W X Y Z
B        B C D E F G H I J K L M N O P Q R S T U V W X Y Z A
C        C D E F G H I J K L M N O P Q R S T U V W X Y Z A B
D        D E F G H I J K L M N O P Q R S T U V W X Y Z A B C
E        E F G H I J K L M N O P Q R S T U V W X Y Z A B C D
F        F G H I K J L M N O P Q R S T U V W X Y Z A B C D E
G        G H I J K L M N O P Q R S T U V W X Y Z A B C D E F
H        H I J K L M N O P Q R S T U V W X Y Z A B C D E F G
I        I J K L M N O P Q R S T U V W X Y Z A B C D E F G H
J        J K L M N O P Q R S T U V W X Y Z A B C D E F G H I
K        K L M N O P Q R S T U V W X Y Z A B C D E F G H I J
L        L M N O P Q R S T U V W X Y Z A B C D E F G H I J K
M        M N O P Q R S T U V W X Y Z A B C D E F G H I J K L
N        N O P Q R S T U V W X Y Z A B C D E F G H I J K L M
O        O P Q R S T U V W X Y Z A B C D E F G H I J K L M N
P        P Q R S T U V W X Y Z A B C D E F G H I J K L M N O
Q        Q R S T U V W X Y Z A B C D E F G H I J K L M N O P
R        R S T U V W X Y Z A B C D E F G H I J K L M N O P Q
S        S T U V W X Y Z A B C D E F G H I J K L M N O P Q R
T        T U V W X Y Z A B C D E F G H I J K L M N O P Q R S
U        U V W X Y Z A B C D E F G H I J K L M N O P Q R S T
V        V W X Y Z A B C D E F G H I J K L M N O P Q R S T U
W        W X Y Z A B C D E F G H I J K L M N O P Q R S T U V
X        X Y Z A B C D E F G H I J K L M N O P Q R S T U V W
Y        Y Z A B C D E F G H I J K L M N O P Q R S T U V W X
Z        Z A B C D E F G H I J K L M N O P Q R S T U V W X Y
```

the information."[43] The three most commonly used keys, as shown in Figure 5.12, are:

1. **Private key**
2. **Public key**
3. **One-time pad**

Private Key

The **private key** system (sometimes called symmetric key, secret key, or single key) uses a single key for encrypting and decrypting information. A separate key is

FIGURE 5.10
Transposition Cipher

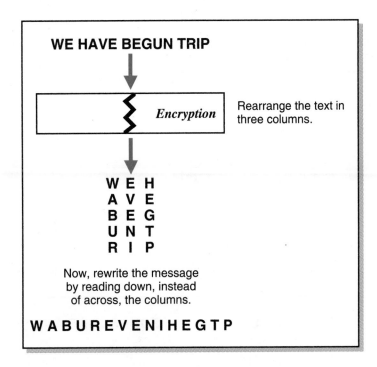

needed for each pair of users who exchange messages; both sides of the encryption transaction must keep the key secret. (See Figure 5.13.) The Data Encryption Standard (DES) algorithm (to be covered later in this chapter) is an example of a private key algorithm.

Public Key

Sometimes called asymmetric key or two key, the **public key** system uses two keys—a public key and a private key. "With a public key encryption system, each user would have a key that did not have to be kept secret. The public nature of the key would not inhibit the secrecy of the system. The public key transformation is essentially a one-way encryption with a secret (private) way to decrypt."[40] (See Figure 5.14.) Within a group of users, each user has both a public key and a private key. The user must keep his private key a secret while the public key is known.

> Private and public keys are mathematically related. If you encrypt a message with your private key, the recipient of the message can decrypt it with your public key. Similarly, anyone can send anyone else an encrypted message simply by encrypting the message with the recipient's public key; the sender doesn't need to know the recipient's private key. When you receive an encrypted message you, and only you, can decrypt it with your private key.[43]

One-Time Pad

In the **one-time pad** system, the key is as long as the message, with no repetition. "This provides for the maximum possible security because there is no way to deduce or recover the key simply by analyzing the ciphertext. The use of the term 'pad' dates from the former practice of intelligence agencies creating such keys and

FIGURE 5.11

The Caesar Substitution Cipher

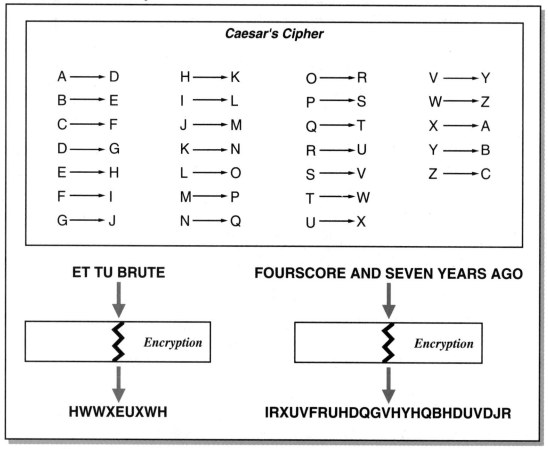

FIGURE 5.12

Types of Keys

- ☑ **Private Key**—Uses a single key.
- ☑ **Public Key**—Uses two keys (one public, one private).
- ☑ **One-Time Pad**—Pad is used *only* once.

publishing them on pads of paper."[33] There are two copies made—one each for sender and receiver. The key is then destroyed by the sender and never reused. (See Figure 5.15.) Sometimes one-time pads are used to encrypt vital diplomatic communications, but they are not practical for most other communications.

FIGURE 5.13
Private Key Encryption/Decryption

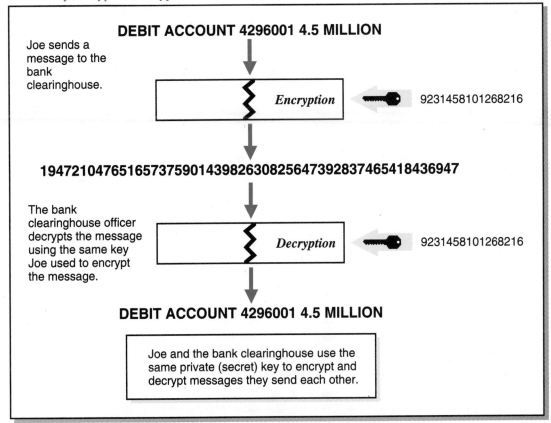

Joe sends a message to the bank clearinghouse.

DEBIT ACCOUNT 4296001 4.5 MILLION

Encryption ← 9231458101268216

1947210476516573759014398263082564739283746541 8436947

The bank clearinghouse officer decrypts the message using the same key Joe used to encrypt the message.

Decryption ← 9231458101268216

DEBIT ACCOUNT 4296001 4.5 MILLION

Joe and the bank clearinghouse use the same private (secret) key to encrypt and decrypt messages they send each other.

THE DATA ENCRYPTION STANDARD (DES)

The **Data Encryption Standard (DES)** is an encryption method published in 1977 by the National Bureau of Standards (now the National Institute of Standards and Technology). DES originally specified the encryption of sensitive government information unrelated to national security. IBM developed DES in cooperation with the National Security Agency (NSA); DES was approved for unclassified information.

This private-key crypto system operates on 64-bit blocks of information and uses a single 128-bit key for both encryption and decryption (which uses 56 independent bits and 8 that may be used for parity checking). The DES was first published in January 1977, as FIPS-Pub-46, which is available from NIST. Also, DES became the standard for ANSI (American National Standards Institute) in 1980. The initial criteria for DES were the following:

- It must provide a high level of security.
- It must be completely specified and easy to understand.

FIGURE 5.14

Public Key Encryption/Decryption

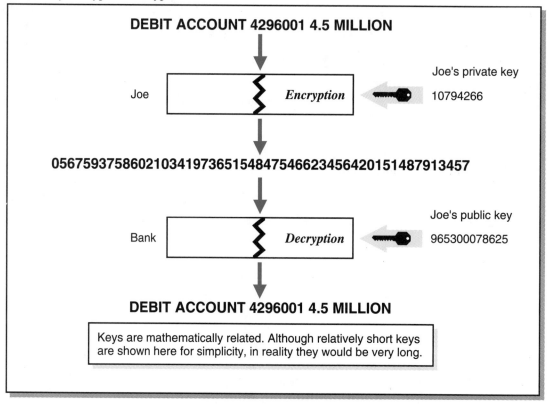

- The algorithm itself must provide the security; the security should not depend on the secrecy of the algorithm.
- It must be available to all users and must be adaptable to diverse applications.
- It must be efficient and economical to implement in electronic devices.
- It must be exportable and able to be validated.

The DES can generate 10 quadrillion different combinations of a 56-bit variable.[20]

Future of DES

"Although the government cannot use the DES to protect classified or extremely sensitive unclassified information, DES products have been very popular in all but the most secret government agencies."[43]

FIGURE 5.15
One-Time Pad

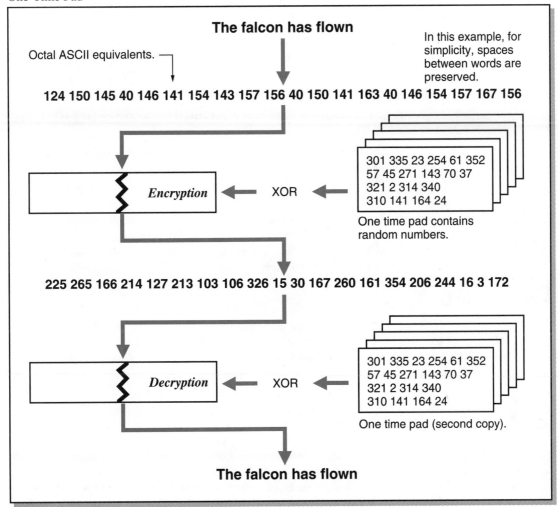

In 1986, NSA announced that, beginning in January 1988, it would not endorse DES-based products as complying with Federal Standard 1027. The agency said it would also recommend that NIST not reaffirm the DES when the standard text came up for review. NSA did say that products already endorsed would continue to be available and would be listed, and that it would continue to provide keys as needed for these products.

Reasons for Not Recertifying

When NSA announced that it would not recertify the DES, the controversy began about the security of DES. One expert summed up this debate as:

The problem with DES is not that it is known, or even suspected, to have been broken; it is just becoming more likely that it could be broken. The extensive debate about the feasibility of an exhaustive key search machine, coupled with the exhaustive analysis of the DES algorithm, has made DES too risky.[40]

Export of the new devices to foreign countries will be prohibited because of national security concerns. This limitation raises serious problems for U.S. companies doing business abroad or having foreign affiliates. Other controversies surrounding DES include these issues:

- Security of communications and data currently protected by DES is in question.

- Some companies have invested substantial sums in hardware that uses DES. Replacing this machinery will be expensive. Other companies that manufacture DES-based devices may have produced merchandise that they will be unable to sell.

- NSA's plan to distribute the devices and the keys constitutes government control of who can use encryption. In addition, if the government retains records of who has received what keys, the government then has the capability to decrypt any intercepted encrypted data, within the private or public sector.[40]

Alternatives to DES

NSA wants its own classified algorithms to be used by both government and industry. Overwhelming support for DES, coupled with the fact that the DES remains technically adequate, caused NIST to recommend to the Secretary of Commerce that the DES be reaffirmed for five more years. (It came up for review and was reaffirmed in 1992.)

GUIDELINES FOR USE OF ENCRYPTION

Information is the single most valuable commodity in our information processing-based society. Yet, as a rule, that information goes unprotected.

There are an estimated 110 million computers in the world today; approximately 50% of them are capable of communicating with other computers via dedicated network and private lines or by the use of massive public telephone services. Due to such capabilities, it has become obvious that there are no longer any physical boundaries between nations and that information has become a global commodity for those who actively seek it.

This information runs the government, military, businesses, and even our own daily lives; it should be protected. We have the technology to ensure the security of our computer systems, and thus the information they contain, but the developers of such security software and programs are "lax, even derelict, in the execution of their fiduciary, legal, and moral responsibilities" to the consumers of their products.[44]

We must press the providers of security mechanisms to use the technology they have. The hackers have the time, the energy, the patience. Shouldn't we expect the same from the people selling us "security" from the hackers? Absolutely.

Susan N. Kesim, Vice President of American Computer Security in South Bend, Indiana, offers these suggestions to prevent an encryption code from being broken:

- Choose a strong encryption cipher.
- Change the key often.
- Segment messages.
- Send nonsense messages in between some of the real ones.
- Encrypt the data during storage as well as transmission.
- Administer regular security checks on employees.
- Properly destroy discarded data (media ribbons, disks, printouts).
- Encrypt or erase hard drives if they are sent for service.
- Keep an unencrypted backup locked in a safe place in case the data is damaged on the disk or contaminated during transmission.
- Keep a personnel policy on file that notes key and password handling.
- Disable all default and utility passwords and accounts.
- Make sure security is handled and checked by more than one person.
- Make sure keys or passwords are stored in an encrypted form.
- Synchronize data transmission times to track whether data is rerouted or slowed by a tap.

In securing data, the best advice of all is to use common sense. By making encryptions easy for authorized users to apply and cost-effective for the company, proprietary data will be less vulnerable to wrongful access.

Security of Keys

The following guidelines will aid in the security of encryption keys.

- Keys in clear form must not exist within the system.
- A separate key should be assigned to each data resource to be protected.
- Keys should be changed frequently. Changes should be made with consideration of cost, convenience, and security.
- Records should be kept of users and their assigned keys.
- Records should be kept of data resources assigned to keys.
- Keys should not be available to persons other than those responsible for generating, setting, and maintaining them.
- The handling of keys should become a physical security problem of easily managed proportions.[24]

Cost vs. Benefit

"Encryption can present some particularly difficult problems for a cost-conscious security manager."[5] Given that encryption almost always reduces the system's response speed and that someone must maintain responsibility for protecting the encryption key, the economics of encryption must be carefully evaluated. When determining whether a proprietary system is to be designed or a commercial product installed, the strengths and weaknesses of each option must be fully analyzed.

The process of introducing cryptography into a computer system should involve the following steps:

1. Define the problem.
2. Conduct market survey.
3. Select the system.
4. Test the system.
5. Configure the system.
6. Install and institute monitoring.[20]

Checklist

The question of whether cryptography is desirable in a given situation is often most perplexing. The following questions should be addressed *before* a cryptographic technique is adopted:

- Has a list been made and approved of the data to be protected and its characteristics?
- Have estimates been made of the financial value of each data item to be protected and of the costs of protection?
- Have the access groups who will be using cryptography been defined, along with their relationship to one another?
- Have physical security levels been evaluated?
- Is it possible to safeguard cryptographic keys?
- Can the system be monitored?
- Is there a simpler solution?[30]

SUMMARY

There are many definitions for the term **encryption**—which means "scrambling" information during transmittal. Other common terms are: **cryptography**, **ciphering**, **encipherment**, **encoding**, **cryptoanalysis**, or **cryptology**.

Julius Caesar was reported to be one of the earliest users of a substitution algorithm to send messages. The Enigma system, developed during World War I and put into use in the 1930s, is one of the most prominent examples of military use of encryption.

The two basic types of ciphers are substitution and transposition. **Substitution** is the ciphering process in which one unit is replaced by another. **Transposition** is the process of changing the order of the units. **Algorithms**, used in the encryption process, are well-defined procedures or a sequence of rules or steps, generally expressed in mathematical terms.

Two well-known encryption techniques in use today are the **Vernam Technique** and the **Vigenere Cipher**.

Keys currently being used when long sequences of complicated code are sent are **private keys**, **public keys**, and **one-time pads**.

The **Data Encryption Standard** (**DES**) is the encryption standard used by the federal government for use of sensitive government information unrelated to national security. DES was released in 1977 after development by IBM and approval by the NSA (National Security Agency). DES was reaffirmed in 1992 and is currently being reevaluated concerning its continued use.

Because information is the most valuable single commodity in the information processing cycle, protection measures aiding in the security of transmitted information are critical. Encryption methods must be fully evaluated before adoption to ensure that the correct method is used. Careful consideration should be given to the security of keys and the cost vs. benefit ratio.

REFERENCES

1. Anderson, John C., "I Spy! Observations on Modern-Day Cryptography," *ISP News*, Vol. 3, No. 2, March/April 1992, pp. 1, 24-26.
2. Andreassen, Karl, *Computer Cryptology . . . Beyond Decoder Rings* (Englewood Cliffs, NJ: Prentice-Hall, 1988), pp. 1-268.
3. Anthansiou, Tom, "Encryption," *Technology Review,* August/September 1986, pp. 57-63, 66.
4. Avarne, Simon, "Cryptography . . . Combating Data Compromise," *Security Management*, October 1988, pp. 39-43.
5. Baker, Richard H., *The Computer Security Handbook,* (Blue Ridge Summit, PA: Tab Books, 1985), pp. 1-281.
6. Bequai, August, *How to Prevent Computer Crime . . . A Guide for Managers* (New York: John Wiley and Sons, 1983).
7. Bosworth, Bruce, *Codes, Ciphers, and Computers* (Rochelle Park, NJ: Hayden Book Company, Inc., 1982).
8. Brown, Nander, "Security and Control of On-Line Systems—Tools and Techniques," *Data Security Management*, No. 84-04-15, 1984, pp. 1-8.
9. Ciarcia, Steve, "Build a Hardware Data Encryptor," *Byte*, September 1986, pp. 97-111.
10. Cole, Gerald D., *Implementing OSI Networks* (New York: John Wiley and Sons, Inc., 1990), pp. 1-336.
11. *Computers at Risk . . . Safe Computing in the Information Age*, National Academy Press, 1991, pp. 1-302.
12. "Data Encryption Standard," National Bureau of Standards, Washington, D.C., FIPS Pub 46, January 1977.
13. "Data Encryption Standard," *NCSL Bulletin*, U.S. Department of Commerce, National Institute of Standards and Technology, June 1990, pp. 1-6.
14. *Data, Fax, and Voice Encryption Equipment Worldwide*, New Canaan, CT: International Resources Development, Inc., 1992.
15. Davies, Donald, "Confidentiality, Integrity, Continuity," *Computer Control Quarterly*, Spring 1986, pp. 28-31.
16. Deavours, C., and Louis Kruh, *Machine Cryptology and Modern Cryptanalysis* (Norwood, MA: Artech House, Inc., 1985).
17. Denning, D.E.R., *Cryptography and Data Security* (Reading, MA: Addison-Wesley, 1982).
18. "Disguising Dollars," *ABA Banking Journal*, January 1987, pp. 77-78.
19. Dror, Asael, "Secret Codes," *Byte*, June 1989, pp. 267-270.

20. Enger, Norman, and Paul W. Howerton, *Computer Security . . . A Management Audit Approach* (New York: AMACOM, 1980), pp. 1-264.

21. "FAX Pirates Find it Easy to Intercept Documents," *Journal of Commerce*, September 7, 1990.

22. Feistel, Horst, "Cryptography and Computer Privacy," *Scientific American*, Vol. 228, No. 5, May 1973, pp. 15-23.

23. "FIPS 140—A Standard in Transition," *CSL Bulletin*, U.S. Department of Commerce, National Institute of Standards and Technology, April 1991, pp. 1-4.

24. Fitzgerald, Jerry, *Business Data Communication* (New York: John Wiley and Sons, 1988).

25. Garon, Tim, "I've Got a Secret," *ISP News*, March/April 1992, p. 21.

26. Glass, Brett, "The Data Encryption Standard: Still Secure?" *Enterprise Computing*, September 9, 1991, p. 71.

27. Glass, Brett, "The Data Encryption Standard: Still Secure?" *Infoworld*, September 9, 1991, pp. 571-576.

28. Goode, W. Mark, "Crypto Standards: A Thousand Points of Contention?" *ISP News*, September-October 1991, pp. 43-44.

29. Gross, Rod, "Encryption's Role in Network Security," *ISP News*, March/April 1992, pp. 21-23.

30. Hutt, Arthur, Seymour Bosworth, and Douglas B. Hoyt, *Computer Security Handbook* (New York: Macmillan, 1988), pp. 1-399.

31. Kahn, David, *The Codebreakers: The Story of Secret Writing* (New York: Macmillan, 1967).

32. Katzan, Harry, Jr., *The Standard Data Encryption Algorithm* (New York: Petrocelli, 1977).

33. Kay, Russell, Ed., "The ISP News Guide to Encryption Terminology," ISP News, March/April 1992, pp. 32-35.

34. Kesim, Suzan N., "Encryption . . .Securing with Ciphers," *Security Management,* October 1988, pp. 45-47.

35. Lsuer, Rudolph F., *Computer Simulation of Classical Substitution Cryptographic Systems* (Laguna Hills, CA: Aegean Park Press, 1981).

36. Meyer, Carl and Stephen, Matyas, *Cryptography: A New Dimension in Computer Security: A Guide for the Design and Implementation of Secure Systems* (New York: John Wiley and Sons, 1982).

37. Mills, Louis, "Four Simple Encryption Techniques Help Ensure Data Integrity," *Data Management*, December 1984, pp. 22-23.

38. "NSA Endorses Xerox Device for Encryption," *Federal Computer Week*, February 26, 1990, p. 19.

39. Parker, Donn B., *Computer Security Management* (Reston, VA: Reston Publishing Company, 1981), pp. 1-308.

40. Pfleeger, Charles, *Security in Computing* (Englewood Cliffs, NJ: Prentice-Hall, 1989), pp. 1-278.

41. Popek, G.I., and C.S. Kline, "Encryption and Secure Computer Networks," *Computing Surveys*, Vol. 11, No. 4, December 1979, pp. 331-356.

42. Rossen, Ken, "Network Security: Just Say 'Know' at Layer 7," *Data Communications*, March 1991, pp. 103-106.

43. Russell, Deborah, and G.T. Gangemi, Sr., *Computer Security Basics* (Sebastopol, CA: O'Reilly and Associates, Inc., 1991), pp. 1-441.

44. Schwartau, Winn, "Information Terrorism Threatens Way of Life," *Enterprise Computing*, September 9, 1991, pp. 72-73.

45. Schwartz, Michael, "Data Authentication," *Data Security Management,* No. 84-03-06, 1984, pp. 1-12.

46. Schwartz, Michael, "Understanding Public Key Encryption—Part I," *Data Security Management*, No. 84-03-04, 1984, pp. 1-16.

47. Schwartz, Michael, "Understanding Public Key Encryption—Part II," *Data Security Management*, No. 84-03-05, 1984, pp. 1-12.

48. Schweitzer, James A., *Computer Crime and Business Information . . .A Practical Guide for Managers* (New York: Elsevier, 1986).

49. Scott, Karyl, "Encryption Schemes Put Safety First," *Data Communications*, March 21, 1991, pp. 17-20.

50. Smid, Miles E., "Integrating the Data Encryption Standard into Computer Networks," *IEEE Transactions on Communications*, Vol. Com-29, No. 6, June 1991, pp. 762-772.

51. "The GOSIP Testing Program," *CSL Bulletin*, U.S. Department of Commerce, National Institute of Standards and Technology, May 1991, pp. 1-4.

52. Wayner, Peter, "True Data," *Byte*, Vol. 16, September 1991, pp. 122-124.

53. Welsh, Dominic, *Codes and Cryptography* (New York: Oxford Press, 1988).

REVIEW QUESTIONS

1. Define encryption and explain how it is used to protect transmission of information.

2. List some of the alternate terms that are used to describe encryption.

3. When were ciphers first used?

4. Explain how cryptology's use heightens during periods of war and conflict.

5. What is the difference between substitution and transposition ciphers?

6. How are plaintext and ciphertext used in message transmission?

7. What are encryption algorithms? How are they used?

8. Explain the difference between the Vernam Technique and the Vigenere Cipher.

9. How are private keys, public keys, and one-time pads used to transmit messages?

10. What is DES? Explain some of the reasons it is now being reevaluated for use.

DISCUSSION QUESTIONS

1. Karl Andreassen, in his book *Computer Cryptology*, states that "there is an increasing interest in cerebral applications of computer software that tax the skill and imagination of the player, and, of this type, none can surpass the challenge of cryptology."[2] How is the introduction of microcomputers, coupled with increasing computer literacy of the general populace, leading to cryptology's popularity?

2. Electronic mail (E-mail) is one of the fastest growing areas of technology, and its use increases daily. Could encryption be used in E-mail to transmit confidential information (military, corporate secrets, personal information)? Would encryption make the transmittal process more cumbersome?

3. Encryption can be used to guarantee the following when data/information is transmitted:

   a. Secrecy or confidentiality

   b. Accuracy or integrity

   c. Authenticity

   Explain each briefly.

4. Government export regulations restrict the freedom of U.S. vendors of DES-based products to sell these products outside the United States. Even software implementations of the algorithm can't be included in international distributions. Is this restriction a constraint to protect the DES?

5. Is it possible to make an encryption technique *absolutely* unbreakable (100% guarantee of its secrecy)? Is it practical to consider this level of transmittal? What factors should be considered?

EXERCISES

1. Locate a journal article or book that reviews some of the early ciphers and why they were used. Briefly summarize these various ciphers.

2. Devise two ciphering schemes—one a substitution and the other a transposition. Encipher your name (first and last) using these two schemes.

3. What are some of the strengths and weaknesses that you can determine to exist with the Vernam Technique and the Vigenere Cipher that would make them both usable and breakable?

4. Select an organization/company in your area (city) and interview a key individual to determine if encryption is used to transmit data and information. What types of information are encrypted?

5. Locate a periodical, article, book, or government publication relating to DES. Briefly review the article and develop a concise listing of the pros and cons of continual use of DES.

PROBLEM-SOLVING EXERCISES

1. Several very reliable encryption techniques can help secure data. One of the easiest schemes is the transposition code, where every character in a record has one and only one replacement character. Complete the following transposition table:

Original Character	Replacement Character
A	G
B	H
C	I
D	J
E	K
F	L

2. Occasionally, double encryption is used to protect data during transmittal. Develop a scheme that could combine two techniques and show the encryption at each step from plaintext to ciphertext to plaintext.

3. The 10 most common sequences in the English language are:

TH	IN	ES	TI
HE	ER	ON	
AN	RE	EA	

Once these are combined with a pattern, the rest of the key is easily broken. Develop a scheme that would ward against these 10 patterns making the information vulnerable. Hint: Vowels are easily broken and often not needed in words to give meaning.

4. Develop a binary ciphering system that uses alternate keys (also using binary) to transmit data/messages.

5. DES uses an 8-byte (64 bit) block to encipher. Develop a ciphering scheme using this 8-byte block.

CASES

Cherokee Products, Inc.

Jacob Leaningtree, president of Cherokee Products, Inc., located in Aldus, Oklahoma, is expanding his company's market of Native American products to include the entire United States, as well as international markets.

Mr. Leaningtree is considering using the fax (facsimile) machine to transmit and receive orders. By using fax, the process is speeded up tremendously, transmissions are not as likely to be misrouted, and the order can be placed immediately. By reducing the information float, inventory is moved much more quickly.

Mr. Leaningtree has been hearing, however, about several instances of competitors breaking into transmissions and gaining valuable information about customers and products. The company is now considering a fax encryption device to protect confidential transmissions.

Some of the factors that should be considered before adopting a fax encryptor need to be fully addressed, such as:

1. How many levels of authority will be needed?

2. Is a key management capability necessary?

3. What restrictions should be applied to user access and password management?

4. Is an audit trail necessary?

5. Can this fax system be interfaced with present software and hardware?

6. What is the cost vs. benefit ratio?

7. What other factors should be considered?

Bank of Shenandoah Valley

Today, banks are transferring large sums of money electronically and facing enormous exposure in the process. The Bank of Shenandoah Valley, located in Roanoke, Virginia, is actively involved in this process. The possibility of funds transfer fraud is prompting many banks to adopt protective measures. The two most common techniques used in the banking industry are encryption and message authentication. Encryption involves the scrambling of messages sent—for example, from a commercial bank to the Federal Reserve Bank. An authenticated message is sent clear—anyone who intercepts it can read it. Tacked on to the message is a related secret code that only the receiving party is capable of decoding.

Most observers say that authentication offers more security than encryption because a key is involved.

The Bank of Shenandoah Valley is considering both options and needs to address the following questions:

1. Are there major differences between the two techniques?
2. What is the volume of transmittal?
3. Are all messages critical or would encryption/authentication be applied only to certain transmittals? If so, which ones?
4. What are the cost factors involved?
5. Which employees should be authorized to transmit?
6. Will these processes slow down operations?
7. What other safeguards should be considered?

SPECIAL CONSIDERATIONS

Chapter 6	**Database Security**	**134**
Chapter 7	**Network and Telecommunications Security**	**163**
Chapter 8	**Microcomputer Security**	**208**
Chapter 9	**Viruses**	**244**

DATABASE SECURITY

LEARNING OBJECTIVES

After studying this chapter, you will be able to:

1. Understand the basic design of databases in use today.
2. List the hierarchy of data.
3. Explain how data is organized in a database system.
4. Understand the advantages and disadvantages of database design.
5. Explain the three types (models) of database design.
6. Contrast the physical and logical design of databases.
7. Explain the security threats to databases.
8. Describe the access control procedures used to protect databases.
9. Describe the methods used internally to protect the information stored in databases from accidental or malicious disclosure.
10. Describe some of the methods used in multilevel databases to limit access and protect information.
11. Define the following terms.

TERMS

- Database
- Database Management System (DBMS)
- Character
- Field
- Record
- File
- Data Entity
- Data Attribute
- Data Key
- Data Structure
- Data Element

- Hierarchical Database
- Network Database
- Relational Database
- Tuple
- Domain
- Logical Access Path (LAP)
- Physical Access Path (PAP)
- Logical Record
- Physical Record
- Schema
- Subschema
- Data Definition Language (DDL)

TERMS (continued)

- Data Dictionary
- Data Manipulation Language (DML)
- Query Language
- Database Security Violation
- Malicious Violations
- Accidental Violations
- Test Data File
- Information Security
- Database Security
- Privacy
- Authorization
- Protection
- System Integrity
- Access Control
- Intentional Resolution
- Information Flow Control
- Semantic Integrity
- Concurrency Control
- Recovery Systems
- Auditing
- Controls
- Field Checks
- Change Log
- Access Control Procedures

- User Authentication
- Integrity Checks
- Audits
- Monitors
- Range Comparisons
- State Constraints
- Transition Constraints
- Boundary Checks
- Precision Checks
- Suppression
- Concealing
- Database Administrator (DBA)
- Multilevel Database System
- Denial of Services
- Partitioning
- Cryptographic Sealing
- Sensitivity Locks
- Front-End Processor (FEP)
- Filtering
- Windows (Views)
- Query Interpretation
- Granularity
- Data Dictionary
- Inference

CREDIT REPORTING NOT ALWAYS AN ERROR-FREE PROCESS

If you've ever learned of an error in your credit reports issued by credit agencies, you'll find it hard to believe that credit bureaus claim a less than 1 percent error rate on their files on consumers. In fact, you're not alone: In a recent *Consumer Reports* survey, 48 percent of credit bureau   *Continued*

CREDIT REPORTING. . .

From page 135

reports contained inaccurate information; nearly 20 percent contained an error of the magnitude that could adversely affect a consumer's eligibility for credit.

One of three survey participants said that third parties had been given access to their credit reports without their permission. Another one in three consumers found the credit report information too difficult to decipher.

Experts estimate there are 400 million individual credit files stored in credit bureau computers; *Consumer Reports'* small sample of 57 consumers suggests the credit bureaus vastly underestimate the error rate and the effect such errors have on consumers.

Now Congress is considering legislation to clean up the credit industry. *Consumer Reports* recommends the following additions to the reform legislation:

1. Consumers should be given, on request, a free copy of each of their credit reports.

2. Information from a report should never be disclosed without the advance written permission of the consumer.

3. Civil penalties should be imposed on bureaus that supply incorrect information.

4. Consumers should be notified in writing when negative information enters their credit files.

5. The bureaus should be required to hire independent auditors to determine the accuracy of their records. The results should be made public.

As reported in "Credit Reports: Getting It Half-Right," *Consumer Reports,* July 1991, p. 453.

In Part I, the physical protection of computers and a brief overview of security were presented. Part II dealt with the software and hardware control features of computer systems and encryption techniques. In Part III, database, network and telecommunications, and microcomputer security will be addressed, as well as viruses—the current plague of systems.

INTRODUCTION TO DATABASES

Before addressing the security considerations of databases, we will review the basic design of databases so that you understand how and where security should be applied.

A **database** is an organized collection of related data. In a database environment, the collection of interrelated files are stored together to reduce redundancy of data and information. A **database management system (DBMS)** is software that organizes data in a manner that will allow fast and easy access to the data. A DBMS is a software program (or set of programs) that creates, manages, protects, and provides access to a database. (See Figure 6.1.)

In a database system, consideration must be given to these questions: What data is to be collected and at what cost (content)? What data is to be provided to which users when appropriate (access)? Where is the data to be physically located (physical organization)? How is the data to be arranged so that it makes sense to a given user (logical structure)?[46]

FIGURE 6.1
Database
Considerations

- ☑ Content
- ☑ Access
- ☑ Physical organization
- ☑ Logical structure

The Hierarchy of Data

For data to be effectively transformed into useful information, it must be organized in a logical, meaningful way. Data is generally organized in a hierarchy that starts with the smallest unit (or piece of data) used by the computer and then progresses into the database, which holds all the information about the topic. (See Figure 6.2.)

A **character** is the basic building block of information in a data hierarchy. A character set generally consists of uppercase and lowercase letters, numeric digits, and special characters such as punctuation marks. A **field** is a collection of characters that form a number, name, or combination of the two. A **record** is a collection of related fields. A **file** is a collection of records. Figure 6.3 illustrates the database hierarchy for an employee file.

Data Organization and Use

A **data entity** is any item, person, place, or thing for which data is collected, stored, and maintained. Examples of entities are employees, inventory, or customers. A collection of data or fields about a specific entity is a record.

A **data attribute** is a characteristic of an entity, such as a Social Security number, last name, first name, date hired, department number, employee number, or salary. Fields are constructed to denote the relevant characteristics or attributes of entities.

A **data key** is a field in a record that is used to uniquely identify the record. Social Security numbers, because they are unique for each employee, are often used as data keys. Many times, more than one key is used for unique identification.[46] (See Figure 6.4.)

FIGURE 6.2

Elements in a Database

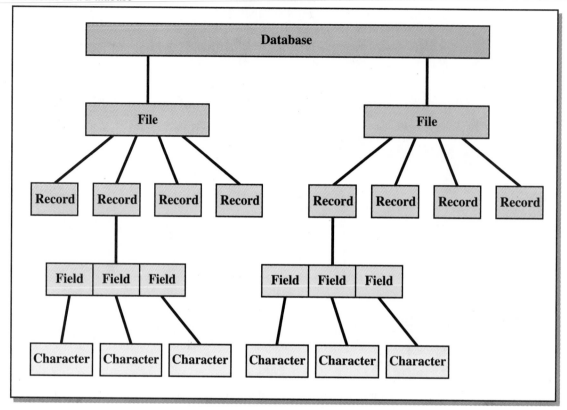

Database Development

"Through the 1980s, most installed information systems were implemented in a crisis environment with a single functional objective in mind. The integration of information systems was not a priority. As a result, many companies are saddled with massive system, procedural, and data redundancies. These redundancies promote inefficiencies and result in unnecessary expenses. Today companies are using database management systems (DBMS) software as a tool to integrate data management and information flow within an organization."[34]

Benefits of a Database

There are many reasons why a company or organization converts to a database environment:

- Reduced data redundancy
- Improved data integrity
- Increased speed
- Easier modification and updating

FIGURE 6.3
Hierarchy of Data

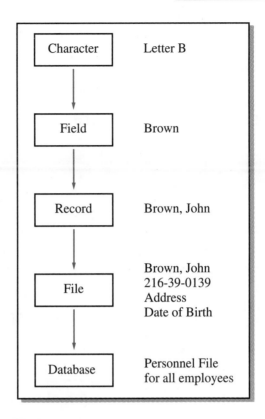

- Data and program independence
- Better access to data and information
- Standardization of data access
- Framework for program development
- Better overall protection of data
- Shared data and information resources[46]

Because organizations have accumulated a great deal of data and information, translating (or processing) data into meaningful information is often difficult in a traditional file organization. The basic structure of an integrated database provides greater flexibility in the types of reports and output that can be produced; it also allows for on-line inquiries.

In a database environment, data is centralized in one location. This "all-the-eggs-in-one-basket" approach fosters better control of information. Generally in a centralized database, security is improved tremendously. The **data structure** refers to the manner in which the data elements and records relate to one another. A database allows for greater coordination of data structures and reduces duplication of data.

Because data in a database are independent of application programs, the programming task is greatly simplified with the conversion to database. **Data elements**

FIGURE 6.4

Attributes, Keys, and Entities

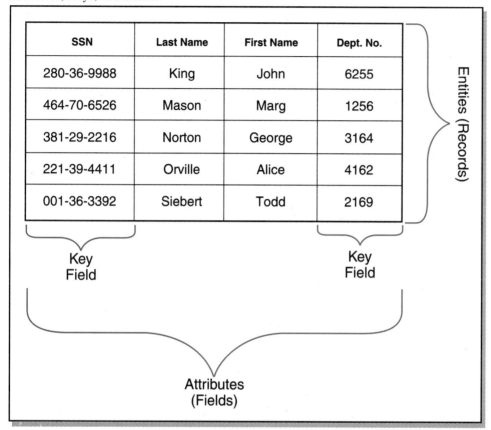

are the smallest logical unit of data used with a particular application. These elements can be added, changed, and deleted from the database without affecting existing programs.

Disadvantages of the Database Approach

As with any technological advance, a few "downsides" accompany the good. Database conversion is no exception. In most cases, the advantages of conversion to database far outweigh the disadvantages. Some key points (or vulnerabilities) to consider with database conversion are the following:

- The relatively high cost of purchasing and operating a DBMS in a mainframe operating environment
- The need for a specialized staff to implement and coordinate the database
- Increased vulnerability to security breaches as more data is concentrated in one system, rather than being spread throughout the organization
- If the database "goes down," the entire system is unavailable, affecting multiple applications

Types of Database Models

Three basic types of database models are in use today:

1. Hierarchical
2. Network
3. Relational

Hierarchical Database. A **hierarchical database** is one in which the data is organized in a top-down or inverted tree-like structure. At the top of every tree or hierarchy is the root segment or element of the tree that corresponds to the main record type. The hierarchical model is best suited to situations in which the logical relationship between data can be properly represented with the one parent-many children (one-to-many) approach. In a hierarchical database, all relationships are one-to-one or one-to-many, but no group of data can be on the "many" side of more than one relationship. (See Figure 6.5.)

FIGURE 6.5
Hierarchical Database
Examples

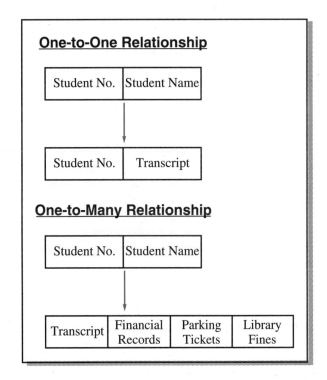

One-to-One Relationship

Student No.	Student Name

Student No.	Transcript

One-to-Many Relationship

Student No.	Student Name

Transcript	Financial Records	Parking Tickets	Library Fines

Network Database. A **network database** is a database in which all types of relationships are allowed. The network database is an extension of the hierarchical model, where the various levels of one-to-many relationships are replaced with owner-member relationships in which a member may have many owners. In a network database structure, more than one path can often be used to access data. "Databases structured according to either the hierarchical model or the network model suffer from the same deficiency: Once the relationships are established between data elements, it is difficult to modify them or to create new relationships."[46] As Figure 6.6. shows, John Smith and Mary O'Reilly are both involved with Report 1, as is Todd Brown. Report 2 involves Mary and Todd, but not John. Hence, there can be more than one "owner."

FIGURE 6.6
Network Database
Example

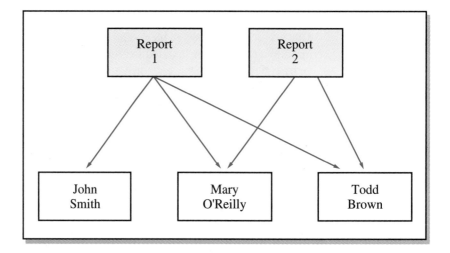

Relational Database. A **relational database** describes data using a standard tabular format in which all data elements are placed in two-dimensional tables that are the logical equivalent of files. In relational databases, data are accessed by content rather than by address (in contrast with hierarchical and network databases). Relational databases locate data logically, rather than physically. A relational database has no predetermined relationship between the data such as one-to-many sets or one-to-one. As long as the tables share at least one common data element, the relational DBMS can link any data element in these tables by generating the desired data elements in a usable fashion.

Using the two-dimensional tables to describe relationships between data, each row of the table, called a **tuple**, represents a record or collection of related facts. The columns in the table are called attributes, and each attribute can take on only certain values. The allowable values for these attributes is called the **domain**. The domain for a particular attribute indicates what values can be placed in each of the columns of the relational table. Once the data has been placed into a relational database, data inquiries and manipulations can be made. (See Figure 6.7.)

Interfacing Between Users and Databases

A DBMS is an interface between application programs and the database. When an application program needs data, it goes to the database. When the application program seeks data from the DBMS, it follows a **logical access path** (LAP). The DBMS then accesses a storage device (e.g., disk or tape) where the data is stored. When the DBMS goes to a storage device such as a disk to retrieve the data, it follows a **physical access path** (PAP). The request is, therefore, logical; the action of seeking is physical.

FIGURE 6.7
Relational Database Example

Student No.	Last Name	First Name	Address				
			Street	City	State	Zip	DOB
1245	Brown	Lisa	124 Elm	Dallas	TX	12460	41770
1398	Green	Robert	169 Oak	Fort Worth	TX	12381	31890
1460	White	Nathan	1460 Black	Houston	TX	14682	31870
1532	Yates	William	1960 First	Kansas City	KS	21690	40691

There are also logical and physical records. **Logical records** are *what* the record contains (e.g., employee number, hours worked). A **physical record** is the actual data stored in a physical storage device. (See Figure 6.8.)

FIGURE 6.8
Logical and
Physical Access

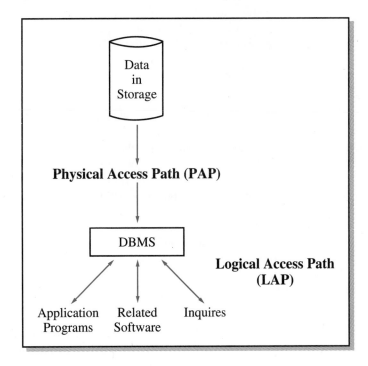

Setting Up the Database

When setting up the database, the relationships among the data to be stored must be defined. A **schema** is a general description of the entire database that shows all the record types and their relationships to one another. A **subschema** shows only some of the records and their relationships. A number of different subschemas can be developed for use by different groups of people. Schemas and subschemas are sometimes called **views**. The schema is the overall view (the way the computer "sees" the database), whereas the subschema is the way the user sees the database. Subschemas can provide additional security as programmers, managers, and other users can see only the records they are authorized to view.

The schemas and subschemas are entered into the computer using a **data definition language (DDL)**, which is a collection of instructions and commands used to define and describe data and data relationships in a specific database.

A **data dictionary** is a detailed description of all data used in the database; it contains the name of the data item, the range of values to be used, the type of data (i.e., alpha or numeric), the storage capacity needed, the person responsible for the file, and the persons who have access. Typical uses for a data dictionary are:

- To provide a standard definition of terms and data elements
- To assist programmers in designing and writing programs
- To simplify database modification

A **data manipulation language** (DML) is a specific language provided with the DBMS software that allows users to have immediate access to and the ability to modify data contained in the database.

A **query language** is a specialized type of data manipulation language that makes retrieving information and manipulating a database easy.

SECURITY REQUIREMENTS OF DATABASES

Now that you have reviewed the general design of databases, let's look at databases from a security perspective rather than just a technical one.

A database system is vulnerable to criminal attack at many levels. Typically, it is the end user rather than the programmer who is often (but not always) guilty of the simple misuse of applications. Thus, it is essential that the total system is secure.

A **database security violation** is the unauthorized reading, modification, or destruction of information stored in the database. The two classifications of security violations are malicious or accidental. **Malicious violations** are those crimes that exploit loopholes in the system, abusing positions of privilege and trust by using data illegitimately. **Accidental violations** include hardware and software failures, human error, or natural disasters.

Threats to the System

Threats to the system come from a variety of sources; thus, the security system must cover as many areas as possible. A partial list of areas and possible procedures or mechanisms that threaten databases includes the following:

1. External procedures
   - Security clearance of personnel
   - Protection of passwords
   - Application program controls
   - Audit

2. Physical environment
   - Secure areas for files/processors/terminals
   - Radiation shielding
3. Data storage
   - Data encryption
   - Duplicate copies
4. Processor software
   - Authentication of user
   - Access control
   - Threat monitoring
   - Audit trail of transactions
5. Processor hardware
   - Memory protection
   - States of privilege
   - Reliability
6. Communication lines
   - Data encryption

To implement these precautions costs money, however. Both startup costs and operational costs are involved. Usually the greater the degree of security, the higher the cost. Thus, it is important to perform a risk analysis of the database to determine what the system is worth and how much should be spent on securing it.

User Beware!

"Warning: Relying blindly on the integrity of your database can be hazardous to your company's health." Companies now rely heavily on personal computers for almost every phase of their operations. But sometimes these same companies put too much trust in the output produced by their computers, allowing even glaring errors to go unnoticed.

One way to help assure the integrity of your database is to perform an audit of the database system. This involves examining the data itself for validity and verifying that data entry forms, reports, queries, batch processes, file maintenance procedures, and custom programs are performing properly.

Some of the ways to accomplish this goal follow:

- Create test data that encompass the full range of possible values that could ever occur.
- Run a manual check in parallel with the system it is intended to replace.
- Check for duplicate records and verify that the system rejects duplicates at the data entry level.
- Examine your data from a different point of view, using new procedures that are not part of your normal processing sequence.
- Verify that your reports are correct.

Miriam Liskin issued this warning to database users in the June 29, 1990 issue of *Personal Computing.*

In just 10 short years, American business has made the transition from fearing and mistrusting personal computers to relying on them in nearly every phase of its operation. For an individual user, the same transition usually takes far less time and often produces a blind trust that can be as counterproductive as mistrusting computers completely . . . Database users need . . . to verify that their software is storing, retrieving, and deleting data as expected; performing calculations correctly; and printing the right records in the right order on reports and listings.[33]

Liskin offers these suggestions to test the database's performance:

- To thoroughly test a new database system, run it in parallel with the manual system it will replace for at least a month.
- To check for duplicate records, sort by the field of interest and scan the printout, or create a custom program that prints potential duplicates.
- For reports, verify that calculated items are computed correctly. You may need to repeat the computations with a calculator.[33]

Test Data Files

Data that are created to test all facets of an information system's operational capabilities are known as a **test data file**. To design a test data file to verify the database, Liskin suggests the following:

- Make sure the data files used to test the database cover the full range of values that could ever occur, and even some that are extremely unlikely, to make certain any trouble spots are anticipated.
- Make the sample files small enough to be manageable, but large enough to be representative.
- Use real, not invented, data.
- For every field, include values that span the permissible range.
- For every field, include "impossible" values to see if the system will react to illegal values.
- Include a full set of matching records from each group of related files.
- Include records that have no match in a related record but should.[33]

Privacy vs. Database Statistics

Many institutions, from businesses and government agencies to hospitals and colleges, maintain large databases containing confidential information concerning employees, customers, patients, or students. The same databases often serve as the sources of statistics that enable regulatory bodies and other groups or individuals to track trends and monitor problems. In such cases, institutions generally release these files only after omitting the names and identifying numbers of the individuals involved.[40]

Many times, the remaining file information can contain enough clues to permit someone to single out a specific individual's record. If a worker happens to know something about another individual (such as college attended, degree, and graduation date), it would not be difficult to search the truncated personnel files to find a record containing matching information and learn this individual's salary or medical history.[40] Kasinath C. Vemulapalli and Elizabeth A. Unger of Kansas State University offer these protection ideas:

- Slightly modifying (or perturbing) the data that a user compiling statistics receives in response to requests for information

- Introducing "noise" into the output data without changing the stored values or unduly affecting the statistics themselves[40]

For example, the computer in response to a query calculates the average value of a certain attribute (such as salary), multiplies this average by a small fraction, then goes through all the records available and randomly adds the calculated product to a record, subtracts the product from a record, or leaves the record unchanged before reporting the query results to the user.[40]

Matching Data to Verify

The Index System, a division of New York-based American Insurance Services Group, Inc., offers a new service aimed at helping self-insured employees, property/casualty insurers, and claims administrators recognize and investigate potentially fraudulent bodily injury claims more quickly. This new system will be able to electronically identify a claimant whose name, address, or Social Security number has been entered into its database during the last five years. The system helps identify claims patterns that may be fraudulent or suspicious. It works by searching for and comparing similar names and addresses on nearly 40 million bodily injury claims in the database. By cross-referencing Social Security numbers, the new system will match past and present claims, and reports will be sent to subscribers via mainframe hookups within two to three days of filing. "Insurance companies and self-insurers will receive more precise data more quickly than ever so they can identify suspicious claims."[15] The size of this database is staggering—1,500 self-insurers, 1,000 insurers and claims administrators, 9 million claims each year, and 2,000 reports to subscribers annually. A database of this magnitude is in definite need of close coordinating and matching of data.

FBI Database Violated

In the January 13, 1992 issue of *Federal Computer Week,* recent happenings with large federal government databases indicate that no system is immune to misuse. "Indictments . . . handed down against insiders who bought and sold confidential information held in Federal Bureau of Investigation (FBI) and Social Security Administration (SSA) computers have prompted agency officials to evaluate how well the government secures its databases."[3] The "information broker" bust was the culmination of an 18-month investigation by the Department of Health and Human Services' (HHS) inspector general's office in Atlanta, Georgia. "Officials said it was the largest case ever prosecuted involving the theft of federal government data."[3] The indictments that were handed down to 18 people from 10 states (including two current SSA employees; a Chicago police officer; an employee of the

Fulton County, Georgia, Sheriff's office; and several private investigators) alleged that the investigators paid for confidential data, including criminal records and earnings histories. "The FBI cannot manage every person in the U.S.," David Nemecek, Section Chief for the FBI's National Crime Information Center (NCIC) stated. NCIC's database contains data on thousands of people suspected and convicted of crimes. Renny DiPentima, assistant Commissioner of SSA's Office of Systems Design and Development, stated that his agency performs 15 million electronic transactions per day—500 per second—and "monitoring the rights and wrongs of these people is a daunting task." Passwords, personal identification numbers (PINs), and audits are routinely used to monitor and control the system, "but despite efforts to tighten the screws on network security, . . . in cases of federal and municipal employees who exploit authorized access, technology and policies can only go so far in affecting human nature."[3]

Tracking the Database Activity

Some possible weak links that should be addressed when evaluating database security include these issues:

- Even though backups are a regular policy, are they tested to determine what has been entered and by whom? Are backups current?
- Is database information screened for correctness and accuracy?
- Is the database constantly monitored for activity by hackers (both inside and outside the organization)?

It is now a well-documented fact that databases are a very valuable corporate asset, and thorough, constant procedures should be in place to ensure their correctness and confidentiality.

DESIGNING DATABASE SECURITY

In *Database Security and Integrity,* E.B. Fernandez, R.C. Summers, and C. Wood[22] offer the following terminology relating to database security:

- **Information security** is the protection of information against authorized disclosure, alteration, or destruction.
- **Database security** is the protection of information that is maintained in a database.
- **Privacy** is a term used broadly for all the ethical and legal aspects of personal data systems—systems that contain information about individuals.
- **Authorization** is the specification of rules about who has what type of access to what information. Access rules for the user are determined by the authorizer.
- **Protection** refers to techniques that control the access of executing programs to stored information.
- **System integrity** is the ability of a system to operate according to specifications in the face of deliberate attempts to make it behave differently. Integrity also is applied to data and to the mechanisms that help ensure its correctness.

- **Access control** is the process of ensuring that information and other protected objects are accessed only in authorized ways.

- **Intentional resolution** controls the actions performed on the data once it is legally accessed.

- **Information flow control** prevents security leakage as information flows through the computer system.[22] (See Figure 6.9.)

- **Semantic integrity** is the part of data integrity concerned with the correctness of database information in the presence of user modifications.

- **Concurrency control** protects data integrity in the presence of changes coming from concurrent programs in execution.

- **Recovery systems** erase and reconstruct changes to preserve integrity.

- **Auditing** is the examination of information by persons other than those who protect the information.

- **Controls** are measures taken to ensure effective management control.

FIGURE 6.9
Information
Flow Control

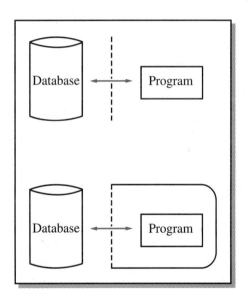

METHODS OF PROTECTION

A database can be monitored and controlled in numerous ways to ensure that adequate protection is achieved. (See Figure 6.10.) Each of these methods will be discussed separately. Bear in mind, however, that in many databases, several of these methods can be used simultaneously for added assurance.

Field Checks

Field checks are tests for appropriate values in a position or action. Possible field checks might be to test for numeric, uppercase, alphabetic, or a given set of acceptable characters. Field checks are designed to check for simple errors as data are entered.

FIGURE 6.10

Methods of Protecting
Databases

- ☑ Field checks
- ☑ Change logs
- ☑ Access control procedures
- ☑ User authentication
- ☑ Integrity checks
- ☑ Audits
- ☑ Monitors
- ☑ Range comparisons
- ☑ State constraints
- ☑ Transition constraints
- ☑ Boundary checks
- ☑ Precision checks
- ☑ Suppression
- ☑ Concealing

Change Logs

By using a **change log**, every change that is made to the system is carefully logged, containing both the original and the new values. If errors are made, the database administrator can correct the mistakes.

Access Control Procedures

Access control procedures ensure that all users who enter data are checked as the data enters the central source (database) to test for accuracy, redundancy, as well as who has authorized access to these elements/data. A procedure should be established to determine who has ultimate authority in the event of two conflicting entries.

User Authentication

All entries and actions should be checked to make certain that the user has authorization for not only the database, but individual files and information. This procedure is known as **user authentication**.

Integrity Checks

Integrity checks ensure that the information is constantly checked for integrity, accuracy, and completeness. If one item is incorrect, it can affect the entire database due to the interconnection of data elements.

Audits

As previously stated, **audits** are performed (either by an internal or external party) to ensure that the system is performing as designed.

Monitors

The structural integrity of the database should be monitored. **Monitors** can check values being entered to ensure their consistency with the rest of the database or with characteristics of the particular field.

Range Comparisons

Range comparisons are designed to check each new value to ensure that the value is within an acceptable range. If the data is outside the range, it is rejected and not entered into the database. Range comparisons can be used to ensure internal consistency and to control allowable data in the database.

State Constraints

If **state constraints** that describe the condition of the entire database are not met, some value of the database is in error. A flag can be used to denote a constraint violation, such as an employee's title. For example, there is generally only one president; if a second president is noted, the state of the data has not been met and, therefore, an error is evident.

Transition Constraints

Before changes can be applied to the database, certain conditions must be met before the transition can take place. Before an employee's Social Security number can be added to denote a new hire, the addition of a new position or the filling of a vacancy must be approved and entered into the database.

Boundary Checks

Data that are entered must fall between two acceptable boundaries; this process is a **boundary check**. Upper and lower boundaries can be visible without revealing the specific records involved. Salary ranges are a common example of boundaries—the amounts of the boundaries can be revealed without disclosing individual salaries of employees.

Precision Checks

Precision checks are aimed at protecting *all* sensitive data while revealing as much nonsensitive data as possible or needed. Precision checks can be set up to show what information can be accessed by queries directly or inferred by a query. Information that can be freely disclosed is also shown. Maximum security levels are also defined.

Suppression

Suppression occurs when, in a query, sensitive data values are not provided and are rejected without response. The query is terminated without any indication of what data/values exist.

Concealing

Concealing occurs when, in response to the query, the response provided is *close to* but not exactly the actual value.

Comparison of Database and Operating System Access

Even though some of the same security precautions used with operating systems are applied to databases, some fundamental differences between operating systems and database security should be considered:

- More objects must be protected in a database.
- The lifetime of the data is normally longer in a database.
- Database security is concerned with differing levels of granularity, such as file, record, or field.
- Operating systems protect real resources. In database systems, the objects can be complex logical structures, a number of which can map to the same physical data object.
- The different architectural levels—internal, conceptual, and external—have different security requirements.
- Database security is concerned with the semantics of data as well as with its physical representation.[22]

Database Administrator

The role of the database administrator should be carried out with total coordination of the entire computer security staff. The **database administrator (DBA)** is an individual responsible for installing and coordinating database management systems. In many cases, the DBA is not one person, but a team of experts. The main function of the DBA is the prevention and resolution of conflicts, along with the identification of possible threats to the security of the database. The DBA estimates the probability that each threat will result in a security violation. "The cost of the potential violation and its impact on the firm's operations must be evaluated. With this information, the DBA can select the type and extent of security procedures required to protect the database."[7] (See Figure 6.11.)

SECURITY OF MULTILEVEL DATABASES

The database models previously described are basically one-dimensional. With the increased use of databases today, a great many installations are utilizing multilevel design. A **multilevel database system** is a database system that supports data having different access classes, where permission to access data is determined by the access class of the user requesting access along with the access class of the data.

FIGURE 6.11
Database Security Methods

Method	Threat	Procedure	Policy
Identification	All	Assigns identity/password	Clears legitimate user
Authorization	All	Grants/revokes privilege	Defines need to know/use
Encryption	Disclosure	Selects algorithm	Protects sensitive data and processing allowed
Audit	Modification	Verifies selected data and database structure	Denotes time, place, and extent of audit

Classification in multilevel databases can be applied at different levels in the database—the relation, tuple, attribute, or element level.

The three basic characteristics of database security to a multilevel database security environment are:

1. The security of a single element may be different from the security of other elements of the same record or from values of the same attribute.

2. Two levels—sensitive and nonsensitive—are inadequate to represent some security situations.

3. The security of an aggregate—a sum, a count, or a group of values in a database—may be different from the security of the individual elements.[39]

Military Security Model

In the military, every piece of information is ranked as unclassified, confidential, secret, or top secret. (See Figure 6.12.) Unclassified is considered the least sensitive and top secret the most sensitive. The military security model is an excellent example of a multilevel database security as individuals are granted access to information on a need-to-know basis (what they need to perform their jobs).

Types of Multilevel Methods

Figure 6.13 shows some of the basic methods used to accomplish the multileveling of databases.

Multilevel models of security provide for an arbitrary assignment of access rights to subjects. Multilevel databases differ from other models described as they

FIGURE 6.12
Military Security
Model

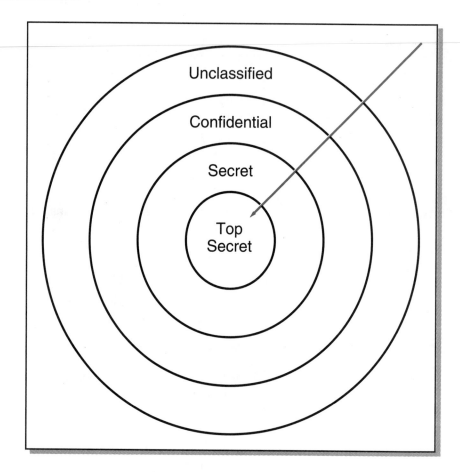

deal with nondiscretionary access control—formal statements about systems security can be made that cannot be made with discretionary systems. "Multilevel models
differ as well in treating not only *access* to information, but also the *flow* of information within a system. Like discretionary models, multilevel models were first
developed for operating systems and later applied to database systems."[22]

Denial of Services. A user at a higher authorization level could have the database
exclusively locked. This is known as **denial of services**.

"This fact must be hidden from the lower level user to prevent a covert signalling channel. However, the lower level user could not access the database if the
higher level user was working with a particular page."[28] Once a covert signalling
channel is exposed, the security of the entire system is vulnerable.

Partitioning. With **partitioning**, "the database is divided into separate databases,
each at its own level of security. This approach is similar to maintaining separate
files in separate file cabinets."[39] However, this approach causes redundancy and
affects accuracy because, with each change, more than one file needs to be updated.
Also, a higher level user may not be able to easily access lower level data; thus,
speed and completeness can be restricted or lessened.

FIGURE 6.13
Methods of Multilevel
Database Security

- ☑ Denial of services
- ☑ Partitioning
- ☑ Encryption/cryptographic sealing
- ☑ Sensitivity locks (labels)
- ☑ Trusted front-end
- ☑ Filtering
- ☑ Windows (views)
- ☑ Query interpretation
- ☑ Granularity
- ☑ Data dictionary enforcement
- ☑ Inference

Encryption/Cryptographic Sealing. The user who accidentally receives sensitive data cannot read it when encryption is used. With **cryptographic sealing**, encrypted check sums are used to determine the authorization and access rights to the data.[22]

Sensitivity Locks. Sensitivity locks combine a unique identifier (e.g., a record number) and the security level. Since the identifier is unique, each lock relates to one particular record. Because many different elements will have the same security level, "a malicious subject should not be able to identify two elements having identical security levels just by looking at the security level position of the lock."[39]

Front-End Processor. A front-end processor (FEP) is a computer through which all input and output travels. Its purpose is to channel and coordinate input and output to a second computer, the back-end processor (BEP). Also known as a guard, the FEP works in the following manner:

1. User identifies self to FEP; FEP authenticates user's identity.
2. User issues a query to FEP.
3. FEP verifies user's authorization to data.
4. FEP issues query to database manager.
5. Database manager performs I/O access, interacting with low-level access control to achieve access to actual data.
6. Database manager returns result of query to trusted FEP.
7. FEP verifies validity of data via check sum and checks classification of data against security level of user.
8. FEP formats data for transmission.
9. The FEP transmits formatted data.[39]

Filtering. The **filter** essentially screens the user's request, reformatting if necessary, so that only data of an appropriate sensitivity level is returned to the user. The filter reformats the query so that the database manager does as much of the work as possible, screening out many unacceptable records. The filter can then provide a second screening to select only data that is authorized for that particular user.[39]

Windows (Views). A subset of a database, a window, contains exactly the information that a user is entitled to access. The **window (or view)** can represent a single user's subset database so that all of the user's queries access only that database. "This subset guarantees that the user does not access values outside the permitted amount, since non-permitted values are not even in the user's database. The view is specified as a set of relations in the database, so that the data in the view subset changes as data in the database changes."[39]

Query Interpretation. **Query interpretation** ensures that the information the user requests from a query is what he/she receives, consistent with the security policy. The integrity of the query must be beyond reproach. There must be no chance for the insertion of malicious code or violations to the system before the query is processed.

Granularity. With **granularity**, data is controlled at the level of the bit, byte, element or word, record, file, or volume (at the lowest level possible). A commonly used level of granularity is the file level, which corresponds closely to the relation level (record level). Often, groups of files are protected as a single object.

Data Dictionary. The **data dictionary** contains the database schema, data conversion algorithms, and characteristics of data attributes. It is vital that data dictionary constraints are strictly enforced. "Since the data dictionary could contain data validity checks which might divulge sensitive information about data value ranges for a database, it should have associated with it a level of trust equal to the highest level of trust for a given database."[28] One solution is to divide the data dictionary into components and protect each according to its relative sensitivity.

Inference. The unintentional compromise or deduction of unauthorized information due to combinations of the possession, known existence, known absence, chronology, and location of authorized information is known as **inference**. The most extreme values are most vulnerable. Protection from inference attacks can include construction of a rule-based semantic layer between the logical database design and the physical implementation of that schema. "It is possible, however, that the performance penalties paid for inference control may make a database management system unusable as an interactive system."[28]

THE FUTURE OF DATABASES

According to the Information Market Observatory (IMO) of the Commission of European Communities, the United States continues to lead the world in the production of on-line databases . . . According to the IMO study, 56 percent of the databases produced and commercially available in 1987 were created in the United States. The EC accounted for just over one-fourth (27 percent), with the remaining 17 percent produced in other developed countries such as Japan, South Korea, Canada, and Australia.[16]

As databases continue to proliferate worldwide, appropriate controls must be instituted to protect the information being collected and stored.

Security and privacy are not incompatible. Privacy involves our right to information about ourselves that is vital to certain of our interests. We may have financial stake in a public utility, and from that someone could ascertain the approximate level of our total assets for purposes of estimating our ability to pay ransom for one of our loved ones; a kidnapping with all its intended agony and likely loss of life can hang in the balance.[2]

Security of databases involves not only the collective maintenance of private, corporate, and government information, but the resultant (and constant) safeguarding of files to ensure correctness and nondisclosure. The task of database protection is, indeed, a formidable (and expensive) one, but all predictions indicate that increased use of databases will continue at a rapid rate throughout the world.

SUMMARY

A **database** is an organized collection of related data, stored together to reduce redundancy. A **database management system (DBMS)** is software that organizes the data used in a database. Data in a database is organized in a hierarchy of **character**, **field**, **record**, **file**, and then recorded into the database. **Entities**, **attributes**, and **keys** are used in a database to describe the data organization and access.

The three types of databases are **hierarchical**, **network**, and **relational**. Interfacing between the user and the database is accomplished by **logical** and **physical access paths** (LAPs and PAPs). **Schemas** and **subschemas** are used to describe the entire database (as seen by the computer or the user).

Security requirements currently being used to protect databases include using **test data files**, matching data for verification, and tracking of database activity.

The design of the database should include the description of **authorization**, **system integrity**, **intentional resolution**, **information flow control**, **semantic integrity**, **concurrency control**, **recovery systems**, **auditing**, and other **controls**.

The basic methods used to protect a database once the design is complete are **field checks**, **change logs**, **access control procedures**, **user authentication**, **integrity checks**, **audits**, **monitors**, **range comparisons**, **state constraints**, **transition constraints**, **boundary checks**, **precision checks**, **suppression**, and **concealing**.

The **Database Administrator (DBA)** is an individual responsible for installing and coordinating the database management systems.

Multilevel databases are systems that support data having different access classes, where permission to access data is determined by the access class of the user along with the access class of the data. Some of the security methods used to protect multilevel databases are **denial of services**, **partitioning**, encryption/**cryptographic sealing**, **sensitivity locks**, **front-end processors**, **filtering**, **windows** (**views**), **query interpretation**, **granularity**, **data dictionary**, and **inference**.

REFERENCES

1. Akl, Selim G., and Dorothy E. Denning, (Rein Turn, ed.), "Checking Classification Constraints for Consistency and Completeness," *Advances in Systems Security* (Norwood, MA: Artech House, Inc., 1988, Vol. III), pp. 271-276.

2. Andreassen, Karl, *Computer Cryptology* (Englewood Cliffs, NJ: Prentice-Hall, 1988), pp. 1-268.

3. Baerson, Kevin M., "SSA, FBI Database Violations Prompt Security Evaluations," *Federal Computer Week*, Vol. 6, No. 1, January 13, 1992, pp. 1, 41.

4. Bagranoff, Nancy, and Mark G. Simkin, "Database Management Systems: Getting the Most from Your Database," *Journal of Accountancy*, May 1988, pp. 122-128.

5. Baker, Richard H., *The Computer Security Handbook* (Blue Ridge Summit, PA: Tab Books, 1985), pp. 1-281.

6. Bradley, James, *File and Data Base Techniques* (New York: Holt, Rinehart and Winston, 1981).

7. Brathwaite, Ken S., "Database Access: The DBMS and the DBA," *Data Security Management*, No. 84-02-12, 1985, pp. 1-8.

8. Berg, H.K., W.E. Boebert, W.R. Franta, and T.G. Moher, *Formal Methods of Program Verification and Specification* (Englewood Cliffs, NJ: Prentice-Hall, 1982).

9. Campbell, John R., "A Brief Tutorial on Trusted Database Management Systems," *Proceedings of 13th National Computer Security Conference*, Washington, D.C., 1990, pp. 553-561.

10. Campbell, John R., "Trusted Database Systems: The Tough Issues," *Proceedings of 13th National Computer Security Conference*, Washington, D.C., 1990, pp. 574-576.

11. Capron, H.L., *Essentials of Computing* (Redwood City, CA: Benjamin/Cummings Publishing Company, 1992).

12. Clyde, Allan R., "Insider Threat Identification Systems," *Proceedings of 10th National Computer Security Conference*, 1987, pp. 343-356.

13. *Computers at Risk . . . Safe Computing in the Information Age*, National Academy Press, 2101 Constitution Ave. NW, Washington, D.C., 1991.

14. "Credit Reports: Getting it Half-Right," *Consumer Reports*, July 1991, p. 453.

15. "Database to Help Curb Fraudulent Claims," *Business Insurance*, June 17, 1991, pp. 29-30.

16. Deans, P. Candace, and Michael J. Kane, *Information Systems and Technology* (Boston, MA: PWS-Kent Publishing Company, 1992).

17. Denning, Dorothy E., Selim G. Akl, Mark Heckman, Teresa Lunt, Matthew Morgenstern, Peter G. Neumann, and Roger R. Schell (Rein Turn, ed.), "Views for Multilevel Database Security," *Advances in Systems Security* (Norwood, MA: Artech House, Inc., 1988, Vol. III), pp. 223-233.

18. Denning, Dorothy E., Teresa F. Lunt, Roger R. Schell, Mark Heckman, and William Schockley (Rein Turn, ed.), "A Multilevel Relational Data Model," *Advances in Systems Security* (Norwood, MA: Artech House, Inc. 1988, Vol. III), pp. 234-248.

19. Enger, Norman L., and Paul W. Howerton, *Computer Security . . . A Management Audit Approach* (New York: AMACOM, 1980), pp. 1-264.

20. Evans, Susan H., *The Computer Culture* (Indianapolis, IN: White River Press, Inc., 1984).

21. Everest, Gordon C., *Database Management* (New York: McGraw-Hill, 1986).

22. Fernandez, E.B., R.C. Summers, and C. Wood, *Data Security and Integrity* (Reading, MA: Addison-Wesley, 1981), pp. 39-53.

23. Fitzgerald, Kevin J., "Security Architecture in an Information Utility . . . Blueprint for Design," *Computer Control Quarterly*, Vol. 10, No. 1, 1992, pp. 21-26.

24. Frenzel, Carroll W., *Management of Information Technology* (Boston, MA: boyd & fraser publishing company, 1992).

25. Garfinkel, Simson, and Gene Spafford, *Practical UNIX Security* (Sebastopol, CA: O'Reilly & Associates, Inc., 1991).

26. Gasser, M., *Building a Secure Computer System* (New York: Von Nostrand Reinhold Company, 1988).

27. Hale, Michael W., "Status of Trusted Database Management System Interpretations," *Proceedings of 10th National Computer Security Conference,* 1987, pp. 340-342.

28. Henning, Ronda R., and Swen A. Walker (Rein Turn, ed.), "Computer Architecture and Database Security," *Advances in Systems Security* (Norwood, MA: Artech House, Inc., 1988, Vol. III), pp. 249-263.

29. Henning, Ronda R., and Swen A. Walker, "Data Integrity vs. Data Security: A Workable Compromise," *Proceedings of 10th National Computer Security Conference*, 1987, pp. 334-339.

30. Herschberg, I.S., "How Secure is Your Database?" *Direct Marketing*, October 1988, pp. 76-77.

31. Hutt, Arthur, Seymour Bosworth, and Douglas B. Hoyt, *Computer Security Handbook* (New York: Macmillan Publishing Company, 1988), pp. 1-399.

32. Jajodia, Sushil, "Tough Issues: Integrity and Auditing in Multilevel Secure Databases," *Proceedings of 13th National Computer Security Conference*, Washington, D.C., 1990, pp. 577-580.

33. Liskin, Miriam, "Can You Trust Your Database?" *Personal Computing*, June 29, 1990, pp. 129-134.

34. Long, Larry, and Nancy Long, *Computers* (Englewood Cliffs, NJ: Prentice-Hall, 1990).

35. Martin, James, *Security, Accuracy, and Privacy in Computer Systems* (Englewood Cliffs, NJ: Prentice-Hall, 1973), pp. 1-626.

36. Nabours, Mac, "Living with Sin: The Security Information Network," *Security Management*, April 1988, pp. 61-72.

37. Nickerson, Robert C., *Computers* (New York: HarperCollins, 1992).

38. Olson, Ingrid, Eugene F. Troy, Milan S. Kuchta, and Brian W. McKenney, "Disclosure Protection of Sensitive Information," *Proceedings of 13th National Security Conference*, Washington, D.C., 1990, pp. 189-200.

39. Pfleeger, Charles P., *Security in Computing* (Englewood Cliffs, NJ: Prentice-Hall, 1989), pp. 1-538.

40. "Privacy Versus Database Statistics," *Science News*, Vol. 140, November 16, 1991, p. 315.

41. Russell, Deborah, and G.T. Gangemi, Sr., *Computer Security Basics* (Sebastopol, CA: O'Reilly and Associates, Inc., 1991), pp. 1-441.

42. Schell, Roger R., and Dorothy Denning (Rein Turn, ed.), "Integrity in Trusted Database Systems," *Advances in Systems Security* (Norwood, MA: Artech House, Inc., 1988, Vol. III), pp. 264-270.

42. Schmitt, Warren, "Information Categorization and Protection," *Proceeding of 13th National Computer Security Conference*, Washington, D.C., 1990, pp. 34-36.

44. Schou, Corey, *Information Security Modules*, Idaho State University, Pocatello, 1991, pp. 1-325.

45. Smith, Gary W., "Going Beyond Technology to Meet the Challenges of Multilevel Database Security," *Proceedings of the 12th National Computer Security Conference*, Baltimore, MD, 1989, pp. 1-10.

46. Stair, Ralph, *Principles of Information Systems: A Managerial Approach* (Boston, MA: boyd & fraser publishing company, 1992), pp. 1-701.

47. Su, Tzong-An, and Gultekin Ozsoyoglu (Rein Turn, ed.), "Data Dependencies and Inference Control in Multilevel Relational Database Systems," *Advances in Systems Security* (Norwood, MA: Artech House, Inc., 1988, Vol. III), pp. 277-286.

48. *The Handbook of Information Security* (San Jose, CA: Arca Systems, Inc., 1991), pp. 1-202.

49. Turn, Rein, "Privacy Transformation for Databank Systems," *Computer Communications Security*, National Computer Conference, 1973, pp. 420-431.

50. Vetter, Linda L., "Relational Database Security," *ISP News*, Vol. 3, No. 1, January/February 1992, pp. 1, 11, 12.

51. Whitehurst, R. Alan, and Teresa F. Lunt, "The Seaview Verification Effort," *Proceedings of the 12th National Computer Security Conference*, Baltimore, MD, 1989, pp. 18-27.

52. Wood, Tim, "A Trusted Database Machine Kernel for Nonproprietary Hardware," *Proceedings of the 12th National Computer Security Conference*, Baltimore, MD, 1989, pp. 11-17.

REVIEW QUESTIONS

1. Briefly describe the hierarchy of data that is input into a database.

2. Explain the differences between a data entity, a data attribute, and a data key.

3. What are some of the advantages and disadvantages of a database?

4. Describe the three different database models—hierarchical, relational, and network.

5. Compare logical and physical access of databases.

6. What are DDL and DML?

7. What are some of the possible threats to databases?

8. What are some of the methods used to protect databases from violation, misuse, or disclosure?

9. Compare a schema and subschema.

10. What is the role of the DBA?

DISCUSSION QUESTIONS

1. In some computer installations, the database administrator (DBA) is in charge of the database but does not necessarily have the authority to override security. The DBA may not have the authority to execute the database; rather, the DBA provides policy analysis and direction of the database. In a nutshell, the DBA may not have the actual power to "run" the database, but instead concentrates on the design and other policy issues. What are the pros and cons of this approach?

2. In *Information Systems and Technology*, a statement says that "On-line databases containing in-depth and timely information are both a *cause of* and *solution to* the growing complexity of conducting business around the world. The international business environment is a twenty-four-hour-a-day arena in

which access to the latest information can shift the competitive balance in the marketplace within minutes."[16] Comment on this statement.

3. Three emerging trends in databases that present related security vulnerabilities are:

   a. Proliferation of on-line databases

   b. Transborder accessibility to on-line data

   c. The structure of on-line databases and related services

   What are some of the security considerations of each of these trends?

4. If databases present such a critical threat to the security and control of information, why don't organizations stay with the "old way of doing things" rather than putting "all their eggs in one basket"?

5. A database presents a unique challenge to a hacker who likes to "snoop around." The database must be protected yet remain free and accessible. Is it possible for a database to be both "open" and "closed" at the same time?

EXERCISES

1. Select an organization/company that has a database system. Interview a key member of their computer center to gain information about what protection mechanisms are employed to safeguard the database.

2. George Orwell, in his well-known book *1984*, comments on "Big Brother is Watching You." Are databases the Big Brother he mentions; why?

3. Locate an article in a computer-related journal that portrays a recent database attack. How could the attack have been prevented?

4. The U.S. Credit Reporting Industry is coming under attack due to its nonregulatory nature. One suggestion to regulate it is to require that all negative information be reported to the consumer. Locate an article that reports this current movement to regulate this industry.

5. Develop a schema and subschema of your student file at your university.

PROBLEM-SOLVING EXERCISES

1. Using Figure 6.3 as an example, create a hierarchy of data for your student file at your school. (Note: Be sure to include all files that your university maintains—e.g., finances, classes, grades, parking tickets, health records.)

2. Using Figure 6.4 as an example, create a listing (using names of your classmates or friends) to show attributes, keys, and entities.

3. Create a hierarchial database to illustrate the members of your family.

4. Create a relational database to illustrate the information contained in your student records at your university.

5. Develop a method of database protection that would limit access to the database to authorized users and, at the same time, not disclose the information if an attempt is made to see if it exists.

CASES

D'Angelo Transportation, Inc.

Nicholas D'Angelo is president of a large trucking company that moves freight for various companies/organizations along the Atlantic coastal areas. This highly competitive industry must maintain accurate records of shipments so that they are not lost or misrouted; these mistakes increase costs and thus decrease profits.

Mr. D'Angelo wants to convert his company's distributed information on shipments (which is currently kept at each of the 10 regional terminals) to a central database in the organization's headquarters in Baltimore, Maryland. Some of the shipments D'Angelo makes are highly sensitive and must remain confidential as they involve military and government agencies.

Mr. D'Angelo wants to assemble each of the 10 regional managers and the key staff members of the company's computer center at the Baltimore headquarters to discuss some of the advantages and disadvantages of the proposed database approach.

1. What are some of the factors that should be considered in the discussion?

2. What are some of the security considerations of the proposed conversion to a database?

3. What training should employees receive if this conversion is implemented?

Tri-City College

Tri-City College, like most educational institutions, maintains a great deal of information in its database about its 5,000 students. Some records are now kept in individual locations for the use of faculty, administrators, campus police, the health center, and the student center activities office.

Joseph Lapton has been hired by Tri-City to bring all information relating to students together in one central database that would be housed in the university's central computer center. Faculty and administrators are concerned about the privacy of this information and the possibility of disclosure. In other words, everyone wants to hold on to their own information in order to protect it and have it easily at hand.

1. What are some of the advantages of Tri-City converting to a central database?

2. What are some of the disadvantages of conversion to a central database?

3. How could the information be "sectioned off" to protect it (so that not everyone has access to all information)?

4. What types of training programs should be organized to ease the transition for faculty and administrators?

CHAPTER 7

NETWORK AND TELECOMMUNICATIONS SECURITY

LEARNING OBJECTIVES

After studying this chapter, you will be able to:

1. Explain the basic concepts of data communication, telecommunications, and networking.
2. Differentiate between analog and digital signals.
3. Describe how a modem is used to transmit signals.
4. List and describe the various communications media in use today.
5. Compare the three major network topologies—star, ring, and bus.
6. Explain the major security risks that are present in a communications environment.
7. List the types of potential network and communications failures.
8. Describe some of the products currently available to secure communications.
9. Describe some of the special communications security considerations present in the emerging applications of connectivity.
10. Define the following terms.

TERMS

- Connectivity
- Data Communications
- Telecommunications
- Teleprocessing
- Networking
- Analog Circuits
- Digital Circuits
- Modem
- CODEC
- Asynchronous Transmission
- Synchronous Transmission
- Handshaking
- Communications Protocol
- Synchronous Data Link Communication (SDLC)
- Systems Network Architecture (SNA)
- International Standards Organization (ISO)
- Open Systems Interconnect (OSI)
- Higher-Level Data Link Control (HDLC)
- Packet-Switched Network
- Wire Cable

TERMS (Continued)

- Twisted-Pair Wiring
- Coaxial Cable
- Fiber-Optic Cable
- Microwave
- Line-of-Sight Transmissions
- Land-Based Microwave
- Satellite-Based Microwave
- Communication Carriers
- Value-Added Network
- Private (Leased) Line
- Switched (Dial-up) Line
- Node
- Network Topology
- Star Network
- Ring Network
- Bus Network
- Local Area Network (LAN)
- Cable Tapping
- Emanations
- Multiplexing
- Inductance
- Parasitic Telephone Transmitter

- Electronic Espionage
- Facsimile (FAX)
- Automatic Teller Machine (A
- Personal Identification Numbe (PIN)
- Electronic Funds Transfer (EF
- Wire Transfer (WT)
- Clearinghouses
- Point of Sale (POS)
- Telephone Bill Paying System (TBP)
- Home Banking System (HBS)
- Passive Wiretapping
- Active Wiretapping
- Cellular Telephone
- Automatic Number Identificat (ANI)
- Caller ID
- Integrated Service Digital Net (ISDN)
- Open Systems Interconnection (OSI)
- Electronic Filing

NEW FOUNDATION COULD MONITOR NETWORK SECURITY

According to the National Academy of Sciences, the United States must establish an Information Security Foundation to improve and monitor the security of nationwide and global networks. If we do not, the Academy warns, legal, scientific, and medical databases will be at risk for security breaches, and the United States could find its sales of computer products to foreign nations seriously undermined.

Continued

NEW FOUNDATION . . .

From page 164

Sixteen Academy experts cited the case of Robert Morris, who was convicted of releasing a replicating program, called a worm, into the global Internet network. About 6,000 computers on the network, on which scientists and government agencies depend to exchange data, were brought to a standstill.

The Morris case is not the only one that concerns the Academy. Unprotected networks have been invaded by pranks, break-ins, and errors. For example, in 1987 electronic thieves changed currency exchange data at a major foreign auto manufacturer, stealing millions of dollars. Hackers came close to stealing $15.2 million from a state lottery fund via its database.

In the case of such security breaches, scientists may be particularly vulnerable because they depend on the free exchange of ideas. However, if their ideas or discoveries are overly publicized (i.e., if unauthorized users can access the data on the network), many journals will reject new papers on their findings. Even patents may be in jeopardy. Other sensitive and vulnerable information includes hospital and insurance medical records.

Majory Blumenthal of the Academy says that, surprisingly, although European countries rely less on networks, they are far ahead of the United States in assuring security by setting strict network guidelines. Because the United States is perceived as "behind the times" in this respect, U.S. computer products could lose European buyers if they fail to incorporate state-of-the-art security devices and protocols. Some Academy members are frustrated by the fact that the U.S. intelligence community has barred the exporting of some of the best products, without explanation. In addition, less sophisticated means of securing data often require export licenses, thus producing another barrier that Academy members recommend lifting.

The proposed Information Security Foundation (ISF) would produce guidelines for the computer industry that would play the same role that building codes play in the construction industry. As the Academy conceives it, the ISF would also foster research on security as well as monitoring computer crime.

As reported in "Computer Cops Needed to Police Networks," *New Scientist,* December 15, 1990, p. 16.

In the last decade, a significant amount of attention has been given to the security and control of computer systems and the data and other valuable resources they contain. The users of early generations of computers were essentially free to use the systems as they wished. In addition, practically anybody with access to a terminal could log on to any early time-sharing system and browse through any of the files on the system. Outside of the defense community and other organizations concerned with national security, there was little, if any, concern about the security of computer systems.

Now that businesses and other organizations have recognized that the information contained in their computers is an extremely valuable resource and, so should be protected as are other resources, dramatic changes are taking place. Most organizations today have an individual or a group that specializes in the protection and control of computer resources. Corporate auditors and CPA firms employ EDP audit specialists, dedicated to examining computer systems and applications software for compliance with prudent security and control practices. Computer vendors and others have developed and successfully marketed security products for practically all models of computers.

Many organizations today are not only storing vast amounts of data and information in databases, but are accessing these large repositories via networks or other telecommunications media. As with databases, networks require special security considerations to protect them from misuse or accidental disclosure. This chapter will briefly discuss the principles of networking and telecommunications before delving into the specifics of security issues.

TELECOMMUNICATIONS AND NETWORKS

Computers and their components need to communicate with each other. "When the hardware is nearby, communication takes place over cables. Longer-distance communication, however, requires other techniques that fall into the area of data communications."[48] In the 1960s, computers were numbered in the tens of thousands. Today, they number in the tens of millions, and information is virtually everywhere. Making this information more accessible to a greater number of people—while protecting the transfer of the information—is the current challenge. Businesses and other computer-using communities are searching for ways to interface, or connect, a diverse set of hardware, software, and databases. **Connectivity,** achieved through data communications, telecommunications, teleprocessing, and networking, is necessary to facilitate the electronic interchange between computers.

- **Data communications** is the collection and distribution of the electronic representation of information from and to remote facilities—in data, text, voice, still picture, graphic, or video format. Data communications usually involves telephone lines, satellites, and coaxial cable.

- **Telecommunications** encompasses not only data communications but any type of remote communication, such as a transmission using a television signal.

- **Teleprocessing** (TP) is the combination of telecommunications and data processing.

- **Networking** is the integration of computer systems, terminals, and communication links.

Communication Concepts

Data communications and teleprocessing are used when it is necessary or more economical to physically separate the processing computer from the source of the input data, the site of the output usage, or the computer user. An airline reservations system is a common example. Reservations systems use data communications equipment and techniques to connect travel agents and airline personnel to a single computer (or set of computers) that is continually recording reservations, answering space availability inquiries, and performing necessary control tasks.

High-capacity cables, capable of carrying hundreds of thousands of characters of information per second, connect computers to high-speed machines such as other computers, disk drives, and tape drives. However, many machines connected to computers operate at much slower speeds. These slower machines are connected to the computer by lower capacity and less expensive cables similar to telephone lines. Direct cable connection becomes impossible at distances of more than one mile and usually becomes inefficient after 2,000 feet. When users miles away are communicating directly with a computer, they are said to be connected via a data communications circuit.

Communication Circuits

There are two types of data communications circuits—analog and digital. **Analog circuits** are signals that use the continuous fluctuations over time between high and low voltage. The voice telephone network typically uses analog circuits capable of transmitting the full range of sounds that the human voice can make. In a similar manner, the hands of a clock can portray the full range of times in a 12-hour period. Analog communication is constantly viable within a predetermined range of frequencies.

Digital circuits use the binary on-off principle to communicate information in digital form. A computer can convert sound into a series of digits that portray the volume, pitch, and other distinguishing characteristics. Another computer at the receiving end then reconverts these digits into sound. Digital circuits are capable of moving more information over a given distance in a given time than analog circuits, and they eliminate the noise distortion problems common to sound-carrying circuits. Digital circuits are replacing analog circuits in the telephone system in many areas. (See Figure 7.1.)

Data are transmitted at the speed of electricity, but one bit at a time. Slower transmission speeds are often used where possible so that slower and less expensive equipment can be used at each end of the circuit. It is also possible to go much faster than 9,600 bits per second, up to several million bits per second on special circuits available from communication carriers.

Transmission errors occur frequently, usually when the communications circuit is momentarily disrupted. These disruptions often destroy some of the bits being transmitted, thereby causing a condition known as a **parity error**. These parity errors are detected by the receiving equipment, which notifies the communications control program in the central computer that an error has occurred. This program takes the necessary corrective action, usually retrying the transmission until error-free data have been achieved.

FIGURE 7.1
Signal Transmission

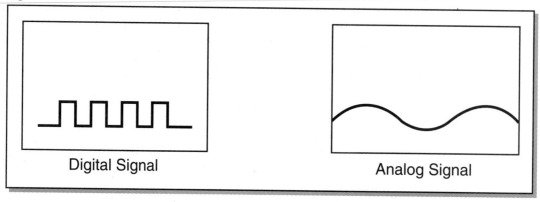

Digital Signal   Analog Signal

Modems and CODEC

Digital information to be transmitted on analog circuits is first converted to analog signals by a special device known as a **modem** (MOdulator-DEModulator), then reconverted to digital information by another modem at the receiving end. The analog circuits are obtained from a common carrier, usually the local telephone company. Data communication circuits may be regular dialed telephone lines or dedicated lines leased from the communications carrier. Leased lines cannot access or be accessed by the dial-up network. Modems are required at each end of both types of lines to perform the necessary digital-analog-digital conversion. (See Figure 7.2.)

When all digital circuits are used for data transmission, a different device is required at each end of the circuit, similar to modems on analog circuits. This device is called a **codec** (COder, DECoder), which codes or decodes the information being transmitted.

FIGURE 7.2
Modem

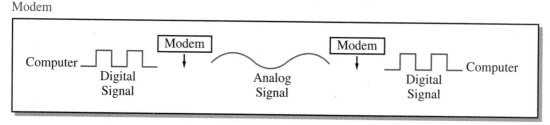

Asynch and Synch Transmissions

The two basic methods for transmitting information are known as asynchronous and synchronous. (These are often referred to as asynch and synch.)

Asynchronous uses a starting bit for information followed at regularly timed intervals by the bits representing a character followed by another start bit, and so on. This is the least expensive and most widely used transmission method for low-speed systems.

The **synchronous** method uses a process called "**hand-shaking**" during which the sending and receiving devices establish a common clocking rate and transmit thereafter at the intervals specified by the clocking rate and without the need for starting bits. The sending and receiving ends are said to be synchronized. Synchronous equipment requires internal clocking and is more expensive, but synchronous transmission does not need the starting bits to separate characters and is faster.

Communications Protocols

A number of communications protocols have been designed for use in the synchronous environment. A **communications protocol** defines the format and characteristics of the data that is transmitted. For example, raw data that is received for transmission is divided into segments or packages, typically of fixed length. Each package of data is enclosed in an "envelope" with a header containing information about the package, such as data length, address of destination, and error detection information, and possibly a trailer specifying other information. The communications protocol rigorously and formally defines the format and content of this packaging and also specifies error handling and other characteristics.

An example of a widely used communications protocol is **SDLC (Synchronous Data Link Communications)**. SDLC was developed by IBM as part of its **Systems Network Architecture (SNA)** which establishes the ground rules and defines the common interfaces for data communication between all IBM-developed computer systems. Because of its wide use, SNA/SDLC is a de facto standard within the data processing industry and has been adopted by other computer and communications equipment vendors as well.

The **International Standards Organization (ISO)** has also developed and published an internationally accepted set of standards known as the **Open Systems Interconnect (OSI)**, a seven-layer communications architecture that is functionally similar to SNA. OSI consists of a suite of protocols that define, or will eventually define, the seven layers. Standards have been published defining the first three layers of the architecture; discussions are ongoing concerning the remaining layers. The second layer, link control, is defined by **HDLC (Higher-Level Data Link Control)**, a synchronous protocol very similar to SDLC.

Perhaps the best known of the ISO/OSI protocols is the X.25 standard, which defines Layer 3, Network Control. X.25 defines networks that are known as **packet-switched networks**. In these networks, messages are divided into packets at the sending site and transmitted one packet at a time. The route from the message's origin to its destination may be a simple, point-to-point routing, or it may be complex, going from node to node (a node is a location on the network that may send or receive messages) before reaching its final destination. As each packet reaches, for example, node A, the packet is switched to the next node along a path to its destination. Depending on the network configuration and availability of links, there may be several possible paths to the destination and each of the individual packets could conceivably take a different path to reach the same final node. For example, to send a message from New York to San Francisco, it may be possible to go through Chicago, Denver, St. Louis, or Dallas. A four-packet message might have one packet routed through each of these locations. In this type of network, packets may arrive out of order in an asynchronous fashion and must be reassembled properly to complete the message. All of the details necessary to handle these and other complex situations are rigorously defined in the protocol.

Other computer vendors have developed similar communications architectures—Digital's DECnet is a good example. In general, however, most of the industry accepts and is implementing either IBM's SNA or the OSI model. Developments over the last few years indicate that these two communications architectures are slowly but surely converging to a common point.

Communications Channel Media

Different media can be used for communications channels. In many cases, a channel can consist of several different media connected in sequence—wire cables, fiber-optic cables, microwaves, or satellites.

Wire Cables. Wire cables are the oldest media. Data is transmitted over a wire cable by sending an electrical signal along the wire (e.g., telegraph). Today, two main forms of wire cables are used: twisted-pair wiring and coaxial cable. **Twisted-pair wiring** consists of two wires twisted together. Most telephone lines use this medium for local voice communications, but it is also used for data communication. Twisted-pair wiring is relatively inexpensive, but its transmission rate is slow compared to other media (up to 80,000 bps). **Coaxial cable,** which consists of copper wire heavily insulated with rubber and plastic, is used with cable television. It is more expensive than twisted-pair wiring but can transmit data at a much greater rate (up to 50,000,000 bps).

Fiber-Optic Cables. An alternative to wire cable, **fiber-optic cables** consist of bundles of glass or plastic fibers. Each fiber is 1/2000 inch thick—about the size of a human hair. Data is transmitted by a laser that pulses light through the fiber. Each pulse represents a bit, so data is transmitted in a digital form. The laser can pulse about one billion times per second, so data can be transmitted at about one billion bps. For long distance communications, fiber-optic cables are less expensive than wire cables.

Most telephone companies use fiber-optic cables for some communications. Because a voice is an analog signal, the voice must be converted to a digital form (bits) for transmission over a fiber-optic cable and then converted back to analog form at the receiving end. Computer communication using fiber optics, however, does not require conversion because computer signals are already in digital form.

Microwave Systems. **Microwaves** are special types of radio signals that are sent through the air from one microwave antenna to the next. Microwave transmissions are **line-of-sight transmissions**, meaning that there must be nothing between the antennas to stop the signal. Both voice and data can be transmitted by microwaves. Microwave systems are very expensive, but they do not require direct cables and can transmit data very rapidly (up to 45,000,000 bps).

Two types of microwave systems are used: land-based and satellite-based.

Land-Based. A **land-based microwave** transmits data from one microwave antenna to the next. Because microwave transmissions are line-of-sight and because of the curvature of the earth, land-based microwave antennas must be no more than 30 miles apart.

Satellite-Based. A **satellite-based microwave** sends data from an earth microwave antenna up to a satellite and then down to another earth antenna. Satellites are usually up about 22,300 miles in space and revolve around the earth at the same velocity as the earth rotates, so they appear to be in a fixed position in the sky. Satellite-based systems can transmit data over a much greater distance than land-based systems.

Communications Carriers

Communications carriers are the companies that supply facilities for transmitting analog and digital information. Several federally regulated companies provide most data communications services in the United States, using the existing voice facilities. The best known are the former telephone operating companies of the AT&T system. The RBOCs (Regional Bell Operating Companies), as they are known, provide the majority of voice and data communications facilities in the United States.

Other communications carriers often specialize in data communications and compete with or supplement the telephone company networks. Western Union, GTE, and MCI are among the better-known competitors licensed to operate as communications carriers within the United States. These and others also supply international data communications services.

Carriers use a variety of technologies including high-data rate microwave facilities, satellites, fiber optics, and radio systems. Most carriers use several or all of these technologies; a single signal may travel over land line, microwave, radio, and land line again before it completes its journey.

Another class of common carriers offers what are known as **value-added networks**. The value-added carriers such as Tymnet or GTE Telenet provide network services using common carrier facilities and specialized data communications equipment. In addition to data communication, these carriers provide other services that add value to simple communications capabilities. Examples include conventional data processing services as well as specialties such as credit card authorization.

A **private line** (or **leased line**) provides a dedicated data communications channel between two points in a computer network. The charge for a private line is based on channel activity (bps) and distance (air miles).

A **switched line** (or **dial-up line**) is available on a time-and-distance charge, similar to a long-distance telephone call. The connection is made by dialing up the computer, then a modem sends and receives data.

Network Topologies

A telephone is an end point, or **node**, connected to a network of computers that route voice signals to other nodes (telephones) around the world. (A node can be a terminal or another computer rather than a telephone.)

A **network topology** is a description of the possible physical connections within a network. The topology—the configuration of the hardware—indicates which pairs of nodes can communicate. The three major topologies are star, ring, and bus. (See Figure 7.3.)

FIGURE 7.3
Network Topologies

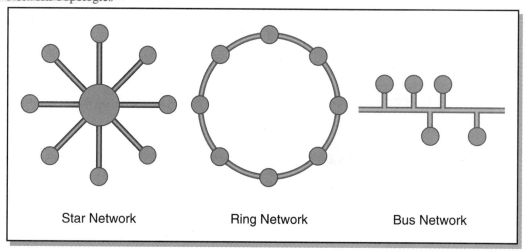

Star Network   Ring Network   Bus Network

Star Network. A **star network** has a central host computer that is connected to a number of smaller computer systems. The smaller systems communicate with each other through the host and usually share the database. Both the central computer and the distributed computer systems are connected to terminals. Any terminal can communicate with any other terminal in the network.

Ring Network. A **ring network** has computer systems that are approximately the same size, with no one computer system as the local point of the network. When one system routes a message to another system, it is passed around the ring until it reaches its destination address.

Bus Network. A **bus network** permits the connection of terminals, peripheral devices, and microcomputers along a central cable. Devices can be easily added to or deleted from the network. Bus networks are most often used when the devices to be linked are physically close to each other.

Local Area Networks

The proliferation of personal computers and intelligent workstations in the office and other workplaces created a need to link these devices to each other and to commonly used peripheral devices. The **local area network** (**LAN**) was developed to meet this and other requirements. LANs are marketed by a wide variety of companies for practically every type of personal computer or workstation. LAN implementations vary considerably from vendor to vendor.

A typical LAN consists of a relatively small number of workstations (usually less than 50) located geographically close to each other in an office or small department. The workstations are connected in a ring manner by communication links, typically coaxial cable or telephone-type lines. One of the workstations, or a special, dedicated file device, usually serves as a file server or primary storage device for the network. Other server devices, such as printers and gateways to mainframe computer systems, may also be connected to the LAN.

Each of the workstations utilizes LAN-specific software to communicate with the other workstations and the servers. Typical functions offered by LAN vendors include electronic mail, file transfer, and backup and recovery facilities. Early LAN implementations paid little attention to security and control of these networks; thus, the data stored on them was quite vulnerable to unauthorized access. More recent LAN products have addressed security and control issues. The somewhat informal nature of LAN use, however, still makes it more vulnerable to security violations than traditional computer systems.

The Ethernet, developed by Xerox Corporation, is perhaps the most well-known example of a LAN. Another popular LAN is the Token Ring Network available from IBM.

SECURITY CONSIDERATIONS

Today, telecommunications are strategic to every nation's progress. Governments are well aware of the economic benefits afforded by an adequate information network . . . There is no question that telecommunications are of major importance for both users and manufacturers of telecommunications as worldwide business expansion demands connectivity. [23]

The decade of the 1980s will probably be remembered in the computer industry as the years of "connectivity." Connectivity is currently being utilized in the form of local area networks (LANs) and wide area networks (WANs). By definition and design, a LAN is a high-speed network covering a limited area (perhaps a campus), connecting terminals, microcomputers, and specialized devices. A WAN is a specialized network that covers a much wider geographical area—sometimes worldwide. Both such networks are connected to the "outside world" via commercial networks through a special interface. Personal computer users connected to a LAN (or WAN) have very powerful capabilities for creating, moving, and storing files; generating graphics; and communicating via message systems with other LAN (or WAN) users; they also have available to them both personal and centralized files.

Widespread Use of Networks

In a 1986 statistical report issued by the National Center for Computer Crime Data, it was pointed out that the growth rate in personal computer network installation had risen rapidly in the previous two years. In 1982, there were 100,000 network workstations; in 1984, 220,000, in 1986, 580,000.

A LAN generally consists of a group of microcomputers physically connected together. A microcomputer must be equipped with the proper hardware, normally an internal circuit board, and software. In all too many instances, however, companies will purchase less-expensive workstations designed specifically for networking. These workstations have a processor that is able to run microcomputer software, but they lack the disk drives or communication parts needed for stand-alone computing. By being connected to the network, these workstations can address and execute software stored on the file server or other workstations and oversee the network operations. In this scenario, it becomes the system administrator's responsibility to act as the security officer of the network.

Networking can offer a company many benefits in cost savings. Conversely, however, due to networking's "inherent openness," securing the data stored throughout the network becomes a monumentally difficult task. "Three years ago, security on micros was never a thought," explained Micke Schwartz, vice-president of Prime Factors, Inc. "Now it's an afterthought."[45] It is imperative, with the widespread use of networks, that security move from an "afterthought to a forethought" *before* the installation of a network is completed. It is naive to think that we can build a security system *after* the network is operational. Security planning must go hand in hand with each step of network planning.

Basically, four main areas should be considered prior to the installation and operation of a network:

1. Network and system design
2. System risk analysis
3. Software protection and security
4. User verification and security

The initial step is to determine the design of the actual network. To properly design the network, the number of users must be estimated and the layout of the workstations must be planned. The more people who will be using the system, the greater the security limitation. This concept is not difficult to grasp when considering that the majority of threats will come from the system users themselves.

Once the number of users has been estimated, the location of these users becomes a primary focus. For the systems design team, determining how the workstations are connected is critical. For the security analyst, special concerns are verifying users to prohibit unauthorized access, maintaining validity of transmitted data to ensure integrity, and preventing corruption of internal data through intentional or unintentional means. It is the security analyst's ultimate responsibility to ensure that the system is designed so as to limit the amount of depravity, both externally and internally.

Total Picture Security

If a firm is to develop a "total picture" security program, the security analyst must consider both inside and outside corruption and violation of the network system. The National Center for Computer Crime Data, in its Computer Crime Census, profiles the typical occupation of computer crime defendants:

- Programmer
- Student
- Input clerk
- Bank teller
- Accomplice
- Unskilled
- Unemployed with access
- Computer executive
- Miscellaneous[10]

As we can readily assess, the threats are real and they come from both inside and outside the organization. If a network is cabled between buildings, then data can be intercepted by electronically monitoring the physical cables running between buildings. The dial-in phone line can be just as easily compromised by using the same means or by the use of a "Trojan Horse" scheme. The phone lines and cable must be shielded so as to limit the amount of data that can be intercepted. Shielded cables alone will not ensure that the data will not be gathered by electronic means by the wrong sources. To thwart these interceptions, data encryption techniques must be used, particularly when data is transmitted from one location to another via phone lines. Encryption should be considered part of the basic system security design, not an "add on." The initial incorporation of encryption not only starts security out "on the right foot," but also prevents large data files from having to be encrypted later—a very time-consuming task. No matter how sophisticated the encryption technique is, however, there are always risks.

To intensify the inherent vulnerability, network users interface with the corporate mainframe and downloading information. This step is very cost-effective for the corporation, but it puts vital information at the point of lowest security within the information system. Much of the data at LAN workstations is stored on floppy disks; this fact alone increases the potential for data corruption because these disks can be copied easily or the data contained can be made incorrect with little effort. These facts point to the need for a good physical layout scheme. Workstations that have the ability—and will be presumed to use—sensitive data should be located in an area that is secured against unwanted intruders. Even though this restriction may present problems for janitorial/maintenance services and MIS visibility, it must be considered crucial if the data's integrity is to be ensured.

System Risk Analysis

Considerable analysis and evaluation must be conducted prior to determining the amount of security features needed for a network. A risk analysis should be undertaken to determine the intrinsic risks involved with the proposed network. Damage to a corporation's computer or data goes far beyond the costs of repairing the system. A thorough risk analysis should include estimates of the impact of events that have not yet occurred, and should point out the potential weaknesses in the system itself.

System risk assessment techniques tend to be time-consuming and not immediately cost-effective. The lengthy statistical evaluation produced by a risk analysis often raises more questions than it answers; thus, the risk analysis is often dismissed for the sake of expediency. This false sense of "safeness," however, could be shattered if a major violation occurred. No matter what the cost and time consumed, a risk analysis should be incorporated into every system's development project and should include all levels of the MIS/SP department. Any point throughout the organization where data is originated or processed should also be included in the analysis to make certain that all the pieces of the puzzle are included. One missing piece makes for vulnerability. The MIS director must work on the premise that security is not the sole responsibility of the security staff—all users must be included.

Many software products on the market today aid in preparing of the risk assessment. Each of these software packages takes a different approach to the risk analysis, yet almost all work in quantitative measures. Many of these products will go help users determine specific areas of risk and provide recommendations based on security standards.

Risk assessment, to be totally effective, must cover all aspects of the proposed network, including assessment of data from origination to execution, workstation placement, and relative importance of the data. For example, if the data stored in Workstation 5 keeps track of departmental office supply usage, then the assigned data security level should be low. On the other hand, if the data is to be uploaded to the corporate mainframe for market research studies, then the data by its own nature will have a much higher level of security. Under the provisions of the Privacy Act, even addresses and phone numbers of employees and clients must be safeguarded against compromise. Also, each time a new workstation or phone line is added to the network, the risk assessment must be reviewed to ensure proper security of the system—an endless process, but a necessary one. Risk assessment is not a one-time process; it must be ongoing and constant to be effective.

Software Security and Protection

Software security and protection are often considered two separate fields. In a mainframe environment, the director of the computer center is often responsible for ensuring that software installed on the system stays on the system and that adequate means of protection to secure their data and programs are offered to the users. In other words, security must be considered when installing software, and protection must be constantly maintained thereafter.

The systems administrator for a network must also be charged with the same responsibilities. One method utilized for network security and protection is to devise a standard operating procedures manual for each user. Along with this manual, the SA should orient each user on the operating procedures and the vital need for security measures. These orientation awareness sessions should go beyond the basics—they should include all parameters that the user will encounter when working with the network.

In a network environment, the workstation is often termed "the file server," and software programs are normally stored on hard disks. The SA is, in essence, offering to share these programs with the other users on the network. Before sharing these disks, the SA should ask the following questions:

1. *What disk or portion of a disk can we share with the user?* The administration can limit sharing to a unit as small and specific as *only* the viewing of a *simple* file within a subdirectory of a disk.

2. *What are the needs of the user?* If the user does not need certain data to perform his or her job, access should be prohibited to prevent "snooping."

3. *Is access to be read only or read/write?* By limiting the user's access to *only* the viewing/reading of a data file or the execution of a program, the SA has reduced the possibility that a user will accidentally erase a necessary file from the disk. At the same time, the user is prohibited from manipulating the data or writing on the file. Thus, basic software protection is maintained.

Password protection schemes can provide a very workable method of limiting user access to *only* those files authorized by the SA. When logging on to the system, the user must enter a predetermined password to gain access. Password protection can be carried down to the individual files and fields within the file.

Another way to prevent access is to "hide" the files or directories on the disk. This is accomplished by changing the program or data file attributes to "hidden" or "read only." The one problem with this method is that the attributes can just as easily be changed back by someone with the proper knowledge; therefore, "write" access must be limited to a small number of people to ensure the integrity of the files.

Software security is best thought of as add-on or value-added security. Through the use of software programs designed internally or purchased externally, added levels of protection, generally not offered by the system itself, can enhance the system's security. It is important, however, to keep in mind that these packages are to be used in addition to local user security steps—not as a replacement for them.

The first and foremost approach, in addition to limiting access, is data encryption. This feature is normally offered by a network vendor as an option or may be offered by a third-party vendor. One of the primary drawbacks of encryption is the initial expense and the amount of time needed to process the data through an encryption technique. The risk analysis, as previously mentioned, should point out the level of risk and indicate whether encryption is needed for a particular application. Obviously, if the information is vital to the corporation, then the investment of money and time must be made.

Encryption cannot always prevent data from falling into the wrong hands. If the data is encrypted and then stored on a floppy disk, the floppy can be easily copied and decrypted at the violator's leisure.

User Verification

User verification is mandatory for networks. If a system administrator cannot determine whether a user is authorized, then the security is for naught. Therefore, adding tiers of protection to the system is vital. The drawback of tiered access is not only added cost, but loss of a user-friendly approach. Once again, however, a determination must be made as to which is more important—convenience or protection.

Using various types of add-on software can facilitate verification protection. Networks that use dial-in lines should use a communications package that demands a log-in procedure. Some products available today use auto-answer modems that work on the premise that, after answering the phone, the system will hang up and dial back the user. This dial-back protection prevents a "Trojan Horse" scheme from gaining access.

Dial-in locations can be limited to certain physical places. For example, a user who logs in is identified by his password as a warehouse worker and may be allowed to call only from the warehouse phone line. For a network that involves many remote users, mandatory call-backs and location checks offer a high level of security.

Audit trails can track user identification by keeping an ongoing record of who logs on, the time spent on the network, which files were copied or deleted, and other records. Each file should have a security profile that is relevant to data being maintained. An audit trail, to be effective, must be constantly monitored and reviewed. The trail, when properly reviewed, can discourage users from attempting to misuse the network or gain access to privileged files. Most users are not going to risk their jobs to be a hacker. All audit trail procedures should be stated in the standard operating procedures manual and reviewed with the user.

Types of Network Failures

Rice Cross and David Yen, in *The Journal of Computer Information Systems,* identify the following types of computer networks that are vulnerable to security failures:

Workstations. Because PCs have their own disk drives, users can download sensitive company data and private software, upload their own versions of documents and spreadsheets; sophisticated users can upload software that can help them defeat security measures already programmed into the network. *Solution: Utilize a diskless workstation.*

Printers. Because printers are connected to the network, they present a threat. *Solution: Procedures that restrict what types of information can be printed on shared printers should be authorized by management and reviewed by employees.*

Cable Tapping. Tapping is making a physical connection to the communications cable being used to transmit data. *Solution: Devices for monitoring network cables are now available. Or use fiber-optic cables because they are more difficult to tap.*

Stray Electromagnetic Radiation. Emanations are electromagnetic signals emitted from printers, CRTs, and computers. Passive intruders can detect these signals with fairly simple devices. *Solution: Purchase "Tempest Certified" computer equipment that uses fiber-optic or shielded cables, or use inexpensive devices that omit electromagnetic signals or emit "jamming" signals to prevent snoops.*

Dial-In Access. Networks with dial-in lines should be equipped with a communications package that will dial back the user attempting to access the system, thus limiting access to the system except from predetermined points. Encryption and data compression also can be used to make data difficult to interpret for unauthorized users.

Employees. Companies are finding that most computer crimes are committed by employees (only 2% of computer fraud comes from outsiders). *Solution: Use the need-to-know rule so that employees have access to the minimum amount of computing resources and data that they need to know to do their job.*

File Locking. File locking capabilities should be installed on the network system to prevent two users from accessing and modifying the same file concurrently.[21]

Communication System Failure

Information being routed from one location to another over communication lines is vulnerable to accidental failures and to intentional interception and modification by unauthorized parties.

Accidental Failures include the following examples:

- Undetected communications errors may result in incorrect or modified data.
- Information may be accidentally misdirected to the wrong terminal.
- Communication nodes may leave unprotected fragments of messages in memory during unanticipated interruptions in processing.
- Communication protocol may fail to positively identify the transmitter or receiver of a message.

Intentional Acts include the following examples:

- Communications lines may be monitored by unauthorized individuals.
- Data or programs may be stolen via telephone circuits from a remote job entry terminal.
- Programs in the network switching computers may be modified to compromise security.
- Data may be deliberately changed by individuals tapping the line (this requires some sophistication, but is readily applicable to financial data).
- An unauthorized user may "take over" a computer communication port as an authorized user disconnects from it. Many systems cannot detect the change. This is particularly true in much of the currently available communication equipment and in many communications protocols.
- If encryption is used, keys may be stolen.
- A terminal user may be "spoofed" into providing sensitive data.
- False messages may be inserted into the system.
- True messages may be deleted from the system.
- Messages may be recorded and replayed into the system (e.g., "deposit $100" messages).[32]

Signal Interception

As stated earlier in this chapter, signals are either analog or digital. Analog signals are voice waves modulated to a high frequency (HF). They are increased in speed from hundreds to thousands of cycles per second, which allows for their transmission over distance. However, analog signals rely on a repeater to amplify the signal periodically.

Digital signals are voice or data vibration signals that are converted into a series of on-and-off pulses—zeros and ones. But both digital and analog telephone calls have to pass through a modem.

Telephone transmissions have four communications paths over which to travel: cable, satellite, microwave, and fiber optics. Cable is the most common and the most secure. However, it is the least economical and practical method for mass transmission over longer distances. Ninety percent of our long-distance telephone transmissions are sent by microwave or satellite for these reasons. Microwave and satellite transmissions can be readily intercepted from the air. Each individual signal is joined with many other signals going to the same region of the country. This process of allowing one line to carry several separate communications is call **multiplexing**.

If you choose a common carrier, rather than a dedicated service, a computer selects the first available route over which to transmit your signal. This allows a less vulnerable circuit to be switched to more vulnerable ones.

A disadvantage of using common carriers is that the responsibility for privacy and security is the burden of the user. The user loses the privilege of controlling the communication security, communication mediums, and the routing. Sharing access, in itself, is another disadvantage of using common carriers. Dedicated service, on the other hand, is a direct line between two entities and has the extra advantage of noise reduction because it does not pass through the normal switching network. Transmitting a message on dedicated lines takes only one to three seconds.

If your signal travels by microwave, it is relayed as a radio wave from tower to tower in intervals of approximately 25-30 miles across the countryside. Microwave is an extremely risky medium because it has a range, or "spills," measuring 12.5 meters in width, and 25 miles between the towers; spills constitute the most vulnerable point of the microwave signal. This signal is easy to intercept with a well-aimed parabolic dish antenna located in an elevated location.

The process is basically the same for satellite transmissions as the signals bounce off the satellite across a wide area. However, because each individual transmission is multiplexed with a vast number of other signals, the chances of interception are reduced.

Fiber optics are cables made up of thin strands of glass that carry light energy and, thus, offer valuable security advantages. Electrical inductance, a serious security consideration, is avoided by using fiber optics because light does not generate a magnetic field. In addition, due to the delicate balance of the fiber-optic network, it is virtually impossible to tap an optical system without detection.

No matter how secure the medium of transmission may be, it is still at risk of having its signals intercepted. Multiplexing helps reduce this risk; however, it does not fully eliminate it. Although the everyday citizen rarely stops to consider it, wiretapping and electronic bugging are common occurrences in today's society. Invasions of privacy could exist in both business and personal lives.

The federal government is also a big user of bugging devices. In fact, a survey conducted by the Congressional Office of Technology Assessment (OTA) reported that as technological advances increase, the government increases its surveillance. The study concluded that government agencies currently use a wide variety of surveillance techniques, such as scanning radio transmissions, monitoring computer use and the use of electronic mail, intercepting cellular telephone calls, and listening to satellite, microwave, and fiber-optic transmissions.

Another vital concern, perhaps more in the business environment than in other areas, is the problem of **inductance**, where the magnetic fields generated by electrical currents can be easily detected by electronic circuitry that is merely close by. This circuitry does not even have to be in direct contact with the wire. Even though the threat of inductance is lessened by circuits carrying a great deal of transmissions, it is still a serious risk worth considering, particularly if the data is critical. Most at risk of inductance are display terminals and dot matrix printers.

Products Available

A wide variety of bugging and tapping devices are available on the market today; their installation does not require a lot of technical knowledge. For example, a passive wiretap accompanied with a voice-activated recorder, available at any entertainment store, is sufficient to intercept a telephone conversation. Another electronic gadget available at any well-known electronic/entertainment store is an FM wireless microphone that fits into the palm of your hand and whose transmissions can be picked up by a regular FM radio. Some products sell for a mere $19.95; for $24.95, you can buy a telephone record control that records both sides of a phone conversation. The telephone plugs into the control, which in turn plugs into a cassette recorder. However, using this device is illegal unless you have the other party's consent because the device does not emit a periodic beep to indicate that the conversation is being recorded.

Generally, bugs can range from wireless microphones, as small as a sugar cube (which costs about $20) to wireless intercom systems the size of a candy box. These will allow anyone within a half-mile radius to listen to any conversation in an office or home. An example of this is a **parasitic telephone transmitter**, a device that does not require batteries or an antenna because it steals power directly from the telephone company (hence the name "parasitic"). This type of device can be received on a regular FM radio, as well as other receivers, up to one-half mile away, depending on its location and size. It cannot be detected unless the device is actually in operation—when the phone is off the hook. There is also a perfectly legal, FCC-registered tap for phones offered by many companies for approximately $25. These devices are only a small sample of what is available on the market today.

Electronic Espionage

As the value of corporate information grows as a crucial strategic asset, security of transmission is increasingly important. **Electronic espionage** is a serious threat to many organizations today. Espionage is stealing information or secrets from another country or company—spying. There is a vast assortment of electronic gadgetry available to aid this intrusion. Whether the goal is the plans for a previously unreleased, hot new product or the company's strategic development plans, electronic espionage occurs quite often and costs the victimized companies millions of dollars each year. However, out of embarrassment and fear, most organizations fail to report the crimes because they are afraid that such a breach of security could tarnish their reputations in the industry. They may also believe that reporting the crime could produce a "textbook" of sorts on how to conduct electronic espionage. Many companies avoid the hassles of reporting the crimes because today's legal system just does not offer sufficient protection for the corporations to warrant either the time or the effort.

Detection and Defense

Because of the size of these bugging devices and the endless ways to hide them, they can be difficult to detect and locate. Because people generally are not aware of communications transmission security, many may have a tap on their phone or in their office and not even realize it. Usually someone with a technical background and expert knowledge, with both the skills and the equipment required for the job, is needed to locate the bugging device.

A wide variety of so-called "counter-intelligence devices" are offered on the market today as a counterattack against the invasion of privacy. There are portable bug detectors for $900-$1,000 that locate hidden transmitters, and there are telephone tap detectors for $1,980. A pocket-sized tape recorder detector is available for $1,950 that detects a cassette recorder operating within a 12 foot radius. And for the increasing number of mobile/cellular phones, there is a mobile phone scrambler that fits into the handset of the telephone. This scrambler, paired with another one in an office or home phone, encodes the conversation as it is being transmitted and decodes it at the other end when it is received. This protection device sells for a mere $890 for *each* scrambler.

Even this relatively short list of the available products demonstrates that there seems to be a protection device for every possible security need. However, remember that these products may or may not perform as promised and should be tested thoroughly. One of the best tools for detecting and locating bugs is the spectrum analyzer. This device, accompanied with a good receiver, can uncover the bug by scanning the RF spectrum for the bug's frequency. A direction-finder is also needed to locate and eliminate all spurious emissions.

One of the most effective security measures against illicit communication interception is encryption. Data encryption provides more security because even if the transmission is intercepted, the data's value is not lost, provided, of course, that the encryption technique is a secure one. Secure key storage with periodic key changes are necessary to increase the chances that an interceptor will not be able to decode the message in the event that it is intercepted. Other protection measures include silent modems, authentication systems, and dial-back connection.

Encryption is an effective security control against inductance as well. Another form of protection is provided by the U.S. government's program, Tempest, which certifies computer equipment as not emitting detectable signals. There are two methods for organizing a Tempest certification: enclosure and emanations modification. **Enclosure** diffuses the signals before they can be picked up by shielding the device in some type of conductive case. To be effective, the entire device must be completely and carefully shielded. Screening and shielding can be extremely expensive, however. Terminals, specially designed to include security screening, cost two to three times more than an equivalent commercial item. Modifying the emanations is accomplished by adding spurious signals to the transmission. Emanations modifications can also be quite costly.

Legal Protection

Unfortunately, our legal system does not offer a great deal of protection against this type of crime. The Electronic Communications Privacy Act of 1986 is an update of the Omnibus Crime Control and Safe Streets Act, more commonly known as the Federal Wiretap Law, designed to extend legislation to some of the latest technology. The earlier 1968 Wiretap Act protects communication over conventional wire telephones. The new law applies to computer and other electronic communication, cellular telephone calls, and stored data. It also expanded the original list of federal crimes for which bugging or tapping could be legally authorized. The new law is more specific about the types of interceptions that are illegal. The law also contains definitions that make it clear that private and public microwave services and cellular telephones are actually protected against interception, but cordless phones are not.

CASES IN POINT

The following examples illustrate some recent occurrences in which networks and telecommunications have failed to remain secure.

Phone Crash Grounds Airplanes

In September 1991, a power shortage at an AT&T switching center left more than one million residents without telephone service for up to seven hours. The victims included hundreds of people stranded at New York airports. Experts stated that the problems could reoccur unless major changes are made in the way the nation's phone networks operate.

Telephone networks run on electricity, and the 93° heat in New York meant additional power was needed to run air conditioners. Con Edison, New York's public utility, asked AT&T to draw electricity for its call-switching computers from AT&T's own diesel-fueled generators. That was a routine request, but then disaster struck.

During the attempted changeover, a sudden surge of electricity knocked out several key devices known as power rectifiers, which failed to switch AT&T's computers from Con Ed's power to AT&T's diesel power. Instead, AT&T's electricity was generated from a battery-operated backup system.

Negligence also played a major role. AT&T admitted that when the attempted power shift took place, no one physically checked to see that the system was operating on the diesel turbines as a routine procedure. Had they checked, they probably would have noticed the system was running on a battery reserve that lasts only about six hours. Audio and visual alarms at the switching center apparently went unnoticed.

The Federal Aviation Administration (FAA) uses AT&T lines to relay air-traffic information from its control center on Long Island to airports from Boston to Washington, D.C.—including New York's three metro airports. With those phone lines down, air traffic controllers couldn't be sure where planes were; neither could the planes' pilots. Planes at the three New York airports couldn't take off; connecting flights at other U.S. airports had to be delayed; dozens of flights were canceled hour by hour.

The major consequences of the New York instance were:

- Airlines such as USAir, Northwest, and American were forced to cancel more than 260 flights over two days.

- Airports in Boston, Philadelphia, Baltimore, and Washington, D.C., were saturated with diverted flights.

- The supersonic Concorde was rerouted to Bradley Airport in Hartford, Connecticut.

- Major banks such as Chase Manhattan had to stay open late because they couldn't transfer funds electronically to Federal Reserve banks.

- Long-distance service for New York was knocked out for several hours. AT&T doesn't know how many calls were affected but said the switching station involved normally can handle 2 million calls an hour.[52]

AT&T used to have thousands of switching centers throughout the country, but advances in computer technology let AT&T handle even more calls with fewer switching centers. Today, there are about 100 such computerized switches—including three in lower Manhattan—that AT&T uses to route calls throughout the country.

One way of preventing a total outage is to spread telecommunication needs among different carriers, minimizing the risk of a total blackout. *Reader's Digest*, as an example, receives about 150,000 calls a month and uses AT&T, MCI, and Sprint.

Other major phone breakdowns that have occurred include the following:

- **May 1988**: A fire at an AT&T switching center in Hinsdale, Illinois, left 35,000 customers without phone service, delaying flights nationwide.

- **November 1988**: A Premier Electric Co. crew accidentally sliced an underground phone cable in Louisville, Kentucky, causing 20,000 customers to lose service.

- **January 1990**: A software problem at a New York signaling center blanked out half the calls on AT&T's nationwide network.

- **January 1990**: An AT&T crew in Newark, New Jersey, sliced a fiber-optics phone cable, affecting thousands in New York and nearby airports.

- **June 1990**: A software bug at Bell Atlantic's C&P Telephone division in Baltimore interrupted local phone service for 5 million customers.

- **June 1990**: A software bug struck in Los Angeles, and 3 million Pacific Bell customers lost local phone service.

- **July 1990**: A bug struck again: 1 million Pittsburgh-area customers lost service for several hours. In an unrelated breakdown, 45,000 Greensboro, North Carolina, customers lost service for 2 1/2 hours.

Tracking Drug Runners Via Satellite

On October 27, 1991, a group of Pakistani Makran Scouts seized 42.3 metric tons of illegal drugs being transported by camel caravan through the same pass in the mountains of Afghanistan and Pakistan that Alexander the Great traversed around 340 B.C. This cooperative effort between the United States and Pakistani governments was hailed in a CNN broadcast on November 11, 1991, as the biggest drug bust in police or military history. The successful drug bust can be attributed to the utilization of the latest satellite imagery, geographic information systems, and international communications networks.

Makran is an area within the Balochistan Province, one of the four provinces of Pakistan, which lies on a principal trade route from South Asia and the Middle East to other parts of the world. Illicit drugs are produced in Northern Afghanistan and the North West Frontier Province and moved primarily through Makran, which now maintains a balance of trade that is 95% illegal. Until now, environmental factors and the isolation of this area made it difficult to install a reliable communications network.

The systems used for this particular enterprise were a combination of satellite imagery technology linked to the latest generation of PC-based geographic information systems (GIS) and informal "bamboo" telegraph networks of verbal information gathered covertly by local operatives. Use of this technology finally enabled the monitoring of the production and distribution of illegal drugs throughout Pakistan and Afghanistan from a Washington, D.C.-based central unit at a personal computer level. The instantaneous transmission of satellite data to Washington from Afghanistan and Pakistan, coupled with consistent reporting from the bamboo network through a communications link with Makran, gave the Washington-based central unit ongoing control over decision making and information dissemination during the operation.

The team of Makran Scouts credited with the success of the raid was composed of 3,000 men divided into twelve strategic units, each unit maintaining daily contact with the central command post based in Turbat, Makran. The information and communications systems used in this project were employed through the United States Agency for International Development (USAID) and remain some of the most complex and technologically advanced information/communications systems in existence today. Despite this fact, these are also some of the most cost-efficient systems in existence as well. It cost the governments of Pakistan and the United States less than $4 million to establish the network, train the personnel, and execute the drug raid. A similar monitoring system, established in the northern provinces of Pakistan for continuous monitoring, cost only $126,000, including all equipment and personnel training. With such efficiency in costs, these systems are not far from being widely utilized on a commercial basis.

The satellite imagery systems are somewhat reminiscent of George Orwell's conception of a "Big Brother" monitoring society, but they have definite advantages for the private sector. For example, satellite imagery can currently be used with local demographic information to pinpoint a prime location for a new school that would offer the most convenient access to a majority of students. Changes in national crop development can be monitored throughout the span of the growing season to regulate consumable markets and facilitate economic forecasting. It won't be long, as a matter of fact, before programs will be developed to enable the monitoring of individuals, for example, to simplify parole programs' monitoring of parolees or to locate missing persons.

The forms of communication and global monitoring currently being utilized are virtually unknown to the general public, except perhaps in the use of geographic information systems to compile information taken from the latest national census. Unfortunately, the little exposure the public has had has been to such useful facts as pinpointing the exact location in the United States on which the country would be evenly balanced, given that each individual counted in the census weighed the same.

Hackers Use Voice-Mail Illegally

Computer hackers in the United States and Canada systematically sold phone records and used local and toll-free 800 numbers to gain access to office voice-mail systems around the United States, allowing them to use the office phones for their own purposes. The size of the intrusion, according to the F.B.I. and Secret Service, could run into tens of millions of dollars in fraudulent phone bills.

Current voice-mail systems are vulnerable because their access codes are relatively easy to crack. Once hackers calling from outside an office break a code and take control of an extension, they have several options—use the line to make long-distance calls (at the company's expense) or leave and receive messages. Some hackers even change the company password, rendering the voice system inaccessible to employees trying to retrieve their messages.

Hackers gain access to corporate voice-mail systems in a variety of ways. The toll-free 800 numbers supplied to corporate and government phone customers by long distance carriers, plus other sensitive phone records, are stored in data banks owned by local phone companies. A hacker using a personal computer and modem can break into these data banks through phone lines and steal customer records showing phone numbers and voice-mail access codes. The hacker can then use this information to manipulate voice-mail systems.

Once inside a system, hackers may use a stolen password to get a dial tone from the company's phone line. They make long distance calls around the world, and the company doesn't find out it has a problem until 30 to 60 days after the intrusion, when it gets its phone bill.

Hackers can also change passwords or create mailboxes in phantom phone extensions.[38]

Waylaid Computer Systems Shut Down Airport

Police are having problems finding witnesses to the fire that waylaid United Airlines' computer systems at O'Hare International Airport. Lt. Thomas Brady of the Chicago Police Department bomb and arson squad said there is a possibility the investigation may turn into a criminal case.

"We know the origin, but not the exact cause," said Brady, who declined to give specifics on the investigation.

The fire, which occurred in a service tunnel, destroyed United Airlines' phone and data cables, shutting down its computers for 12 hours. The staff was forced to record all flight information—including arrivals, departures, and reservations—manually. United canceled 110 flights.

United's communication cabling is all in one place, which may indicate poor planning. American Airlines said that its operations have alternative data switches, with more than one set of communication cables. "If one communications node is knocked out, another is available," said an AA spokesman.

Ray Hipp, president of Comdisco, the Rosemont, Illinois, disaster recovery firm, says that airlines handle backup internally and don't use third parties.[58]

DOD Computers Hit by Hackers

The government faces increased levels of risk for information security because of greater network use and computer literacy, and greater dependence on information technology in general. For years hackers have been exploiting the security weaknesses of systems attached to the Internet, an unclassified network composed of more than 5,000 smaller networks nationwide and overseas that is used primarily by government and academic researchers. Their techniques have been publicized in hacker bulletin boards and magazines, and even in a bestseller, *The Cuckoo's Egg,* written by Clifford Stoll. Hackers, however, continue to successfully exploit these security weaknesses and undermine the integrity and confidentiality of sensitive government information.

Between April 1990 and May 1991, computer systems at 34 Department of Defense (DOD) sites attached to the Internet were successfully penetrated by foreign hackers. The hackers exploited well-known security weaknesses—many of which had been exploited in the past by other hacker groups. These weaknesses persist because of inadequate attention to computer security, such as password management, and the lack of technical expertise on the part of some system administrators.

DOD officials, however, are still unable to determine the full scope of the problem because security measures for identifying intrusions are frequently lacking. At many of the sites, the hackers had access to unclassified, sensitive information on such topics as military personnel—personnel performance reports, travel information, and personnel reductions; logistics—descriptions of the type and quantity of equipment being moved; and weapons systems development data.

Although such information is unclassified, it can be highly sensitive, particularly during times of international conflict. For example, information from at least one system, which was successfully penetrated at several sites, directly supported Operation Desert Storm/Shield. In addition, according to one DOD official, personnel information can be used to target employees who may be willing to sell classified information. Further, some DOD and government officials have expressed concern that the aggregation of unclassified, sensitive information could result in the compromise of classified information.[12]

SPECIAL COMMUNICATIONS SECURITY CONSIDERATIONS

As the use of various traditional communications continues to grow, so too has the introduction of new devices, such as facsimile (fax), Caller ID, E-mail, voice mail, and open systems environment (OSI). Each of these new communication applications increases the speed of information transfer, but also presents new avenues for security breaches. As each new device is utilized by a growing number of users, the potential for accidental disclosure or intentional misuse also grows.

Fax

Facsimile (fax) allows users to transmit text, graphs, blueprints, and other pictures and drawings electronically from one location to another via telephone lines. The effect is similar to putting a document in a copy machine at one location and getting a copy at a distant location. The pictures are converted to data by the sending of a fax, transmitted to the receiving fax, and restored to picture format before being printed by the receiving machine. Some fax transmissions are computer-to-computer via fax boards.

Fax machines have been in use for more than four decades, with the first commercial fax being marketed in 1966. This machine required six minutes to transmit one page that was hardly readable, and the cost of the machine was outrageous ($10,000 or more). With technology, the prices dropped below $2,000 and the speed of transmission increased to 10 seconds per page. This led to a sudden "fax revolution." Currently there are approximately 4.2 million fax machines in use worldwide; the United States accounts for about 2 million of these.

As a result of the expansive fax technology, the fax machine has become a necessity for efficient business communications. Its dominant feature is convenience; however, the fax does pose a potential problem for security. Direct marketing, sending unsolicited fax messages, which have come to be known as "junk fax," is a growing problem as well.

As the use and reliance on fax machines to transmit important information increases, so does the risk of interception as eavesdropping equipment can be used to make crucial and confidential information easily accessible. An additional problem with security involves the open access to fax machines. Many fax machines are simply installed in mailrooms, which are obviously very open areas. One chairman of a company claims that unauthorized access to facsimile transmissions is one of the greatest hidden liabilities to the integrity of business operations. Before fax messages are delivered to the designated recipient, anyone may have access to them in an open area. With fax messages so easily accessible, there are immense opportunities to gain abundant information about a company's operations. This idea of "faxpionage" is a serious threat, but it is frequently ignored by many companies. Messages are often faxed that would otherwise be mailed and stamped "private" or "confidential." When traditional mail is used, the sender expects that the correct recipient will receive the message without any interference or problems. Fax messages have gained an "unclassified" type of reputation, one that needs to be reversed so that fax messages will be treated as private communications.

The following methods of fax protection are recommended:

- **Monitored rooms.** All fax machines should be installed in rooms that will be constantly and carefully monitored.

- **Reduction of "openness."** If a room cannot be monitored at all times, the fax machine should be put in a separate, small room.

- **Limited access.** Only those with access privileges should be allowed to pick up messages.

- **Separate machines.** Those companies involved in receiving and sending confidential information should install separate machines in the offices of individuals with valid access privileges.

- **Internal memories.** Unauthorized reception may be prevented by using fax machines with internal memories that allow messages to be addressed to individual electronic mail boxes. The recipient must then supply a security code to get a printout of the message.

- **Encryption.** A code should be used so that only the recipient of the message is able to understand (decrypt) the message.

- **Smart cards.** Access should be limited to those who have properly coded smart cards.

As the use of fax machines increases and the technology advances, the security issues will become more important. As with many other devices, the first concern with the fax is its efficient and effective use; unfortunately, security is an afterthought. Many corporations believe that because significant losses caused by a lack of security with fax transmissions have not been reported, there is no real problem.

One problem with faxes that has received more attention than the security issue is the transmission of junk faxes—unsolicited and unwanted messages. Along with the "fax revolution" came the marketing plan to use the fax machine as a means of advertising. Now, many businesses are demanding protection from such unsolicited advertising. Currently a war is being waged between those in favor of legislation concerning the uses and abuses of fax transmissions and those who oppose such legislation in favor of self-regulation.

Junk faxes are the cause of the following problems:

- Additional, unnecessary electricity expenses
- Additional, unnecessary telephone costs
- Wear and tear on the fax machine
- Waste of paper (The recipient must pay for the paper even though he or she does not want the message.)
- No means of disconnection
- Waste of significant personnel time as employees separate the legitimate messages from the junk
- Busy fax lines (While receiving unwanted messages, the company is unable to send or receive messages that may be crucial.)
- Invasion of privacy which occurs when marketers gain access to employment data and use that information to target their advertising

Possible solutions to fax security problems include the following:

- **Safeguarding of fax numbers.** Businesses should not give their numbers to fax directories and should be selective when distributing their number. Many businesses no longer print their fax number on letterhead, business cards, and other similar items.
- **Call-screening function.** Machines are being manufactured and devices are being developed with a call-screening function to reject calls that are not on an approved list of incoming numbers.
- **Lock-out devices.** These devices simply refuse to accept random messages.
- **Software.** Software has been developed that ignores unwanted callers so that only those wanted messages will be printed, thus providing an automatic screening.
- **Legislation.** Many people feel that there is a need for legislation to ban unsolicited fax advertising or impose strict regulations is needed. This idea has become a heavily debated topic and will continue to be an issue as restriction of free trade becomes a paramount concern.

Many have come to believe that some type of legislation is necessary to prevent or at least curb unsolicited fax messages. The following are a few examples of why some form of regulation is needed:

1. Mario Cuomo, the governor of New York, was forced to wait for a memo from the state consumer protection chief concerning nuclear power plants because his fax machine was busy receiving a three-page menu from a local sandwich shop.

2. *The New York Times* banned all unsolicited news releases sent by fax. *The Times* was receiving so many unsolicited messages that their reporters were having trouble receiving important, needed information from their sources.

3. A major problem has developed with fax directories. For example, many companies have come to detest Mr. Fax (a seller of fax machines and supplies), which has created a database of at least 500,000 fax numbers. Companies are enticed to provide other companies' fax numbers as Mr. Fax offers prizes in return.

4. One company reported receiving a 100-page unsolicited document.

These examples reveal an obvious need for some type of legislation, regulation, or restrictions to be imposed. Those in favor of regulation have a variety of negative comments about fax advertising. Some feel the ad is not worth the paper on which it is printed. Today's audience expects creative and sophisticated advertising, standards that plain, black-and-white faxes cannot match. In addition, many add that faxes are simply a wasted effort—after all, few people other than the secretary or mailroom worker who sorts and delivers the fax receipts will see the ad. Others claim that all other forms of advertising have some form of regulation; therefore, faxes should not have the privilege of avoiding some restrictions. But the biggest problems are the simplest ones: fax advertising is annoying, and the recipient is forced to absorb its cost.

Connecticut was the first state to enact a law establishing penalties for unsolicited fax transmissions in May 1989, proposed by state representative Richard Tulisan. As more states band together in opposition and limitation, it is hoped that such actions will lead to nationwide regulation, but interstate cooperation is necessary for any real success with regulations and their enforcements.

A variety of proposals within various states are concerned with the type of regulations and restrictions that legislation should enforce; some range from fines up to $20,000 or jail terms up to six months for multiple offenses. However, the Telephone Advertising Regulations Act, an attempt at a reasonable form of legislation, seems to be receiving the most attention. This bill is not designed to impose unreasonable burdens on marketers or ban the legitimate use of fax technology. The bill does intend to protect the privacy rights and safety of the public. The bill would require the Federal Communications Commission (FCC) to create a national list of customers who do not want to receive commercials or advertisements on their fax machines. Telemarketing firms would then be barred from sending messages to numbers on this list, and the FCC would set and impose penalties for violations. Those numbers not on the list would be protected only if they were unlisted. This bill would impose no cost on consumers; instead, marketing firms would be required to pay for access to the list.

ATMs and EFT

An **ATM** (**automated teller machine**) is a system or machine that allows customers to make a wide range of financial transactions, such as deposits, cash withdrawals, cash advances, account transfers, bill payments, and balance questions or inquiries. Most automated teller machines are directly on-line with the bank's main computer, with access to the machine controlled by the use of a magnetically encoded card and a **personal identification number** (often called simply a **PIN number**). Two types of ATMs are generally used—free-standing or wall-mounted. Free-standing machines are placed away from the bank's physical location and are common in retail stores, food stores, and office buildings. These machines provide convenient banking in convenient locations. This convenience, however, comes at the price of extra security risks. Wall-mounted ATM machines are physically a part of the financial institution and provide transactions after normal banking hours or for individuals not wishing to wait in line inside the bank.

Electronic funds transfer (**EFT**) is used by banking and financial institutions to transfer money electronically from point to point. ATMs and EFT are frequently confused as they are often administered by the same bank, but these two terms represent separate operations. EFT is the act; the ATM is the mechanism used. EFT moves funds through the telecommunication/computer medium by using electronic signals instead of checks (the medium of the past) to transfer funds.

Automated teller machines are the most popular form of electronic funds transfer devices. At the present time, automated teller machine transactions account for 70 percent of all electronic funds transfer transactions, largely due to the increase of available ATMs for public use.

With the vast increase in the number of automated teller machines and their use, ATM theft and fraud have also increased over the last few years. However, although automated teller machines account for the majority of all electronic funds transfer transactions, they represent, in monetary terms, the smallest segment of EFT. This fact can be attributed to the relatively low dollar amounts involved in each case of ATM fraud.

There are three main ATM security problems at this time—physical attacks as a bank customer uses an automated teller machine, unauthorized use of access cards and PIN numbers, and the manipulation of financial data. Of the three, physical attacks are the most common form of ATM security violations. These attacks occur when someone uses force to rob the ATM user after he or she makes a transaction. Many states, due to a rapid increase in ATM attacks, have established the following requirements:

- Lighting must be a minimum of 3.5 watts per square foot within a 40-foot radius of the ATM.
- Landscaping with a 40-foot radius cannot exceed 40 inches in height.
- No structure of land can obstruct the vision of an ATM user within a 20-foot radius.
- Surveillance cameras must record all transactions within 20 feet of the ATM at all hours of operation.
- If three or more crimes occur at an ATM in one year, its hours of use will be restricted to 6 a.m. to 7 p.m. daily.

Users of ATMs should follow these precautions when using ATMs:

- Never approach an ATM if you have any doubts, fears, or concerns.
- Never count your money at the ATM location site.
- Never approach an ATM if you see suspicious people near or around the machine.

PIN numbers should be chosen by the user with the following cautions:

- Don't pick numbers that can be associated with you.
- Don't pick numbers in sequence (e.g., 1234 or 1111).
- Don't use a birthday or telephone number.

The automated teller machine should be equipped with many built-in security features to help protect both the bank and the individuals using the ATM, such as:

- Limiting the amount of each transaction.
- Limiting the number of transactions in a given time period.
- Alerting the bank (and possibly "eating" the card) if any inappropriate activity is discovered or limit is reached.

Bank cards and PIN numbers, although often abused by many people, will continue to be used because of their low cost and convenience. Because of the relatively small amounts of money affected at one time, security measures such as eye scanners, voice recognition units, and fingerprint analysis, although tested, are still too expensive to implement at each ATM site.

Electronic funds transfer currently takes on six distinct forms.

1. Wire transfers (WTs)
2. Clearinghouses
3. Automated teller machines (ATMs)
4. Point-of-sale systems (POS)
5. Telephone bill-paying systems (TBPs)
6. Home banking systems (HBs)

The first two forms (WTs and clearinghouses) are accessed by the Federal Reserve System and the banking industry, while the other four forms are accessed mainly by the average consumer.

- **Wire Transfer. Wire transfer (WT)**, one of the oldest uses of electronic funds transfer, was originally set up by the Federal Reserve System for the transferring of funds between the Federal Reserve and the banking industry. Currently the WT system is used by 800 banks and accounts for approximately $360 billion in transfers of funds each day.
- **Clearinghouses. Clearinghouses** are used to sort and route electronic transactions between corporations and banks. One use for a clearinghouse is the processing of direct deposits. A company's payroll is recorded on a magnetic tape that is released to a clearinghouse. The clearinghouse sorts the payroll by

bank and then transfers the transactions to each separate bank. One example is the Clearing House Interbank Service (CHIPS), which services more than 90 banks and processes more than $120 billion daily.

■ Point-of-Sale System. A **point-of-sale system (POS)** is set up like an ATM system and grants access to your bank account. The POS card, used in place of a credit card and writing checks, will directly subtract the cost of an item from your bank account. This process eliminates writing checks to pay for items or services. Using a POS system can eliminate the risk of bad checks and resulting losses because the POS card can check the amount of money in a person's account; the retailer knows before the sale whether the customer has enough money to purchase the item.

■ The Telephone Bill-Paying System. The **telephone bill-paying system** uses a touch-tone telephone. A subscriber to this system would call his/her bank and use the phone to respond to prompts made by the bank's computer. The bank's computer uses speech synthesis to "speak" with the user and guides the user through the process with a series of questions or directions. For example, the computer may state, "Press 1 for account balance, press 2 to transfer funds, press 3 to end session." The computer guides the user through the series of prompts until the transaction is completed. TBPs have simplified bill payment: no longer is it necessary to pay a bill by mailing a check.

■ Home Banking System. The **home banking system** (HBS) is variation of the TBP. Through the use of home computers, people are now able to bank, shop, and carry out financial transactions from the privacy of their homes.

Even though electronic funds transfer is widely available to the public, it is mainly used by the banking industry and the federal government. There are more than 5 million home computers, but only about 44,000 people now use their home computers for financial transactions. At first the computer industry thought that HBS would catch on and banking fees would eventually drop, but HBS has not been widely accepted by customers. Many people still are reluctant to buy products through a computer. The future holds promising signs for HBS as many people now use television shopping networks to purchase products. The next move seems to be heading toward home banking systems. The results for ATM usage are equally mixed for the near future. Even though the number of ATMs in the United States is growing, only a small percentage of all ATM transactions transfer funds from one account to another; most simply supply cash. Similarly, the use of point-of-sale terminals for credit and check verification, and eventually direct transfer of funds to the stores, is in its infancy.

On the other hand, more than $480 billion in funds are transferred daily by the banking industry. Clearly the Federal Reserve system and the banking industry are the major players in EFT. EFT systems hold great potential for consumer use, but it is unlikely that things will change in the near future.

EFT basically functions with four elements:

1. The terminals at which instructions are entered into the system and messages are received

2. The computers that carry out the instructions

3. The telecommunications lines or networks that join terminals and computer

4. The software or programs that guide the working of the system

Because of the many different avenues that criminals can take to violate an EFT system, many different types of security measures have to be incorporated to protect the EFT system.

Three primary areas are of concern with regard to telecommunications security:

1. Line noise, which destroys the meaning of information transferred on a communication circuit

2. Intentional interception by an unauthorized person

3. Disconnection of circuits for either short or long periods of time

All of these threats to security are especially crucial to EFT systems. If line noise is too great, then the integrity of the entire EFT system has been destroyed since none of the data can be trusted for accuracy. If data is intercepted by unauthorized persons, the security of the network is again compromised. If frequent disconnections occur, then the EFT network is simply inoperable.

A major threat to the telecommunications element of EFT comes from interception and fabrication of transactions. When a transaction is intercepted, the perpetrator is just listening in but does not interrupt or modify the transaction being sent. The term for this type of intrusion is **"passive" wiretapping**. Fabrication occurs when the perpetrator in some way modifies a transaction, or the perpetrator sends a fabricated transaction. This type of intrusion is called **"active" wiretapping**.

It is unlikely that an intruder would tap a long-distance phone or trunk line to access an EFT system. To extract one transmission, the intruder would have to know exactly which cable contained the transmission and exactly which channel on the cable carried it. However, it would be much more likely that an intruder would attempt to wiretap or eavesdrop close to the source or the receiving end of the transmission. In the case of EFT systems, this point could be a phone line coming directly out of a particular ATM.

Some might argue that there is little danger in a transaction that was intercepted, because the perpetrator has only listened in on the transaction, not modified it. The problem here is that information is being divulged. By intercepting transactions, the perpetrator is able to collect account numbers, account balances, how much money is being transferred, where the money is being transferred, and possibly the secret PINs (personal identification numbers) used to verify and secure people's accounts.

If a perpetrator is able to intercept a message, it is possible that the perpetrator can also fabricate transactions. After collecting information concerning people's account numbers, account balances, and PIN codes, a perpetrator has the information needed to fabricate a transaction.

One example of a telecommunications threat is the tapping of an ATM's communications link between the ATM and the bank's computer. With the tap in place, the perpetrator can then wait for several days or weeks, collecting copies of the transactions sent along the communication link. After collecting enough information, the perpetrator can now send fabricated messages. By fabricating messages, the perpetrator could transfer funds out of people's accounts into an account he or she controls or withdraw cash from the ATM. It is very difficult to protect against telecommunication penetrations because of the many different connection points in a nationwide network. All it takes is one wiretap on any of the connection points to compromise the security of the network.

A case that occurred at the Japanese Hokkaido Bank typifies this type of security breach. An engineer for another Japanese company had access to the bank's on-line computer system. He built equipment based on a simple tape recorder and effectively wiretapped the bank's telecommunications lines. He then used the intercepted information to develop counterfeit bank cards that he used to illegally withdraw funds. After being caught by the authorities, he agreed to show them how he had perpetrated the crime. Police were amazed at how ridiculously easy it was to commit such a crime.

In 1978 in Los Angeles, California, a thirty-two-year-old systems analyst transferred $10,200,000 to his account in New York. How did he do it? He walked into the wire transfer room of Security Pacific Bank of Los Angeles and observed the process in which users were assigned identification codes. Then he read a sheet posted on the wall showing all the identification codes to be sent to the specific bank. He then transferred money to his account. What made it so easy was that he had access to the wire room and knew the bank's security procedures, allowing him to transfer the money when the computer system was most vulnerable.

Cellular Telephones

Wireless communication, most specifically the cellular telephone, is adversely affecting the human population in four major ways:

1. A feeling of invasion of privacy, especially because laws are not keeping pace with the new technology
2. A high price tag on wireless technology when there is little justification for it
3. A change in the social customs due to the presence of this new technology
4. A change in the worker and the way in which work is accomplished

Cellular telephones are telephones that operate by transmitting signals to and from cellular receiving stations distributed throughout a metropolitan (or restricted) area.

Because cellular telephones use a form of radio transmission, scanners tuned to the appropriate frequencies will be able to hear both sides of a cellular telephone conversation. Currently, well over one million Americans already own scanners that can be used to intercept the frequencies used by cellular telephones. People who want to eavesdrop on callers using conventional mobile phone service can do so on known frequencies. Those involved in this eavesdropping range from people who listen to the conversations out of curiosity to hackers who listen to gather information for malicious reasons. Unauthorized monitoring of cellular telephone conversations is an obvious invasion of privacy to those who are being overheard. The United States Electronic Communications Privacy Act (EPCA) of 1986 prohibits the interception of radio communication such as those involving cellular telephones. One problem is that obtaining the evidence of eavesdropping that would be punishable under the ECPA is difficult.

For those worried about people eavesdropping on their cellular conversations, car phone scramblers that jumble the signal of a phone call are now available. When this encryption is used with a cellular telephone, it will either scramble audio content or modify the sound frequencies used.

Cellular telephones offer convenience and ability to communicate from a mobile unit, but they present a considerable expense:

- The average monthly bill for cellular phone subscribers is $183.95.
- Car phones cost $100-$500 plus installation fees.
- A transportable phone costs $100-$700.
- A pocket phone can cost $400-$1,500.
- A cordless phone can cost $50-$200.

From a sociological and psychological view, the spread of wireless technology may diminish human contact while disrupting social customs.

With this new technology, people are split into the "haves" and "have nots." In addition, some people view cellular phones as a hazardous distraction. Most everyone has seen someone driving down the highway during the morning rush hour holding a cellular telephone handset pressed to one ear in one hand and a cup of coffee in the other. These people appear to steer the car with their knees.

Cellular phone technology has a profound effect on the worker and the ways in which the worker performs his duties, as employees now have the ability to work anywhere and anytime. Cellular phones have become the new tool of workaholics, those who work compulsively and put their job above such human needs as eating and sleeping. An increase in technology and an increase in workaholism correlate very strongly. Consequently, the wireless revolution is transforming the way business is being done. The goal of the telecommunications industry is to have a telephone in everyone's pocket by the end of the decade. The wireless phone is becoming an indispensable business tool for many companies.

As this new technology marches on, cellular phones will become smaller, lighter, and cheaper. Street lights, basements, subway tunnels, homes, and offices will house small base stations that will transmit and receive calls over the airwaves. The base stations will be as universal as fire hydrants, allowing people to use tiny pocket handsets to make calls from a subway train, the ballpark stands, or an underground garage. Telephones will be cordless handsets about the size of a cigarette pack and will cost less than $100. To activate them, people will insert a credit card coded with an electronic identification number.

All indications predict that cellular telephone use will continue to increase. If users transmit private or confidential information by cellular telephones, they should take precautions. During the early 1980s, NSA sponsored a development program that resulted in the Secure Telephone Units (STU) III terminal. The STU-III looks like a typical telephone but provides end-to-end security between any two STU-III devices, even those manufactured by different vendors. Three U.S. vendors are authorized to make the devices: AT&T, General Electric, and Motorola. The STU-III utilizes current analog telephone communications but secures the speech signals by digital security techniques. There are also cellular STU-III terminals that provide "end-to-end" security between any two cellular STU-III terminals or between a cellular mobile terminal and a fixed terminal. The STU-III terminals provide protection against what is considered a high level of threat; they are most often used in government and defense-related settings.

Cordless telephones are available to consumers and are used in many homes and offices. Cordless phones, like cellular phones, offer convenience and mobility, but also have a security privacy concern. Some commercial cordless telephones protect conversations between a handset and its base station. Unprotected cordless telephones usually have a range of approximately 100 feet, and conversations can sometimes be picked up by an identical base station located in a neighbor's house. The "protected" telephones typically use a simple coding system, with a number of user-selected codes, to prevent someone from possibly listening to a conversation. They also protect against someone making a long-distance telephone call from a handset outside a residence. These telephones provide protection against what is considered a low level of threat.

ANI and Caller ID

When the regional telephone operating companies (the "Baby Bells") were separated from long-distance carriers (e.g., MCI, Sprint, AT&T), the operating companies had to develop a means by which to provide the long-haul carriers with information on the originating telephone number. This data, referred to as **automatic number identification** (**ANI**), was originally intended only to be used for billing purposes. But the carriers were quick to realize the revenue potential of this new product, and they promptly began offering ANI as a fee-based service.[14]

ANI is currently being used for cable TV pay-per-view services and direct-mail catalog marketers. ANI is generated independently by the telephone company as a way of physically identifying the cables over which the call is originated. The ANI code is unique to each telephone circuit and cannot be altered by the caller.

ANI is currently available from more than 90 percent of the phone lines in the United States, and it is increasing daily as small, independent phone companies convert their rural exchanges to equal access offices. ANI is obtained by subscribing to the service with a long-distance carrier. The ANI code can be provided "in band" over the 800 or 900 telephone networks, or "out of band" over the ISDN (integrated service digital network). A third alternative, **Caller ID**, is available on a more limited, local-calling basis.

Caller ID, a service offered within local calling areas by local telephone operating companies, allows a subscriber to view the originating telephone number of an incoming call before the call is answered. The adaptation of ANI to accommodate Caller ID is a simple task.

As a dial-in security medium, Caller ID has its limitations as it can only be forwarded from within a local calling area. *ISP News* summed up the issues this way:

The legal controversy stems from an individual's right to privacy, namely the display of that person's unpublished telephone number against his or her wishes. On an ANI basis, right to privacy is not an issue since, as the theory goes, the called party is paying for the phone call and accordingly has the right to know the source of the call. The issue becomes more complicated on a Caller ID basis, but recent court rulings have upheld the telephone companies' rights to offer Caller

ID if a free call blocking service is also offered. Call blocking prevents the forwarding of Caller ID information on a line-by-line basis. This development has been viewed as positive by the ANI dial-in security industry, and rapid expansion of the number of states allowing Caller ID is anticipated.[14]

It's relatively simple to apply ANI or Caller ID to dial-in security. The phone numbers of all authorized callers are entered into a database. When the screening device receives an access attempt, the calling number is stripped off, a comparison is made to the database, and the device grants or denies access. If access is granted, the caller can be subjected to a second level of security (e.g., a PIN number), or he or she can be passed through directly to the modem. If denied, the unauthorized call can be terminated or the caller can be given a message. The key features of ANI/Caller ID systems, from a security perspective, are that unauthorized callers never obtain access to a modem or hear a modem tone, and that a permanent record of all callers' telephone numbers, authorized and unauthorized, is maintained.[14]

ISDN and OSI

Integrated service digital network (**ISDN**) is a set of standards whose adoption provides additional telephone capabilities without scrapping existing telephone lines. In a global economic environment, the pressure to improve a country's business infrastructure will come, not from within, but from outside the country, as more and more businesses depend on international trade to survive. ISDN is a critical element of an effective business infrastructure for integrating business functions such as intelligence, marketing, manufacturing, and research and development.

ISDN can connect different processing devices, such as mainframes, PCs, PBXs, terminals, telephone handsets, fax machines, central office switches, and even control panels of security systems. It integrates computer and telecommunication technologies and, more importantly, allows for the infusion of intelligence into networks. ISDN also provides high-speed multimedia transmission over telephone wires. It was first defined by the CCITT, the United Nation's committee for data and voice communications, in 1972. Since then, there has been an ongoing effort to improve the technology. Organizations like IEEE and COS (Corporation for Open Systems) have also joined the effort to define ISDN standards.

The transmission structure of an ISDN link can be made up of three channel types: B (64 kbps), D (16 to 64 kbps), and H (384, 1536, or 1920 kbps). The B channel transmits text, voice, and graphics information. The D channel carries control information related to the B channel information to ensure the robustness of information exchange. The H channel provides high bit rates for services such as facsimile, video, high-speed data, and time-division multiplexed information streams. Different combinations of these three channel types can be packaged to satisfy specific user needs. For example, two commonly used packages are Basic Rate Interface (BRI) and Primary Rate Interface (PRI). BRI consists of two 64 kbps B channels and one 16 kbps D channel and is designed to satisfy individual users and small offices. The PRI consists of 30 (in Europe) or 23 (in the United States, Canada, and Japan) 64 kbps B channels and a 64 kbps D channel. It is useful for companies with greater capacity requirements.

ISDN can support both local and long distance communication. Today, ISDN can be integrated into any micro, mini, or mainframe computer. Therefore, for new users of telecommunication technology, ISDN is an inexpensive way to satisfy their internal needs since only regular telephone wires are needed. But the more important use of ISDN is to serve, in the future, as the backbone of information expressways, providing access to any individual or business in any location throughout a nation or even the world. Gateways can be provided to any wide or local area network to complement many private networks already in use by business, hence enhancing connectivity among business partners.

Open Systems Interconnection (OSI) is a model for guiding the development of standards for communications between networked systems, regardless of technology, vendor, or country of origin. OSI standards are the result of efforts of two major international standardization groups—the Consultative Committee for International Telephone and Telegraph (CCITT) and the International Standards Organization (ISO).

> The OSI Reference Model provides a layered set of network services. In many ways, it is analogous to the layers of an operating system. In both cases, the top layer provides services that are directly of concern to the applications programmer. Similarly, the bottom layers provide detailed control of hardware devices such as disk controllers for the operating system or local area network (LAN) interface boards for the networks. In both cases, the intermediate layers provide increasing levels of service, with one layer building on and adding to the services of the layer below it.[17]

With both ISDN and OSI, a concern for security and control should be addressed. The five major components of security and control should include the following:

1. **Configuration management** maintains an awareness of the physical and logical topology of the network, including the existence of components and their connectivity.

2. **Fault management** maintains an awareness of the current up/down status of each switching component and interconnection in the network and the current activities with regard to restoring any faulty units.

3. **Performance management** maintains an awareness of the current and previous performance of the network, including statistical parameters such as delay, throughput, availability, packets per second, bits per second, and number of retransmissions.

4. **Security management** maintains an awareness of who is using the network and validates access to resources within authorized constraints based on need to know.

5. **Accounting management** maintains an awareness of how and by whom network resources are being used.[17]

Electronic Filing

Electronic filing—the process of transmitting data to the Internal Revenue Service over the telephone by the use of computer—is used in conjunction with computerized tax return preparation. Some restrictions must be followed in order to qualify for the magnetic media process. Depending on the circumstances, most find these limitations to be minimal. This project involves three years of research that overcame many obstacles. The pilot project was designed to test the concept of electronic filing with three main goals:

1. To verify that the plan could work in the real world
2. To verify that the market would actually participate in the idea
3. To see if refunds were actually processed faster

The designers also drafted an operating policy that consisted of the following factors:

- Electronic filing would be voluntary.
- The IRS wouldn't provide a telecommunications network.
- Electronically filed returns would have to contain the same information as paper returns.
- An electronic return wouldn't be considered filed until the IRS recognized and acknowledged the data as processible.

All of these procedures worked to make the pilot an incredible success. In the first year, the Internal Revenue Service processed approximately 25,000 returns with an error rate of 15 percent lower than that of the manual filing process. This success encouraged the Internal Revenue Service to expand its project each year until 1989, when it was offered across the country.

The electronic method, on the other hand, is both time and money conscious as processing an electronic return costs just $.03 and reduces the steps involved. The restrictions with electronic filing cause some inconvenience because the return must include a refund and the tax return must be sent by a qualified preparer. This preparer must have a computer and compatible IRS software as well as a modem. The preparer must also be eligible and have permission from the Internal Revenue Service after a successful transmission test. The taxpayer must also file Form 8453 (U.S. Individual Income Tax Declaration for Electronic Filing) to request permission to file electronically. This form contains the taxpayer's signature, a necessity for electronic filing.

The steps for magnetic media filing are the following:

1. Return is completed and given to a qualified preparer.
2. The tax return is transmitted electronically by modem to the IRS mainframe where magnetic tape is made.
3. The tape containing the tax return is loaded and verified for accuracy. The IRS also notifies the taxpayer of receipt of the return.
4. The mainframe processes the tax return.
5. The tape is sent to the Martinsburg, West Virginia, Computer Center for posting.
6. The tape, with refunds, is sent to the Treasury Department where checks are disbursed.

As seen through the last years of this project, it has many advantages. A major plus involves the cost savings. The government can reduce its processing costs by 90 percent when filing electronically instead of manually, which will save the U.S. government approximately $200 million by 1995. With the transfer of the tax return from one computer to another, human error drastically declines. The qualified preparers must examine for errors and discrepancies before transmission, which reduces the IRS's need to check returns. No data entry is required at the IRS Service Center so increased efficiency and reduced error occur. Additionally, there is little chance that the return can be lost in the mail as receipt of filing is given and the refund arrives in approximately two to three weeks, either in the mail or by direct deposit.

Numerous security risks can be found throughout the projects as most IRS information is labeled confidential. In the manual process, many people have the chance to look over the tax return, but the electronic method lowers the chances of others seeing this information by computer transmission. The lack of paper copies can also be a danger as all the IRS backups could be lost or destroyed. For this reason, the IRS is considering a contingency procedure where filers would still need to send paper copies of previous electronic returns. The IRS has also established a physical break between the receipt of electronic data and all subsequent processing—this separation of duties serves as an important safeguard. The ability to tap the modem transmitting a return is quite feasible; consequently interrupting, intercepting, modifying, and even fabricating the transmission of the return could occur. In the future, preparers should have passwords as well as other authenticity checks to verify their existence. Because the IRS is a government agency, few breaks in its security are disclosed and little evidence can be found to verify this critical information. Controls over viruses and worms must also be installed even though the preparers are screened. A bug of this nature could virtually shut down the IRS computers and cause serious problems. Security is a top priority for the government, especially the Internal Revenue Service. New procedures and controls are constantly being established to ensure higher security.

THOUGHTS ABOUT THE FUTURE

From their earliest history, computers have been used to keep records and exchange information. The 1970s witnessed the rise of communication-based computer systems and networks. The 1980s subsequently witnessed a continuation of the computer revolution. Among these 1980s contributions were:

- A growth in the use of personal computers, thus increased vulnerability
- An expansion in the number and size of networks
- The growth of information services
- A higher-level integration of data services in such industries as insurance, banking, travel, entertainment, law enforcement, commodities/securities exchange, medical service, and others

Regardless of what direction the computer revolution takes, networks are a burden as well as a blessing. Our data can be transported at speeds unheard of in the 1960s, but data vulnerability is our newest potential threat. Our challenge comes, therefore, from both technology and from human frailty. We must ensure that the "information in the hands of many" is protected at all cost. Only then will we have a functioning network, not a nightmare of unequalled proportions.

SUMMARY

Computers and their components need to communicate for speed and accessibility. **Connectivity** facilitates the electronic interchange between computers via **data communications**, **telecommunications**, **teleprocessing**, and **networking**. Signals are transmitted by either **digital** or **analog** circuits via a **modem**. The two basic methods for transmitting information are **asynchronous** or **synchronous**.

Various media are utilized for communications channels—**wire cables**, **twisted-pair wiring**, **fiber-optic cables**, **microwaves**, and **satellites**. The three most often used **network topologies** are **star**, **ring**, and **bus**.

Communication presents security vulnerabilities that must be addressed—network failures, **eavesdropping**, **dial-in access**, **file locking**, **communication system failures**, **signal interception**, and **electronic espionage**.

As the use of communications increases, new applications are introduced and present unique security risks—**Fax**, **ATMs** and **EFT**, **cellular telephones**, **ANI** and **Caller ID**, **ISDN** and **OSI**, and **electronic filing**.

REFERENCES

1. Alter, Steven, *Information Systems . . . A Management Perspective* (Reading, MA: Addison-Wesley, 1992).
2. Alzua, Roy, "The Problem of Communications Fraud," *Access*, 2nd Quarter, 1989, pp. 8, 24-26.
3. "An Introduction to Secure Telephone Terminals," *CSL Bulletin*, U.S. Department of Commerce, Technology Administration, National Institute of Standards and Technology, March 1992.
4. Axner, David H., "Access and Authentication: Key Issues in Network Security," *Business Communications Review*, March 1992, pp. 61-66.
5. Baker, Richard H., *The Computer Security Handbook* (Blue Ridge Summit, PA: Tab Books, 1985), pp. 1-281.
6. Bequai, August, "The Rise of Cashless Crimes," *USA Today*, January 1986, pp. 83-85.
7. Berman, Alan, "Security in a Dial-Up Environment," *Data Security Management*, 84-04-17, 1985, pp. 1-8.
8. Berry, S.L., "FAXPIONAGE!" *Security Management*, April 1991, pp. 59-60.
9. "Beware! FAX Attacks," *ABA Banking Journal*, June 1990, pp. 57-60.
10. BloomBecker, Jay, *Commitment to Security*, National Center for Computer Crime Data, Santa Cruz, CA, 1989.
11. Bonner, Paul, "Keeping the LAN Manager Honest," *PC Week*, Vol. 4, No. 9, March 3, 1987, pp. C1, C6-C8.
12. Brock, Jack L., "Hackers Penetrate DOD Computer Systems," Testimony before the Subcommittee on Government Information and Regulation, Committee on Governmental Affairs, November 20, 1991, pp. 1-5.
13. Burnham, David, "Flaws Are Cited in Treasury Computers," *The New York Times*, February 13, 1986, p. D5.

14. Callahan, Michael, "Who's Calling? ANI, Caller ID, and Dial-In Security," *ISP News*, January/February 1992, p. 42.

15. Capron, H.L., *Essentials of Computing* (Redwood City, CA: Benjamin/Cummings Publishing Company, Inc., 1992).

16. Cole, Gerald D., *Computer Networking for Systems Programmers* (New York: John Wiley and Sons, 1990).

17. Cole, Gerald D., *Implementing OSI Networks* (New York: John Wiley and Sons, 1990).

18. "Computer Cops Needed to Police Networks," *New Scientist*, December 15, 1990, p. 16.

19. Cook, William J., "Paying the Bill for Hostile Technology: PBX Fraud in 1991," *ISP News*, September/October 1991, pp. 34-37.

20. Craig-Van Collie, Shimon, "Banks That Master the Cat's Cradle of Communications," *Bankers Monthly*, June 1989, pp. 69-73.

21. Cross, Rice H., and David Yen, "Security in the Network Environment," *The Journal of Computer Information Systems*, Fall 1991, pp. 4-12.

22. Danca, Richard A., "EPA Juggles LAN Security, Access," *Federal Computer Week*, January 27, 1992, p. 34.

23. Deans, P. Candace, and Michael J. Kane, *Information Systems and Technology* (Boston, MA: PWS-Kent Publishing Company, 1992).

24. Enger, Norman L., and Paul W. Howerton, *Computer Security . . . A Management Audit Approach* (New York: American Management Association, 1980), pp. 1-264.

25. "FAX Security: Curse of the Wrong Number," *Canadian Business*, August 1989, p. 68.

26. Fernandez, E.B., R.C. Summers, and C. Wood, Database Security and Integrity (Reading, MA: Addison-Wesley, 1981), pp. 39-53.

27. Fitzgerald, Kevin, "The Management and Control of Downloaded Data," *Computer Control Quarterly*, Vol. 9, No. 3, 1991, pp. 22-29.

28. Forcht, Karen A., "The Path to Network Security," *Security Management*, Vol. 31, September 1987, pp. 152-157.

29. Frenzel, Carroll W., *Management of Information Technology* (Boston, MA: boyd & fraser publishing company, 1992).

30. Gantz, John, "Largely Ignored Network Security Could Prove Ticking Time Bomb," *Infoworld*, May 8, 1989, p. 34.

31. Gillooly, "Internet Working Poses Big Problems About Security," *Computer Control Quarterly*, Vol. 9, No. 4, 1991, pp. 25-26.

32. *Guideline for Automatic Data Processing Risk Analysis*, FIPS Pub. 65, U.S. Department of Commerce, National Bureau of Standards, Springfield, VA, 1979.

33. Hains, David, "LAN Security: Are You Taking It Seriously?" *Computer Control Quarterly*, Vol. 10, No. 1, 1992, pp. 7-11.

34. "How to Put the Q-T on Your FAX," *Canadian Business*, August 1989, p. 17.

35. Hutt, Arthur E., Seymour Bosworth, and Douglas B. Hoyt, *Computer Security Handbook* (New York: Macmillan, 1988), pp. 1-399.

36. Jan, J.K., C.C. Chang, and S.J. Wang, "A Dynamic Key-Lock-Pair Access Control Scheme," *Computers and Security*, Vol. 10, 1991, pp. 129-139.

37. Johnston, Robert E., "Solving the Mystery of Network Security," *Access*, 2nd Quarter, 1989, pp. 16-18.

38. Keller, John J., "Hackers Open Voice-Mail Door to Others' Phone Lines," *The Wall Street Journal*, March 15, 1991, p. B1.

39. Keough, Howard, "Safety Pins," *Security Management*, February 1988, pp. 33, 34, 37.

40. King, Julia, "Secure Communications Stuck in Certification Mire," *Federal Computer Week*, November 4, 1991, p. 40-41.

41. Long, Larry, and Nancy Long, *Computers* (Englewood Cliffs, NJ: Prentice-Hall, 1990).

42. Manheim, Marvin L., "Global Information Technology . . . Issues and Strategic Opportunities," *International Information Systems*, Vol. 1, No. 1, January 1992, pp. 38-67.

43. Martin, James, *Security, Accuracy, and Privacy in Computer Systems* (Englewood Cliffs, NJ: Prentice-Hall, 1973), pp. 1-626.

44. Mase, Mike, "Your Number Please? Gains New Meaning," *Telephony*, April 14, 1986, p. 60.

45. Mullen, Jack B., "A Checklist for Online Terminal Security," *Data Security Management*, 1986, pp. 1-16.

46. Murray, William H., "To Dial or Not to Dial," *ISP News*, March/April 1992, p. 57.

47. Nabours, Mac, "Living with Sin: The Security Information Network," *Security Management*, April 1988, pp. 61-72.

48. Nickerson, Robert C., *Computers* (New York: HarperCollins, 1992).

49. Pfleeger, Charles P., *Security in Computing* (Englewood Cliffs, NJ: Prentice-Hall, 1989), pp. 1-538.

50. Press, Jim, "Secure Transfer of Identity and Privilege Attributes in an Open Systems Environment," *Computers and Security*, Vol. 10, 1991, pp. 117-127.

51. Russell, Deborah, and G.T. Gangemi, Sr., *Computer Security Basics* (Sebastopol, CA: O'Reilly and Associates, Inc., 1991), pp. 1-441.

52. Schneidawind, John, and Mark Land, "The Big Hang-Up . . . Phone Crash Grounds Airplanes, Raises Anger," *USA Today*, September 19, 1991, p. B1 (27-28).

53. Schweitzer, James A., *Computer Crime and Business Information . . . A Practical Guide for Managers* (New York: Elsevier, 1986).

54. Sieber, Ulrich, *The International Handbook of Computer Crime* (New York: John Wiley and Sons, 1986), pp. 1-276.

55. Stair, Ralph M., *Principles of Information Systems: A Managerial Approach* (Danvers, MA: boyd & fraser publishing company, 1992), pp. 1-701.

56. *The Handbook of Information Security* (San Jose, CA: Arca Systems, Inc.), pp. 1-202.

57. *Trusted Network Interpretation*, National Computer Security Center, July 31, 1987.

58. "United IS Failure May Be Arson," *Information Week*, December 21, 1987, p. 15.

59. Wells, Ken, "Lithuanians Find They're Missing One Perk of Red Army Occupation," *The Wall Street Journal*, November 11, 1991, p. B1.

60. Wright, Benjamin, *The Law of Electronic Commerce . . . EDI, FAX, and E-Mail: Technology Proof, and Liability* (Boston, MA: Little, Brown & Co., 1991).

61. Zachary, G. Pascal, "Computer Viruses May Win Military Role," *The Wall Street Journal*, November 11, 1991, p. B1.

62. Zachary, G. Pascal, "Software Firms Keep Eye on Bulletin Boards," *The Wall Street Journal*, November 11, 1991, p. B1.

REVIEW QUESTIONS

1. Briefly explain the term connectivity and how it is achieved.
2. Compare analog and digital circuits.
3. How are modems used in communications?
4. What is a communications protocol?
5. Describe several communications media.
6. What are the three major network topologies? Describe each.
7. How can encryption be used in communications to secure transfer of information?
8. Explain the use of user verification.
9. What are some of the common network failures?
10. Describe some of the various special communications applications and their accompanying security risks.

DISCUSSION QUESTIONS

1. Electronic messaging in the form of electronic mail (E-mail), voice mail, and bulletin boards provides a free and open exchange of ideas and correspondence. What are some of the possible misuses or security risks associated with each of these systems?
2. Security in data communications and networking is an extension of the security concern of a single computer. The same basic concerns apply, but the components are geographically distributed. Three problems need to be addressed when this dispersion takes place: security policy, security mechanisms, and assurances. Briefly discuss each of these.
3. In a networking or communications environment, two major factors to be addressed are confidentiality and integrity. What are some of the factors that apply to each of these?
4. The high concentration of information stored in electronic devices, along with the increasing dependence on computers, make computer sabotage and espionage particularly dangerous for businesses. The objects are the tangible computer facilities as well as the intangible information and data containing computer programs and other valuable information. What are some of the objects, both tangible and intangible, that could be the targets for these efforts?
5. Networks are subject to two types of security threats—passive and active. How can network managers control these intrusions?

EXERCISES

1. Locate a periodical article that discusses an application of communications and relate some of the possible security risks that might be present with the application.
2. Interview someone in your area who is employed in the communications industry to gain some insight into the security that is present in his or her position.

3. Discuss the following statement: "Communications has opened up a multitude of new opportunities and the security will first have to 'catch up' with the technology."

4. Interview a key employee at a local bank concerning how ATMs and EFTs are secured.

5. Observe an ATM machine in your area and list some of the possible security vulnerabilities that are present.

PROBLEM-SOLVING EXERCISES

1. Preventing the interception of data during transmission from one site to another is virtually impossible. Other means must be used to make the data of as little use as possible. Describe three methods that could be used to achieve this goal.

2. Machine identification is needed where telecommunications switching equipment is used. Some switching mechanisms occasionally make a wrong connection and have no means of recognizing their error. What are some of the design flaws that could cause this problem? How can they be corrected?

3. Design a topology (ring, bus, or star) and indicate how and where security control procedures would be added to protect the network.

4. Network security is an old problem that "came with the territory" of connectivity. Users, in many cases, are still paying "lip service" to the issue as they interconnect. Design a basic orientation session for communication users that would convince them that each of them is responsible for the security of the entire system.

5. People communicate best when they speak the same language. Machines communicate better when they operate under consistent standards and formats. How would stricter standards make security of communications easier to enforce and administer? What are some of the drawbacks to stringent standards?

CASES

Aston Enterprises

John Aston, president of Aston Enterprises, has been hearing a lot of talk about communications and how it speeds up the transfer of information. Aston Enterprises is an independent geological survey company, located in Altus, Oklahoma. The geologists are out in the field most of the day and call back to the headquarters via cellular phones located in their vehicles.

Mr. Aston is considering adding mobile faxes, additional cellular phones, cordless mobile phones, beepers, and voice mail so that the flow of information will become a constant two-way process. In the highly competitive geological survey business, speed is of paramount concern. Mr. Aston, however, has not considered the security risks involved.

List several factors he should consider to lessen the possibility of accidental or malicious breach of information flow both to the field and from the field.

Paxly Products

Paxly Products is a mail-order business, headquartered in Brockton, Massachusetts, specializing in computer components. Customers are able to phone or fax their orders for speedy service. Paxly is growing increasingly concerned with the accuracy of its ordering process: if the wrong components are added to a computer system, damage or malfunction can occur, resulting in irate customers and possible legal action.

What are some of the verification procedures that could be added to Paxly's present system to ensure accuracy and reliability of information transfer?

MICROCOMPUTER SECURITY

LEARNING OBJECTIVES

After studying this chapter, you will be able to:

1. Explain the growth and use of microcomputers in today's computing environment.
2. Describe the different types of microcomputers (PCs) currently being used.
3. Compare micros to minicomputers, mainframe computers, and supercomputers.
4. Describe some of the techniques used to provide security for micros.
5. Describe some of the internal data security controls used to protect data and machines from malfunction and damage.
6. List the basic elements of a microcomputer security plan.
7. Explain some of the security issues involved in a micro-to-mainframe link.
8. Describe password protection schemes and their application to the microcomputer environment.
9. Explain some of the security elements of special microcomputer applications, such as downloaded data, multiuser micros, on-line terminals, and workstations.
10. Define the following terms.

TERMS

- Personal Computer (PC)
- Laptop Computer
- Notebook Computer
- Workstation
- Microcomputer
- Minicomputer
- Mainframe Computer
- Supercomputer
- Multiuser Capabilities
- Master File
- Transaction File
- Backup File

- Grandfather-Father-Son Principle
- Hidden Files Technique
- Dial-Back System
- Password
- System-Generated Password
- Challenge Response
- Shared-Use System
- Multiuser System
- Removable Media
- Fixed Media
- On-Line System

RED FACES AT THE BRITISH MINISTRY OF DEFENCE

Shortly before Christmas 1991, an unknown thief stole a laptop computer from a British Royal Air Force (RAF) officer's car as he visited a used-car showroom. On the laptop were stored files describing the battle plans for the recapture of Kuwait from the Iraqis. The British Ministry of Defence's luck held, however, when the laptop was returned anonymously and it became apparent that the information had not fallen into Iraqi hands. The incident raised some disturbing questions, including whether the information had been properly encrypted.

As the story broke via the British tabloid press, the MoD refused to discuss the incident, which led to the conclusion that the department had failed to encrypt the data stored on the laptop's hard disk. If that is true, the MoD is left with egg on its face, because inexpensive encryption programs are readily available and very reliable.

On such program, which sells for around $200, contains programs to recover files inadvertently erased, repair corrupted data, and wipe out files completely so they can never be reconstructed. This package also inclues a program that protects confidential files by coding them with the data encryption standard developed by IBM and the U.S. government.

The user simply chooses a random password. Thereafter all text stored in the computer can be read only after the password has been reentered. The encrypted text does not even show up on the computer's normal file listing. The only hope of cracking the code would be to use the supercomputers operated by the National Security Agency at Fort Meade near Washington, D.C., or at British Government Communications Headquarters at Cheltenham, England.

As reported in "Defense of the Data," *New Scientist,* January 19, 1991, p. 17.

In the previous two chapters, the basics of database and telecommunications security were discussed. As many computer systems are configured with microcomputers used as stand-alone computers or linked to mainframe databases and telecommunications networks, this chapter will discuss the special security considerations that microcomputers present. To fully address the microcomputer environment's security problems, this chapter will cover the current trend toward micros; discuss the unique characteristics of the micro environment, security considerations, the link between micros and mainframes, special micro applications (databases, telecommunications, on-line systems, multiusers, and workstations), password considerations; and briefly review micro-based security products.

MICROCOMPUTER PROBLEMS AND SOLUTIONS

With the widespread use of microcomputers has come an equally widespread knowledge about computer systems in general. There are many more people today who know how to invade private data files. There are also more "beginners" who are likely to have accidents that irretrievably destroy the data they are working with. Thus, the importance of planning and policies that minimize or prevent data loss from theft, accident, or equipment failure cannot be over stressed.[60]

Today, most offices within corporations are equipped with a desktop PC that not only has all the corporation's information within a user's reach, but also is more vulnerable to security risks due, quite simply, to its open location. Add to this risk the number of computer systems that use MS-DOS, the most prevalent operating system for the PC. MS-DOS is well known and easily understood, even by some of the most inexperienced computer users. Corporations are working to promote security in their systems, but they must rely on the users to uphold the systems and follow security procedures. Users agree that companies need security, but they may feel that security is a burden that makes using the computer too much of a bother.

Many large corporations are stepping up their security efforts. Amoco's system will not allow a person to log on from outside. Instead, the computer will dial the authorized user back on authorized phone lines only. Amoco's PCs will not allow modules to be executed if they have been altered in any way. Moore Business Forms feels that security is important, but only to the degree that it does not hamper the end user's ability to use the system. Baxter Healthcare Corporation feels that cooperation between the security personnel, system developers, and end users is the key to a successful security system.[33]

The bottom line is that, as long as humans are involved with computers, security mechanisms must be in place to protect the systems. Users must understand the importance of security measures for the measures to work.

Growth of Computing

From very limited beginnings, organizational computer use has grown substantially. Examination of statistics prepared by International Data Corporation, a firm whose business is to study the computer industry, shows quite readily how widespread this growth has become:

- About 100,000 computer instructions are executed every second in the United States for *each* of its citizens.

- *Each* U.S. citizen "appears" in 10 to 1,000 databases.

- Most homes have 2 to 10 "computers" (microprocessors).

- In the next five years, a trillion dollars will be spent in the United States on information processing.

- The information processing industry is growing at a yearly rate of 16 percent with many sizable sectors growing at rates of 25 percent or higher.

- At present growth rates, by the early- to mid-1990s, there will be more than one intelligent office "gadget" for every white-collar worker.[14]

During their rapid growth stage in the 1980s, computers proliferated in organizations of all sizes and types. Often, as this growth proceeded, security was not considered in the system design stages. This attitude may well have resulted from management's lack of understanding of the functioning of their computer systems and their peculiar vulnerability to novel forms of fraud and malicious interference.

414s Strike

How microcomputers could be used to enter large mainframe systems and databases came to light during the summer of 1983 when seven youths ranging in age from 15 to 22 in the Milwaukee, Wisconsin area gained access to computers spread across the United States and Canada. The 414s (as the group called themselves, after the Milwaukee area code) penetrated such vital computers as:

- Memorial Sloan-Kettering Center in New York City

- A bank in Los Angeles

- A cement company in Montreal, Canada

- An unclassified computer at a nuclear weapons laboratory in Los Alamos, New Mexico

The Sloan-Kettering Caper and hit movie (1983) *War Games*—the story of a young computer buff who nearly sets off a nuclear war when he accidentally gets into one of the Defense Department's most sensitive machines—have focused attention on a serious question: How to safeguard information stored inside the computer.[15]

The potential for fraud is awesome. The American banking system alone moves more than $400 billion between computers every day. Corporate data banks hold consumer records and business plans worth untold billions. Military computers contain secrets that, if stolen, could threaten the United States' security. Microcomputers link many of these machines to the telephone system, thus enabling them to communicate with other computers and with users in remote locations.

Undetected Breaches

Experts in the industry estimate that nearly 95% of all computer security breaches go undetected, and agree that the fastest-growing problem area involves the personal computer. Elaborate security measures are now routinely employed on mainframe computers, but PCs afford an almost unregulated opportunity for problems. The current trend of integrating PCs into networks and mainframe systems now makes once secure data vulnerable.[60]

The information that magnetic media contains is easily copied, and PC media are readily transportable and therefore easy to conceal and steal. An entire hard disk of data can be loaded onto a tape cartridge that slides easily into a jacket pocket.

Moreover, the PC can change a minor theft of office equipment into a major disaster if the pilfered equipment happens to be a computer containing the only copy of a company's records.[60]

Growth Will Continue

James A. Schweitzer, in *Computer Crime and Business Information,* states that:

> Computers are changing our society and the world. We do not have to understand these effects to use computers, but if we wish to use them carefully and manage our business information resources efficiently, we must understand the business implications of the information age . . . The ready availability of computing power, through local computers and their connections to high-power central computer services, suggests that personal use of computing as both tool and playmate will continue to increase rapidly.[48]

THE MICROCOMPUTER ENVIRONMENT

Over the years, people have developed various terms to describe computer systems including personal computers, workstations, minicomputers, mainframe systems, and supercomputers. Most people have a good idea about the size and purpose of these systems, yet providing precise definitions is sometimes difficult. Computers are usually classified by characteristics, speed, size, and cost. (See Figure 8.1.)

A **personal computer** (**PC**) is a single-user computer that sits on a desktop or can be carried around by the user (such as a laptop or notebook computer). A **laptop computer** is small enough to be portable but big enough to have a very large capacity—most weigh about 15 pounds. A **notebook computer** is a computer that is small enough to fit into a briefcase but big enough to carry out sizable computing tasks—most weigh 6 to 7 pounds.

FIGURE 8.1
Classifying Computers

Characteristic	Personnal Computer	Workstation	Minicomputer	Main Frame	Super computer
Processor speed	8-16 MHz	16-32 MHz	1-2 MIPS	10-60 MIPS	1-10 GFlops
RAM size	16-64 MB	16-64 MB	32-128 MB	32-256 MB	Greater than 256 MB
Physical size	Desktop	Desktop	Closet-size	Automobile-size	Room-size
Cost	$100 to over $5,000	$5,000 to over $20,000	$2,000 to over $100,000	$50,000 to over $5 million	$5 million to over $20 million

Source: Ralph M. Stair, *Principles of Information Systems* (Boston, MA: boyd & fraser publishing company, 1992), p. 79.

Workstations are powerful single-user computers that can be used for complex data analysis and design work. (Both personal computers and workstations are often called **microcomputers**.)

Minicomputers are larger computers typically shared by a department in a company for processing transactions, accessing corporate databases, and generating reports.

Mainframe computers are even more powerful and are used to process large volumes of on-line transactions and to generate reports from large databases.

Supercomputers are specialized high-speed computers used for lengthy calculations rather than for processing transactions or generating reports.

With the continuous stream of developments in computing, the classifications are a moving target. Personal computers and workstations were originally aimed at different audiences, but they are starting to converge as the underlying technology becomes cheaper and more powerful. These computers have also started to incorporate **multiuser capabilities** as a means of coordinating and integrating the work performed by various members of the same department or groups.

According to Carroll W. Frenzel in *Management of Information Technology,*

> The personal computer puts the power of yesterday's mainframe on the desks of millions of today's workers . . . The growth in numbers of PCs rivals their growth in power. It is estimated that the total number of PCs installed in 1992 will be nearly 93 million and that nearly half of these will be part of a local area network. U.S. purchases of microcomputers in 1987 were 40 percent of the $26.7 billion CPU market.[20]

SECURITY OF MICROCOMPUTERS

A few years ago, most computing was done on mainframes housed in secured rooms. Security of this closed environment was a matter of "watching one spot." As micros, networks, workstations, and distributed computing grew rapidly, security became a widespread dilemma. The basic principles of security are somewhat universal, but applying them to micro and distributed environments presents some unique challenges.

Security problems on PCs or micros are more serious than on mainframe computers for two reasons, one related to people, and one related to hardware and software.

- **Lack of Sensitivity.** People often do not understand the security risks associated with the use of personal computers.

- **Lack of Tools.** The tools—hardware, software, and combinations of these—are fewer and less sophisticated than those used in the mainframe environment.[41]

James A. Schweitzer, in *Computer Crime and Business Information,* offers the following points relating to the unusual risks that microcomputers present:

1. Because of its power and especially its communications capabilities, the personal computer and its information storage elements present severe vulnerabilities.

2. These vulnerabilities are made worse because the personal computer is frequently used outside the normal, controlled business environment—where rules cannot be enforced and where unknown persons often can have access.

3. Nevertheless, if carefully used with the controls and software available for the purpose, the personal computer can provide better security than that typically provided to information on paper.[48]

When PCs are first introduced to users, a few basic rules should be reviewed. If PCs are already in place, the same rules should be applied. Only when *all* micros are secured can the vulnerability of the entire system be lessened. (See Figure 8.2.)

FIGURE 8.2
Micro Security Policies

☑ Back-up procedures
☑ Disaster recovery plans
☑ Software theft and misuse
☑ Threat from within
☑ Physical security
☑ Encryption and passwords

Backup Procedures

The practice of using effective backup procedures *regularly* should be followed. Making copies of diskettes and tapes should be as routine as booting up the system. Since storage media can be damaged by fire, electrical fluxes in the power lines, mechanical failure, misuse, and operator error, backups are vital to an ongoing operation. The recommended practice is to have one copy available on the premises for immediate backup and one stored off-site. Many organizations follow the grandfather-father-son method of backups that has been applied to the mainframe environment for years. (See Figure 8.3.)

FIGURE 8.3
Floppy Disk Backups

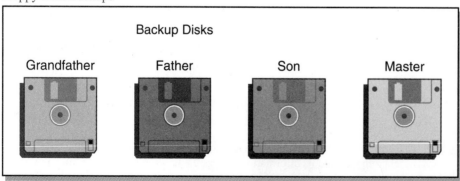

Backup Disks

Grandfather   Father   Son   Master

A **master file** is a permanent group of records stored on computer media. These files must be updated to make them current and accurate before transaction processing begins.

A **transaction file** is a computerized file that contains business transactions or activities that affect the master file. Transaction files should be checked closely for accuracy.

Backup files are copies of critical or essential files that are routinely run along with the original. Some organizations require backup copies to be placed in a centralized and secure location, ensuring that key data and programs will not be lost due to poor individual backup practices.

The **grandfather-father-son principle** of backups means that several backups of all critical master files will be made.

Disaster Recovery

High on the list of accidental causes of data loss in the small system environment are disk damage and disk erasure. The floppy disk is susceptible to damage resulting from cigarette smoke, staples, magnets, and moisture. Even when the hazards are known, it is not always easy to avoid them; consequently, a disaster recovery system should be set up to work in conjunction with the backup procedures. Some of the common hazards found in today's PC environment are:

- **Sunlight.** Sunlight can be extremely damaging to a disk. Avoid placing disks on window sills.

- **Static electricity.** Place a non-static rug in the computer area to avoid static build-up.

- **Cigarette smoke.** Smoke can interfere with the reading of data by the head.

- **Ballpoint pens.** The pressure of the ballpoint pen can scratch the disk surface. Use only felt markers.

- **Heat.** Do not place the disk on or near any hot surface as warping can occur.

- **Power surges.** Use a power surge protector to avoid loss of data. This is extremely important in certain metropolitan areas of the country where "brown-outs" are common.

Software Theft and Misuse

Some industry commentators maintain that, when dealing with the software theft issue, vendors like licensing software rather than selling it outright, for that appears to give them greater control over the product's use. Users generally feel that once they have purchased the software, the vendor should make allowances for backup copies instead of "blowing up" the program when the user tries to make a copy. Both software vendors and corporate users are finding ways to deal with the question of illegal copying by drafting site licenses for multiple machine users (agreements that allow the copying and use of programs on large numbers of machines for a price that is less than the number of copies times the retail price per copy) and employing various software-protection systems. Organizations are formulating new policies for employees' use of software, issuing corporate-wide policy statements, and restricting program disks to a limited number of employees to avoid any infringement violations.

The Threat from Within

According to James A. Schweitzer in *Computer Crime and Business Information,* "The primary risk to business information today is its theft or misuse in written form by employees."[48] Most information today exists in some tangible form— paper, disks, microfilm, etc.—and is, therefore, quite portable. The ready availability of copying machines makes the task of running an extra copy quite simple. Many times, employees run copies merely as a backup but, all too often, illegal copies are made and used for illegal purposes or financial gain. Establishing a formal corporate policy on employee ethics and loyalty should become a part of the organization's security program. Many employees may not fully understand how vulnerable information becomes when multiple copies are run, are left on file cabinets or desks, and are easily accessed by those intent on misusing that information.[48]

Physical Security

The ideal location for storing the microcomputer is in a locked room, an area not accessible to all employees. The layout of the office or the sheer number of micros in an organization may prohibit this type of security. In that case, securing micros at the users' desks is the next option. Susan Foster, writing in *Computer Decisions,* suggests the following:

- Using computer tables with lockable covers that enclose the entire system, allowing for dust protection as well as physical security
- Bolting the CPU to the desk and limiting access by a key to unlock the casing.[18]

Some additional ways to protect your hardware and other devices include the following:

- Use a surge protector for all computers. A fluctuation in power supply can lead to memory loss or destruction of the entire system.
- Leave room around your computer. Air circulation is important to keep your computer cool.
- Static electricity can also damage equipment. Keep humidity above 40 percent when possible. If static is a problem, use the available products, such as anti-static mats for both your computer and floor.
- If you notice a burning smell, unplug your computer. **Never** use a fire extinguisher!
- Use covers to protect computers and keyboards from dirt and food spills when not in use.
- Never eat or drink near your computer. Liquids and food crumbs should be kept away from sensitive equipment because they can damage electrical connections.
- Don't smoke near your computer. Cigarette smoke can cause particle buildup that damages computer components.

- Clean the monitor regularly. Spray the cleaning agent on a soft cloth, not directly on the screen. If available, use anti-static cleaning solutions or pads designed for screens. If you have an anti-glare screen, follow manufacturer's instructions for cleaning it.

- Never force a diskette into a drive or force a drive handle.

- Be sure there are no loose labels on diskettes. Cover adhesive remaining from previous labels with a new label.

- Dust and dirt damage disk drives. Never place 5 1/2″ diskettes on work surfaces; always put them directly into their protective sleeves. Also, keep your computer area clean and dusted.

- Back up your hard disk and all important diskettes. Remember that computers are only machines, and machines can fail.

Encryption and Passwords

A very effective method of limiting access to only authorized personnel is encryption, a method of garbling and compressing data to make it unreadable to anyone without a translation command code. Although, until recently, encryption has been used almost exclusively when information is transmitted to and from remote locations, it is a practical and effective way to safeguard any important data.

Password protection may also be used to limit access to only those users able to supply the password. If the user fails to enter the correct password after several tries, the system shuts down. Indexes containing passwords should be kept locked in a separate location, away from the work site; access to the index should be limited. As an additional security measure, passwords should be changed regularly. The following ten computer security solutions cover the basic problems of microcomputer security:

1. Establish formal policies concerning the use of corporate computers—both at the office and at the employee's home.

2. Make each user accountable for the device, the software, and vital information being used.

3. Provide facilities for the safe storage of media.

4. Instruct *each* user in the minimum care requirements of the computer.

5. Develop a backup plan and recovery procedures.

6. Provide rules and facilities to protect sensitive information in shared environments.

7. Locate off-premises storage for critical data.

8. Adopt standard naming conventions for all data and files to avoid "losing" data when someone cannot remember a file name.

9. In unsecured areas, provide a means to protect the computer and its components from unauthorized use or theft.

10. If the computer is connected to dial-up communications facilities, ensure that sensitive data is protected from unauthorized users.

"The personal computer user must be required to use all the security features available and must be motivated to follow good security discipline."[48] This means:

1. Using passwords of at least six characters in length, not easily guessed by others and changed at least every 90 days.
2. Using encryption or file lockwords when a fixed disk is in use. Again, these should not be easily guessable.
3. Locking away floppy disks when not in use.
4. For high-value information, using encryption when appropriate, in such situations as strategic planning, funds transfer, product research, financial reporting, and market research strategy applications.
5. Avoiding printouts where unnecessary and destroying unneeded printouts by shredding or burning. Avoid using the wastebasket for *any* sensitive information.
6. Protecting tapes, disks, microfilm, and other documents when en route or in public places.

INTERNAL DATA SECURITY

There are certain things that a user can do to ensure that data are protected without having to worry about elaborate forms of security. These simple steps can be followed to ensure that basic data security has been carried out.

Rules of the Road

1. When the computer is not in use, the user should not leave the terminal on with confidential data being displayed on the monitor.
2. When running a confidential application, the user should not leave the terminal unattended in an unlocked room.
3. The user should not store disks in unlocked drawers or in the cabinet on which the terminal is kept.
4. The user should remove all disks from the disk drive when the terminal is unattended or not in use.
5. The user should create backups for all critical data files in case they are destroyed or they are rendered inaccessible.

The user may follow these simple techniques to create his or her own informal form of data security. These techniques, however, are not fail-proof. An intruder may still break through these simple measures; that is where the issue of external data security comes in.

Accident Security

If a user is unfamiliar with computers, potentially destructive commands should be kept out of reach. Store them on separate diskettes. If someone accidentally FORMATS an important diskette, or performs FDISK on a hard disk, all the files will be deleted! If write-protect tabs are used on source diskettes, the danger of erasing a needed diskette is eliminated.

Commands

For good security, all potentially destructive programs should be kept out of reach. Everyone who uses the computer doesn't need to back up and format diskettes, and certainly no one needs to perform FDISK on the hard disk.

As in any business situation, there is always the possibility of accident. If precautionary measures should be taken, the following guidelines establish a hierarchy of access to MS-DOS commands.

1. **Standard security.** The following commands can be used by all employees:

CHKDSK	FC
COMMAND	MEMTEST
COMP	TREE
DIR	VER
DIRCOMP	VOL
DISKCOMP	

2. **Good security.** Employees responsible for maintaining backups on diskettes will need:

BACKUP	FORMAT
COPY	REN
DISKCOPY	VERIFY

   These commands will allow selected employees to copy all data and programs and, if careless, to erase data and program diskettes.

3. **High security.** Employees familiar with the computer can use the following commands:

ANSI	EXE2BIN	MODE	RESTORE
ASSIGN	EXIT	MORE	RMDIR
BREAK	FIND	PATH	SET
CHDIR	FOR	PAUSE	SHIFT
CLS	GOTO	PHDINIT	SORT
CTTY	GRAPHICS	PRINT	SYS
DEBUG	IF	PROMPT	TIME
ECHO	LINK	RECOVER	TYPE
ERASE	MKDIR	REM	

4. **Ultra security.** The following commands should have the highest security and should be used only by authorized personnel:

   DISKTEST

   EDLIN

   FDISK

In the Windows environment, the Program Manager, Print Manager, File Manager, and Control Panel allow you to perform some of these functions by clicking on icons.

Naming Conventions

Naming conventions should be established to organize files. If users are using their hard drives for more than one type of activity (say accounting and word processing, or inventory, accounts receivable, and accounts payable) they can start their organization by storing each type of file in a separate directory.

One way to organize naming conventions is by application, department, or project. All files having to do with the same thing should have a common set of characters in their filenames. Wildcard characters can be used to view and manipulate these names.

Another naming convention is to use each person's name. Use a particular character or characters in the filename to specify the person who created or is responsible for the file.

Hidden Files

The user has the ability to hide files, directories, and characters from view by using **hidden files techniques**.

Hiding a file or directory simply means that the user can't see any part of the file or directory name. Hiding characters allows the user to see part of the file or directory name, but keeps one character invisible. The advantage of this technique is that an intruder will not be able to access what cannot be seen unless he or she knows about the MS-DOS feature. To access the feature, all the user has to do is enter a standard notation within the file or directory name or as the file extension. Another feature provided by most operating systems is an alarm and lockup procedure, executed by a command, that locks up the keyboard and causes the alarm bell to ring on the terminal. To stop this function the user would have to press the CONTROL, DELETE, and ALT keys all at the same time. The advantage of this feature is that it startles the intruder and may force him or her to abandon the machine. The techniques of the MS-DOS operating system offer the user an inexpensive way to secure data because the system is used by a good percentage of the microcomputing environment.

Categorizing Data

Data and information should be routinely categorized to determine the security level needed to protect the privacy and credibility of those particular files. Some of the security options for different types of personal data are shown in Figure 8.4.

THE THREATS TO MICROS

PC Magazine, in its April 28, 1987 issue, named the following invaders as threats to a micro (PC) environment:

- **The Corporate Spy.** The same guy who tried to jimmy your file cabinet may now be after your electronic data. While espionage tends to target trade secrets, certain PC areas such as financial, personnel, and bookkeeping are likely targets for professional snoops.

FIGURE 8.4

Security Options for Different Types of Personal Data

Type of Personal Data	Security Option

Routine
  meeting announcements
  course catalogs
  organizational charts
  simple correspondence
	■ No Security Needed

Sensitive
  salary ranges
  seniority
  EEO summary reports
  changes to policies
	■ Password protection
	■ List of data files
	■ Disks locked up
	■ Keyboard or system
	  lock up

Proprietary Personal
  union negotiating strategy
  layoff plans
  individual salaries
  individual performance
  ratings
	■ Password protection
	  with encryption of
	  data files
	■ Disks locked up
	■ Keyboard or system lock up

- **The Hacker.** It's the hackers who have the stick-to-itiveness to keep on trying until a system is penetrated. Some relish decoding the ultimate encryption scheme; others plant undetected Trojan Horses in working programs; others decide to borrow the company Enhanced Graphics Adapter (EGA) because it'll do wonders for their PC clone at home.

- **The Innocent Incompetent.** Hard disks and a mysterious DOS environment make navigating the PC waters tough going for the uninitiated. An accidental disk format, a copy of a file overwriting an existing but different file, and a file wipeout are all potential dangers.

- **The Unhappy Employee.** A recent survey indicated that more computer crime is committed by insiders than by outside hackers. Ignorance, listed as the number-one motivation, was only slightly ahead of maliciousness and revenge.

- **The Thief.** Office crime was, is, and will always be a problem. Unprotected, out-in-the-open PCs are a thief's wish come true.[1]

These same threats are present in mainframe computing, but the sheer multitude of machines and users make the vulnerability of the PC environment much more significant.

DEVELOPING A MICRO SECURITY PLAN

The degree of security to be applied to a microcomputer system will depend on the importance and sensitivity of the applications involved. The security aspects of a microcomputer system operation should be reviewed before the system is developed and installed. The primary objective of a security review is to assure that adequate protection is provided and to make security recommendations. Any individual charged with the responsibility of reviewing the adequacy of data processing security must consider microcomputers.

Developing a good security review plan is essential to conducting an effective, efficient, and economical security review. The identification of security risks becomes the basis of determining the adequacy of security. Competent data processing systems personnel and the internal audit staff should be asked for their opinions as to what security measures to install and what policies and procedures should be followed. If the microcomputer system has been operating and little thought has been given to security considerations, then a review of the operation should be conducted to assess the adequacy of the security methods.

Two primary objectives of a security review of a microcomputer system are:

- **To assure that adequate protection is provided.** This process consists of identifying resources requiring protection, determining the types of potential hazards, assessing the adequacy of protective controls and devices, and continually testing the security measures. The computer system may supply and control operating data critical to the success of the rest of the organization. A review of the protection provided assures that the security measures are adequate.

- **To make security recommendations.** It is impossible to foresee and prevent all potential dangers; therefore, it is necessary to periodically review the security of an operation and to recommend control improvements. These recommendations may include methods and procedures for the microcomputer operation, creation of a disaster recovery and backup plan, establishment of various physical measures, and so on.

Checklists can be used to probe the critical areas associated with microcomputer security planning, management, and control. They will help determine the actual security condition rather than the perceived circumstances. These checklists can be used either in the planning stage or after the installed equipment has been operating. They cover the principal points of management concern relative to security.

A sample microcomputer security checklist that could be expanded as needs and applications increase is offered in the *Computer Security Handbook, Third Edition.*[29]

1. Are data copied to new diskettes and tapes before age and wear cause loss of or damage to the files they contain?

2. Are backup copies of each system's disk contents made promptly and regularly?

3. Are diskettes kept away from magnetic fields?

4. Are external storage labels written on with a soft tip pen?

5. Are food, beverages, and smoking materials kept out of microcomputer areas?

6. Are contaminants avoided, such as dust, hair, fingerprints, and chemicals?

7. Are all devices covered with static-resistant covers when not in use?

8. Is contact with exposed portions of tape or disk media avoided?

9. Do maintenance contracts guarantee timely repairs?

10. Is a carbon dioxide fire extinguisher kept near each microcomputer?

11. Does insurance cover damage from users' negligence; losses on or off the premises; damage caused by power surges, fire, or water; cost of replacing data and programs; business interruption expense; and loss of or damage to equipment rented, borrowed, or leased?

12. Is owner's identification etched on equipment covers?

13. Is the equipment housed in a locked cabinet, or locked or firmly attached to fixed adjacent furniture?

14. Is the microcomputer's incoming telephone number kept confidential?

15. Is a password required for initiating use of the microcomputer if it contains sensitive data?

16. Are passwords never displayed or printed as they are used?

17. Are logs maintained that record each attempt to use the microcomputer?

18. Must a user be called back via telephone at stored telephone numbers before access is granted?

19. Are passwords changed frequently?

20. Is encryption used to protect highly sensitive data?

21. Are menus designed to prevent use by someone unfamiliar with the system?

22. Are sensitive files kept from appearing in on-screen listings?

23. Are users trained to be security conscious?

24. Is a written security policy circulated and enforced?

25. Has a security coordinator been appointed with responsibility for data resources?

26. Are penalties for security infractions clearly stated and imposed when necessary?

27. Are proper procedures and security rules audited regularly?

Source: *Computer Security Handbook, Third Edition,* by Arthur E. Hutt, Seymour Bosworth, and Douglas B. Hoyt, 1994, John Wiley & Sons, Inc.

ESTABLISHING A MICRO-TO-MAINFRAME LINK

Microcomputers have quickly become an integral artery of the corporate information processing resource. It's now time for the next logical step—linking microcomputers to mainframes. Before rushing to implement these essential corporate links, DP managers must consider user needs and security issues and develop a comprehensive "action plan."[58]

User Needs

Micros are used to:

- Emulate a terminal
- Download major portions of a mainframe-resident database
- Execute microcomputer-resident programs with data downloaded from mainframes
- Execute a microcomputer-resident program using data obtained at the microcomputer level
- Upload data obtained by processing at the microcomputer level to the mainframe database
- Upload and download data to/from other microcomputers
- Execute programs maintained at the mainframe level
- Upload programs from the microcomputer to the mainframe for processing

The proliferation of microcomputers has transformed the person on the other end of a mainframe link from a DP professional or a data input/retrieval clerk to almost any employee. What was once a highly technical and rather recondite area now requires only minimal computer sophistication.[58]

Security and Control Issues

The new microcomputer environment requires DP managers to address:

- Risk evaluation
- Ease-of-use trade-off
- Responsibility
- Computer literacy
- Implementation

Action Plan

An action plan must be designed to address the following issues:

- Obtaining full support and involvement from top management
- Identifying exposures connected with applications subject to the micro-mainframe link
- Selecting controls appropriate to the identified exposure[58]

Linking microcomputers to mainframes has some advantages, such as:

- The extraction of data by the microcomputer directly from the mainframe will eliminate the time and effort users take to reenter this information on the microcomputer.
- Some programming support work can be shifted from the SP department to the users.

- Microcomputer processing often provides faster response time, shorter training time, and lower cost than mainframe processing.
- Microcomputers can be used as front-end and back-end processors.
- Micros can replace data entry terminals.
- Micros have extensive editing, formatting, and computing capabilities.
- Mainframe programs can be downloaded to the micros, thus reducing the load placed on the mainframe.
- Peripherals as well as data files can be shared.[7] (Figure 8.5 shows the microcomputer-to-mainframe connection.)

FIGURE 8.5
Microcomputer-Mainframe Connection

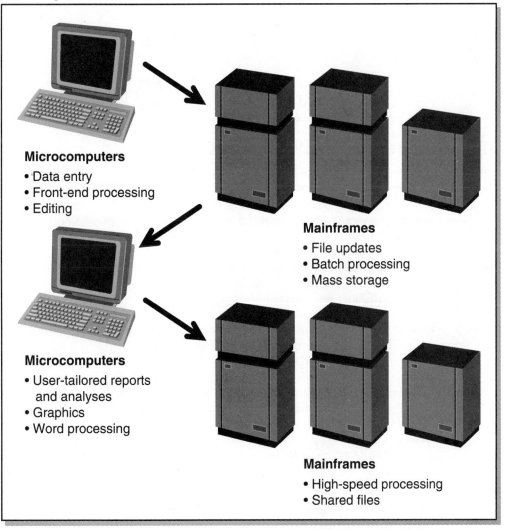

Microcomputers
- Data entry
- Front-end processing
- Editing

Mainframes
- File updates
- Batch processing
- Mass storage

Microcomputers
- User-tailored reports and analyses
- Graphics
- Word processing

Mainframes
- High-speed processing
- Shared files

Along with these advantages of linkage should go a few cautions in order to ensure the security of the entire system. Some of the security considerations should include controls over communication lines, such as:

- Frequent telephone number changes (with unlisted numbers)
- Mainframe operator intervention for making the connections
- Validation of a terminal identification code
- A **dial-back system** in which the mainframe contains a table of authorized telephone numbers and will call back to the microcomputer when queried
- A limited time frame during which transmissions from remote locations are accepted[8]

File transfer controls should include methods that ensure the security of files being uploaded or downloaded.

> One of the best methods to control a microcomputer's effect on data stored in the mainframe is to allow only downloading of data files needed for manipulation and storage on the microcomputer. This requires the user to have read-only access.[8]

> In many cases, the microcomputer connection may necessitate granting write access authorization to mainframe resources . . . The main concern here is the integrity of files resident on the mainframe. They must be secured against alteration or overwriting when files are uploaded from the microcomputer.[8]

PORTABLE MICROCOMPUTER SECURITY

By its very nature, a portable computer (laptop or notebook) has little or no physical security. It is designed to be small, lightweight, and portable, all of which make it an easy target for theft. Also, very few laptops have any type of keyboard lock or any other device to prevent unauthorized use. It is generally in the possession of one person, who is traveling to customer sites, airports, and hotels, so the security controls in place are few to nonexistent.

PC manufacturers are beginning to address the unique security needs of portable PCs using a variety of hardware and software techniques. "For the most part, these features are designed to keep precious corporate data away from prying eyes in the event the machine is lost or stolen."[19] These features include a power-on password, a keyboard password, a disk lock/screen blank function, diskette boot control, and a cable lock to secure the portable to a disk.

Portables are now available that use a Data Encryption Standard (DES) coprocessor that automatically encrypts all hard disk files. Data going to and from the floppy drives, serial ports, fax/modem, and printer port can also be encrypted.

Portable PCs that connect to networks pose a more difficult security challenge. Many IS managers rely on the same security packages that are used to secure desktops rather than one of the many utility packages designed specifically for stand-alone PCs.

Figure 8.6 offers some suggestions for security of portable computers.

PASSWORD PROTECTION

The only way to guarantee that a system has not been tampered with is to make absolutely certain that no one with motive to do so is given access into the system.

FIGURE 8.6
Portable Computer
Security Guidelines

☑ Carefully observe and access surroundings whenever using or transporting devices.

☑ Make devices inconspicuous in public places (including physical security tokens or locks).

☑ Always lock devices when unattended.

☑ Use reliable password schemes.

☑ *Always* treat information as highly confidential!

The Lap-Gard

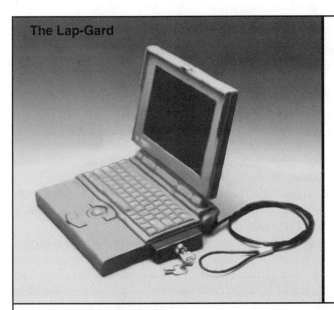

Disk Drive Security Lock

Portable computers, such as laptops and notebooks, are particularly vulnerable to theft. The Lap-Gard Security Device incorporates a Disk Drive Security Lock to prevent unauthorized users from taking electronic files from the secured laptop. The lock also has a vinyl-coated, flexible steel cable which can be looped around a desk leg or other fixed structure. This offers double protection for the security of the electronic files stored on the laptop as well as for the physical protection of the unit itself.

Courtesy of: COMPU-GARD, INC., Seekonk, Massachusetts

The hacker or observer must be cut off before he or she gets into a system. The headlines may read "mastermind of a rogue computer program has brought the country's foremost scientific information network and all connected PCs to their knees." How did this person do such damage?

The most efficient and widely used method to prevent access to a computer system, whether personal computer or mainframe, is the password. A **password** is a secret sequence of characters used to authenticate a user's identity, usually during a log-in process.

What Is Password Protection?

Whenever a would-be user wants to use the services of a computer system of any type, the computer system must be sure that this would-be user is not using a false identity. This job is initially given solely to a password system, which has two enormous responsibilities:

- *The password must identify who is currently authorized to use the system.* This is accomplished by the use of a file or list of valid individuals and, in some cases, the status of each person or file (to be used in limiting access to some programs or files).

- *The computer system must adequately use a system of controls to provide authentication of the would-be user.* If properly protected, this method of security can make a system invulnerable.

Who Needs Password Protection?

Password protection is the most commonly used form of security in computer systems. However, if this protection is not utilized correctly, it causes the systems to be vulnerable to attack.

Any computer system that involves interaction with a user is open to a breakin which could cause much damage and require costly restoration. Whether the system is a personal computer, a multiserver computer, or even an automatic teller machine, passwords are a vital part of providing identification and authentication before allowing a would-be user into the system.

Passwords play a big part in safeguarding the files kept on the hard disk drive inside the personal computer. It is microcomputers that will impact the security of office operations. It is this security that ensures the owner that files have been protected. The micro equivalent of a large computer terminal session is initiated by turning on the CPU. Once inside, the micro provides the user with menus and help text to access files. The personal computer literally is at the user's mercy.

Problems in Password Application

Basing the security of a computer system on a password protection system puts a large responsibility on the user. It is totally up to the user to keep the password process secret. With some systems, users are given the opportunity to use a password of their choice so that they may remember it easily. With this freedom comes several problems; one problem is the simple fact that passwords can be figured out. Surveys are done on outdated passwords, and the results prove to be ironic. Most passwords that are used can be figured out by knowing a characteristic related to the possessor, the possessor's nickname, a spouse's name or girlfriend/boyfriend's name, or the address of the possessor. All of these facts are not difficult to figure out or obtain by asking an unsuspecting coworker or friend of the PC user.

"Password guessing is the Achilles' heel of modern computer access-control systems . . . password-guessing attacks pose an increasingly serious threat, because growing numbers of relatively computer-illiterate people are added to the rolls of users each day, and these novice users often receive inadequate guidance in the choice of difficult-to-guess passwords."[61] The basic theory with passwords is that "if you know the secret password for an account, you must be the owner of that account."[46] Passwords are still, due to their simplicity, the authentication tool of choice.

In addition to the relatively easy "guessability" of passwords, users are often careless when dealing with their passwords. Some of the most obvious places where passwords are found are on the inside of the cover of the computer system instruction manual, under the keyboard, under the mousepad, on the screen, in the pen drawer, or taped to a favorite coffee mug.

The static password is a further problem that plagues computer systems. Eventually, computer systems do become obsolete, and often passwords that have never been changed are found by searching through used hard drives. Or, a password may have been shared and an employee leaves the company to work for a competitor. Passwords that do not change are eventually figured out, and soon the system is helpless.

Creating Good Passwords

A good password has these six characteristics:

1. **Composed of letters, digits, and other characters**, so that the base alphabet for an exhaustive attack is large.
2. **Long**, so that there are many possibilities for an exhaustive attack.
3. **Not a common word or name**, so that a dictionary attack will fail.
4. **Unlikely**, not a characteristic related to the possessor, such as a spouse's name or a street address.
5. **Frequently changed**, so that even in the event of someone's guessing it, the period of vulnerability is short.
6. **Not written down**, so that it will not be found by outsiders.[41]

Figure 8.7 indicates some considerations on types of weak passwords. Figure 8.8 gives some hints on protecting passwords.

Establishing Policies

An effective password policy normally has the following characteristics:

- Passwords are sufficiently long to reduce the possibility of guessing.
- The user selects passwords after initial access to the system; thereafter, the system forces changes in passwords on a periodic basis, typically 60 days.
- A password entered at a terminal is not visible on the screen or to other people in the area, nor is it visible on any printed paper coming from the terminal.
- Password holders are periodically indoctrinated about the secrecy of their passwords and their responsibility for safeguarding passwords.
- Safe password administration is required, including elimination of or imposition of strict security on password lists, frequent password changes, separation of duties in the administration of passwords, accountability for the safety of passwords, and background investigation of those people in high positions of trust who administer passwords.

FIGURE 8.7
Types of Weak
Passwords

☑ Regular words (house)
☑ Reversed words (esuoh)
☑ Regular capitalization (Program)
☑ Reversed capitalization (margorP)
☑ Unconventional capitalization (maRGoRP)
☑ Acronyms (NASA)
☑ Acronyms with periods (U.S.A.)
☑ Abbreviations (etc.)
☑ Words followed by a period (USA.)
☑ Special abbreviations with interspersed periods (Ph.D.)
☑ Special abbreviations with unusual capitalization (MHz)
☑ Common first names (John)
☑ Common last names (Smith)
☑ Possessives (John's)
☑ Verb conjugations (see, sees, saw, seen)
☑ Plurals (computers)
☑ Legal words (litigation)
☑ Geographical words (New York)
☑ Biographical words (Washington)
☑ Medical words (lipoprotein)
☑ Technical words (modem)
☑ Scientific terms (clone)
☑ User-IDs (Brownly)
☑ Reversed user-IDs (Ylnworb)
☑ Words constructed with only one character (aaaaa)
☑ Keyboard scales (asdfghjkl)
☑ Cultural icons (Superman)

- Password lists stored in the computer and used for authorization purposes are encrypted. As soon as a password enters the computer from a terminal, it is immediately encrypted and compared against the master password in encrypted form only. This feature reduces the exposure of actual passwords in the computer.

- Time delays are imposed on terminal users so that repeatedly attempting to use unauthorized passwords requires discouragingly large amounts of time. Also, an individual is not allowed to input an incorrect password more than three times before being disconnected.

- The system displays a banner message during the logon process warning users and potential intruders about trespassing.

FIGURE 8.8
Password Schemes

☑ Passwords should be six to eight characters in length. The longer the password, the harder it is to guess.

☑ Don't use obvious characters (e.g., your name, college, city, state, name of program).

☑ If allowable, use a space, nonalpha character, or capital letter somewhere in the password.

☑ Keep your password secret—do not write it down and post in an obvious place.

☑ Change your password often.

☑ Don't embed your password in a macro or template routine.

- All system logon attempts, successful or not, are recorded, or logged, by the computer system. The computer then analyzes the data files produced and produces exception reports indicating deviations from normal use that might indicate attacks on the system.

- Procedures are established for imposing alternative methods of security when the password system and the computer equipment supporting it fail to function properly.

- Sanctions are clearly known by password holders, and violators are punished.

System-Generated Passwords

System-generated passwords are generated randomly by the system, rather than relying on the user's selection. Some systems allow the users to view several random choices. The inherent danger of system-generated passwords is that they are often too hard to remember so users write them down and then the algorithms used to generate the passwords become known.

One method of computer-generated passwords uses a random word-generator program to form pronounceable syllables and concentrate them to create words. Examples of this method include: qua-va, ri-ja-cas, te-nort, oi-boay. The advantage of using a password like this is that people will be less inclined to write down the password if they can pronounce it.

Challenge Response

Aside from a password that must be memorized or written down, another form of password protection exists. The computer can be given an algorithm that can be used to authenticate users after identification has been made. In this type of situation, called a **challenge response**, the host sends a message m and the user replies with $E(m)$. Both the message m and its encryption $E(m)$ may be obtained (by observation or by line tapping), but this loss will not reveal the encryption algorithm. This type of system is particularly beneficial to users who work in a high security environment in which the password must be changed frequently. Memorizing a new password daily is much more cumbersome than remembering how to respond to one challenge-response system periodically.

Cost Considerations

When deciding how to implement the optimal password scheme, one must consider the costs involved in design and adaptation. The costs of a given password scheme are those incurred by the intruder as well as by the protector. The protector must consider the costs in conjunction with the value of the information that must be protected. Every byte of data is put together to construct the information that resides in a computer system. That information has a value to it, and protection should be implemented proportionally with that value. For example, an elaborate password system on a stand-alone personal computer that contains only the names of employees would *not* be a system requiring maximum security. However, a similar system that stored production specifics or employee wages may require heavy security. Some general rules relating to cost versus benefit of passwords are as follows:

1. Simple password for identification: cost of software, systems performance, storage.
2. Changeable passwords: cost of software, updating lists and storage, systems performance.
3. Password transformations: cost of software, cost of random lists and storage, systems performance, computational cost.
4. Magnetically encoded cards with constant or changeable passwords: cost of terminal to read/write, cost of software, systems performance.

A Simple Rule to Follow

Any password protection is not enough to ensure security. When L equals password lifetime, R equals guess rates, and G equals the total number of guesses that can be made during a password's lifetime where

$$G = R \times L,$$

the solution is simple. Make the rate small and the lifetime short! Educate password possessors of their responsibility and use a protection scheme that will adequately protect your system. A password protection system may be the system's best defense against intruders. What they cannot see will not be accessed!

SECURITY OF SPECIAL MICRO APPLICATIONS

As micros are applied more and more to an organization's overall computing system, some unique applications are introduced that challenge the mainframe or stand-alone micro security plans.

Downloaded Data

When data is downloaded from the mainframe to the micro, both avoidance and deterrence security options should be implemented and strictly followed. They include the following:

- Policies that require staff to sign agreements that they will respect the confidentiality of the information they handle
- Deterrence policies for disallowing staff to use floppy disks not marked with the company logo
- Detection policies, such as records of what information exists on storage media and who is responsible for its safety
- Detection controls in extremely sensitive areas requiring staff to submit to a personal search on entering and/or leaving the building or floor/office
- Avoidance policies of diskless workstations that will reduce the exposure to the file server disks and tapes.[16]

Figure 8.9 shows some of the ways preventive controls are applied to PC-based systems. Effective security that will control unauthorized access by a microcomputer user should require:

- Personal identification
- Identification authentication
- Access-protected audit trails
- Message authentication codes where communication is concerned
- Encryption of messages[16]

Multiuser Micros

An increasing number of organizations are using distributed microcomputers to accommodate multiple end users. The primary issue to be addressed in this ever-growing end-user environment is "Do not spend more money protecting your assets than you would lose if the assets were lost or destroyed." The growing use of distributed, micro-based systems means that data security procedures should be simple, yet effective and cost-efficient. Questions to be addressed in multiuser micro environments are:

- Who is permitted to log on to the system?
- After logon, which files and programs should be available to each user?
- Which system functions should be at the disposal of each user?
- Should data be exchanged between systems with different security files?

Shared-use and multiuser systems are often thought to be the same, but there is, in fact, a distinct difference between the two configurations.

A **shared-use system** is a system that is used by more than one person, but *not by more than one person at a time*. A **multiuser system** is a system that *can be used by more than one person at a time*.

If the system is to be shared by several users, and not all users will have the necessary clearances and need-to-know for all information that will ever be processed or controlled by the system, the possibility of using a removable-media-only system should be investigated. With this type of system, information can be removed and locked away to prevent compromise.

FIGURE 8.9
Preventive Controls

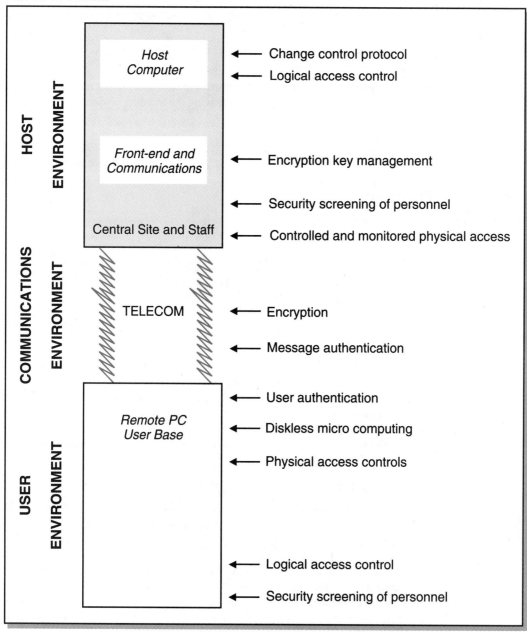

If a system with fixed media is used, any information that is stored may be accessible to all users of the system and should, therefore, be restricted in some appropriate manner. Users with different clearances or need-to-know, or both, should be able to access the system appropriately.

Removable media is any magnetic media used for the storage of information that is designed to be frequently and easily removed by the users. Examples of removable magnetic media include floppy disks, removable hard disks, and magnetic tapes.

Fixed media (non-removable media) is any magnetic media that is used for the storage of information that is not designed to be regularly removed from the system. Examples include fixed or Winchester disks (or nonremovable hard disks).

On-Line Terminal Security

On-line terminal systems are used primarily for two functions:

- To update data files through a batch or real-time system
- To provide on-line programming capabilities to the installation

An **on-line system** consists of the specific programs that communicate with the terminals and pass the information to on-line applications programs.

Some of the specific security risks that should be reviewed for on-line systems are discussed below.

Physical Access. The user area in which the terminals are located should be inspected to determine whether access is limited. The following controls should be checked:

- Terminals should be within management view and under its control.
- Keyboards should be able to be deactivated by keylocks or controller sign-off.
- Terminals should be locked at the end of each workday.
- A log of employees who are assigned terminal keys should be maintained.
- The card access system should have a logging device to track entry and exit.
- A log of employees who are authorized to use the terminals should be maintained.
- An individual's terminal key should be collected on termination of employment.
- If possible, the locks should be changed periodically and new keys issued.
- Operator IDs and passwords should be rotated or changed periodically.
- The responsibilities for maintenance and security of IDs, passwords, and keys should be assigned to a senior employee or department manager.
- Management should review terminal activity reports daily.
- Employees should be alert for unauthorized personnel in the area and report questionable terminal access to their supervisor.
- Prospective employees should be screened to a relevant work performance level.

Access Control. The on-line system should include an access control scheme to control operator access. The sign-on scheme should:

- Identify the user with a unique identifier (e.g., an employee number and password, which should be used to track operator activity)

- Verify operator identity by matching the sign-on key and password with internal files or tables

- Authenticate the user's access authority by matching each transaction to internal authorization tables, which should define the transactions, files, and programs that the operator is authorized to use

Authorization Controls. The on-line security system should correlate users to resources in an authorization table to control user functions. The following procedures are recommended:

- Transactions should be assigned to users.

- Transactions can be assigned to terminals if additional security is needed.

- The system should check each transaction entry.

- The system should abort attempts to perform unauthorized transactions.

- Repeated invalid attempts should result in the disability of the terminal and user sign-on key.

Message Information Requirements. The message control system should require and validate the following information in each message:

- **Terminal address.** This should be compared with the one in the terminal table. If an address is not in the table, the message control system should ignore the terminal, deactivate the line, or deactivate the network. In addition, it should sound an alarm or print a diagnostic message at the master terminal.

- **Operator ID, password, and transaction code.** These should be verified against the appropriate authorization files.

- **Sequence number.** Each message should contain a sequence number to account for all messages. The sequence number is used in recovery to determine the last message processed.

- **Supervisor's approval code.** The use of approval codes is based on the sensitivity of the transaction. If a required approval code is absent, the transaction should be rejected.

Although the growing use of microcomputers in on-line systems has increased the risks and exposures to loss for all companies, the specifics of security vary from organization to organization.

Workstation

The total collection of microcomputer equipment, physically located in one place, that makes up the resources meant to be used by one person at a time are known as **workstations**.

Developing security for workstations has been a rapidly increasing concern for companies that employ micro networks across the country. While the industry is producing the most secure networks ever seen, they are still hard pressed to keep up with the growing knowledge of would-be computer criminals and saboteurs. Today's companies need not only be concerned with theft; they must also guard against the increasing threat of virus infection. When installing secure micro work-stations on the network, companies have two options:

- Add new, secure workstations
- Add security features to existing workstations

The current spread of computer virus infection has led many companies to upgrade their LAN (local area network) security, particularly for micro-based systems. The damage that can be caused by a virus is many times greater in a network than it is for a single PC. The virus can quickly spread to every computer that logs onto the network file server, making the debugging process very difficult. Not only must the network server be debugged, but each individual workstation must also be checked for infection.

One of the major methods used to create a secure workstation is to install one without a floppy disk drive. This type of workstation is extremely effective in avoiding the insertion of viruses into the system and makes it difficult for individuals to copy and remove data. The major question with these types of workstations is how the computers boot up—a network boot server is the most logical solution. The boot file that the workstations use can be configured to allow specific use by different users. Configuring a customized boot file makes it easy to set up PCs as special function devices, such as remote access servers, electronic mail gateway servers, fax servers, and print servers. All of these available options make the use of workstations without floppy disk drives increasingly popular.

Several options are available to consumers who wish to upgrade the security of their network workstations without buying all new PCs, and at a relatively low cost. One solution would be to acquire certain specified software to increase security, but the focus is generally on hardware solutions. One option is to equip the computer with a remote-boot network interface card and no floppy disk drive. To accompany this, however, the computer casing must be locked to avoid removal. It is also advised that the consumer purchase high-quality locks, because many of the lower-priced models can be opened with the same key. A simple solution for a system that retains the disk drive would be to purchase a floppy disk drive lock. These locks look and work much like the locks that are used on phone dials. Other locks are designed so that the entire computer cannot be removed from the station.

There are many effective ways to secure a workstation that retains the floppy disk drive, as well as the hard drive. The driving force behind this concept is the "something known and something owned" philosophy. This type of configuration will require the user to provide a password as well as some type of security clearance card to access the workstation. The advantages of this type of system are that the user can gain access through several different workstations while still allowing the manager to control the information and data to which users have access. This may seem like an ideal solution to the security problem, but the cards are often lost and must be reissued, thus leaving active cards in circulation. One solution to this problem is the use of personal credit cards instead of standard issue security cards.

SUMMARY

The **microcomputer** "explosion" in our offices has refocused information security on the business environment.

Corporations must ensure that business users and management are security aware, and that management is prepared to develop and enforce policies and procedures that support this awareness. It has also called for development of new products and new combinations of products. Most importantly, it has called for the development of products that have a long-term view of security including the integration of **mainframe** security and field support considerations.

Today the downloading of information is broadly uncontrolled. With a better understanding of our information and its uses, management will be able to take advantage of the products that are currently available to provide security to the intelligent **workstation** environment.

Some of the policies that should be introduced into the microcomputer environment are **backup** procedures, disaster recovery plans, deterrence concerning software theft and misuse, threats from within, physical security, need for encryption, and **passwords**.

Internal data security of micros should include techniques to protect data and equipment, accident security, commands to protect programs, naming conventions, **hidden files**, and categorizing data.

The two main objectives of a security review of a microcomputer system are to assure that adequate protection is provided and to make security recommendations.

In establishing a micro-to-mainframe link, some issues that should be considered are the needs of the user, security and control issues, action plans, and the advantages and disadvantages of linkage.

Passwords are often used to protect microcomputers. Some of the issues to be addressed are who needs password protection, problems in password application, creating good passwords, protection of passwords, establishing password policies, **system-generated passwords**, **challenge-response** systems, and the cost of passwords.

Some of the special applications that are often associated with micros include downloaded data from mainframes, **multiuser micros**, **on-line terminals**, and workstations.

The issues of data security that have plagued the mainframe computer environment for a long time must now be considered in the microcomputer environment as well. Many of the issues, such as personnel selection, disaster and fire protection, and user authorization definitely have a role to play in the micro environment, just as they did with mainframes. The micro environment, at the same time, introduces a new set of problems and considerations that are unique to this rapidly advancing technology. We must readjust our "security thinking" to keep pace with all the changes that technological advances introduce and be prepared to apply this new knowledge concerning security and control in our operation so that users are aware of the breaches and accidents that they will face. With knowledge and awareness, we will be ready to face the challenge.

REFERENCES

1. Aarons, Richard, and Robin Raskin, "Security Strategies . . . Hardware Protection for PCs," *PC Magazine*, April 28, 1987, pp. 105-120.
2. *Advisory Memorandum on Office Automation Security Guidelines*, NITSSAM Compusec, pp. I-87, National Telecommunications and Information Systems Security, January 16, 1987.

3. Aiken, Dick, "Common-Sense Security for Laptops," *ISP News*, July/August 1992, p. 45.

4. Baker, Richard H., *The Computer Security Handbook* (Blue Ridge Summit, PA: Tab Books, 1985), pp. 1-281.

5. Bosen, Bob, "When Passwords Are Not Enough," *ISSA Access*, 4th Quarter, 1989, pp. 12-19.

6. Buttross, Thomas E., and John C. Walley, "What You Need to Know About Microcomputer Security," *The Practical Accountant*, Vol. 23, June 1990, pp. 94-95.

7. Callis, Melinda, and Sheryl Skolnik, "Establishing the Microcomputer-Mainframe Link," *Data Security Management*, Auerbach Publishers, Inc., 1985, No. 84-04-19, pp. 1-8.

8. Callis, Melinda, and Sheryl Skolnik, "Securing and Controlling the Microcomputer-Mainframe Link," *Data Security Management*, Auerbach Publishing, Inc., 1985, No. 84-04-20, pp. 1-8.

9. Christie, Linda Gail, "Putting a Lock on Microcomputer Security," *Interface Age*, November 1983, pp. 92-93.

10. "Computer Security Guidelines for Microcomputer Users," Lawrence Livermore National Laboratory, January 1985.

11. *Computers at Risk . . . Safe Computing in the Information Age*, National Research Council, National Academy Press, National Academy of Sciences, Washington, D.C., 1991, pp. 1-302.

12. "Defense of the Data," *New Scientist*, January 19, 1991, p. 17.

13. DerFler, Frank J., "Driving Without a Disk: Booting Up Diskless PCs," *PC Magazine*, Vol. 10, No. 17, October 15, 1991, p. 218.

14. Dickson, Gary W., and James C. Wetherbe, *The Management of Information Systems* (New York: McGraw-Hill Book Company, 1985).

15. Elmer-DeWitt, Phillip, "The 414 Gang Strikes Again," *Time*, August 29, 1983, p. 34.

16. Fitzgerald, Kevin, "The Management and Control of Downloaded Data," *Computer Control Quarterly*, Vol. 9, No. 3, 1991, pp. 22-29.

17. Forgione, Dana, and Alan Blankley, "Microcomputer Security and Control," *Journal of Accountancy*, Vol. 169, June 1990, pp. 83-84.

18. Foster, Susan, "On Maintaining Security and Sanity," *Computer Decisions*, July 1983, p. 70.

19. Francis, Bob, "How to Secure a Portable PC," *Datamation*, March 15, 1992, pp. 43-44.

20. Frenzel, Carroll, *Management of Information Technology* (Boston, MA: boyd & fraser publishing company, 1992), pp. 1-565.

21. Gallegos, Frederick, and Thomas R. Halsell, *Data Communications Management*, Warren, Gorham, Lamont, Inc., New York, NY, 1991, pp. 1-14.

22. Georgas, Nora, "12 Portables for the Road," *PC Magazine*, March 29, 1988, pp. 93-95.

23. Graham, Carol, and Melanie Freely, "Keep Your Data Secure," *Datamation*, October 1, 1990, pp. 97-99.

24. "Guidelines for Developing Personal-Computing (PC) Policies," Commonwealth of Virginia, Department of Information Technology, COV Guideline 86-1, 1986.

25. Hansen, James V., and Marshall B. Romney, *Internal Auditor*, December 1987, pp. 44-47.

26. "Hardware Maintenance," *Information Technology*, Bulletin, Winter 1991.

27. Highland, H. J., "Access Control for Micros," *Computers and Security*, Vol. 9, No. 8, pp. 671-672.

28. "How to Protect Your Hardware," *Computing and Communications Services News*, August/September 1990.

29. Hutt, Arthur, Seymour Bosworth, and Douglas B. Hoyt, *Computer Security Handbook*, 3rd ed. (John Wiley & Sons, Inc.), pp. 1-399.

30. Johnston, R.E. "The Best Answer Yet," *Infosystems*, September 1986, pp. 44-47.

31. Kingston, Peter, and Martin Goulbourn, "Security for Microcomputers Used in Small Business," *Computers and Security*, 1987, pp. 68-72.

32. Kochanski, Martin, "How Safe Is It?" *Byte*, June 1989, pp. 257-264.

33. LaPlante, Alice, "Guarding Their Turf," *Infoworld*, September 9, 1991, pp. 560-564.

34. "Laptop Security: A High-Stakes Game," *ISP News*, July/August 1992, p. 46.

35. "Managing Microcomputer Security," *FTR Technical Library,* Port Jefferson Station, New York, pp. 1-1, 1-15.

36. Moulton, Rolf T., *Strategies and Techniques for Preventing Data Loss or Theft* (Englewood Cliffs, NJ: Prentice-Hall, 1986).

37. Mullen, Jack B., "A Checklist for On-line Terminal Security," *Data Security Management*, Auerbach Publishers, Inc., New York, NY, 1986, pp. 1-16.

38. "Multi-user Micros Pose Greater Security Threat," *Data Management*, April 1987, pp. 42-43.

39. Mutter, John, "On-line on the Road," *Publishers Weekly*, January 13, 1989, pp. 50-51.

40. Norton, John C., "A Systematic Approach to Microcomputer Security," *Personnel*, Vol. 63, June 1986, pp. 8-12.

41. Pfleeger, Charles P., *Security in Computing* (Englewood Cliffs, NJ: Prentice-Hall, 1989), pp. 1-538.

42. Poor, Alfred, "Playing It Safe: Securing Options for Existing Workstations," *PC Magazine*, Vol. 10, No. 17, October 15, 1991, p. 227.

43. "Products for Laptop Protection," *ISP News*, July/August 1992, p. 46.

44. Przybylowicz, Edwin P., Excerpt from *Business Week*, 1989 Special Issue, p. 67.

45. Reithner, Robert, Excerpt from *Business Week*, 1989 Special Issue, p. 179.

46. Russell, Deborah, and G.T. Gangemi, Sr., *Computer Security Basics* (Sebastopol, CA: O'Reilly & Associates, Inc., 1991), pp. 1-441.

47. Scannell, Ed, "Apprentice 2.0 Adds Security, Archiving, Compression Features," *Infoworld*, October 21, 1991, p. 19.

48. Schweitzer, James A., *Computer Crime and Business Information* (New York: Elsevier Science Publishing Company, 1986), pp. 1-195.

49. Seymour, James, "Special Report: The Software Outlaws," *Today's Office,* August 1984, pp. 21-34.

50. Sparks, James G., "Minicomputer Operating System Security—RSTS/E," *Data Security Management,* Auerbach Publishers, Inc., New York, NY, 1984, pp. 1-16.

51. Spencer, Cheryl England, "Data Safety," *Macworld*, January 1990, pp. 142-149.

52. Steinauer, Dennis, "Securing Your PC—Which Security Devices You Should Select," *Government Data Systems*, September/October 1986, p. 41-52.

53. Steinauer, Dennis D., *Security of Personal Computer Systems: A Management Guide*, NBS Special Publication #500-120, January 1985.

54. Stephenson, Peter, "Personal and Private," *Byte*, June 1989, pp. 285-288.

55. Taylor, Thayer, "How the Best Sales Forces Use PCs and Laptops," *Sales and Marketing Management*, April 1988, pp. 64-74.

56. "Technical Evaluation—Plug-In Security Cards," *Computer Fraud and Security Bulletin*, Elsevier Scientific Publishers, Inc., Vol. 10, No. 8, 1988, pp. 6-12.

57. Walsh, John, "Designs on a National Research Network," *Science*, February 19, 1988, p. 861.

58. Ward, Gerald M., "Securing a Micro-Mainframe Link Demands Detailed Action Plans," *Data Management*, December 1984, pp. 20-21.

59. "Weeding Out Risky Passwords," *Science News*, Vol. 140, November 16, 1991, p. 315.

60. Whitmyer, Claude F., "Computer Security: Some Problems, Some Solutions," *The Office*, December 1987, p. 27.

61. Wood, Charles Cresson, "To Guess or Not to Guess," *ISP News*, Vol. 1, No. 3, September/October 1990.

62. Wood, Patrick, "Safe and Secure," *Byte*, May 1989, pp. 253-258.

REVIEW QUESTIONS

1. Why has the use of microcomputers grown so rapidly in the last few years? What are the projections for future growth?

2. Briefly describe the classification of micros (PCs) laptops, notebooks, workstations, minicomputers, mainframes, and supercomputers.

3. What are some of the unique security problems of micros?

4. Why are backups necessary in the micro environment?

5. What are some of the methods used to protect hardware and other physical devices?

6. Briefly describe some of the internal controls used to protect data, information, and programs.

7. What are some of the unique security risks associated with a micro-to-mainframe link?

8. Describe some of the basic rules of password selection.

9. What are some of the advantages and disadvantages of system-generated passwords?

10. What are some of the inherent security problems associated with downloaded data, multiuser micros, on-line terminals, and workstations?

DISCUSSION QUESTIONS

1. Today's desktop microcomputers will be increasingly linked to mainframes, databases, and telecommunications networks through high-speed data links, giving users access to a wide spectrum of information. The challenge to the users, their managers, and organizations is to use this capability to full

advantage. Will we lose control of the invention that has fallen into the hands of so many?

2. The microcomputer explosion that has occurred in the last 10 years is unprecedented in the computing environment. This rapid growth has introduced many advantages and availability of computing to more users, but it has also introduced new-found security and control problems. Why was it not possible to easily transfer mainframe security principles to the micro-based environment?

3. One of the dangers found in microcomputer environments, due to the down-sizing of hardware, is the portability of the equipment—it is easy to remove quickly. What are some of the ways to physically secure equipment (micros, monitors, printers, and circuit boards) that would be unobtrusive, yet effective?

4. "True security may not exist on small computers as they are excellent copying devices and should not contain any confidential data." Comment on this statement.

5. "Passwords are merely a patchwork solution and cannot truly protect a micro-computer system." Comment on this statement.

EXERCISES

1. Select an organization/company that is using micros to process information. Interview a key employee to determine what types of security measures have been applied to their micros.

2. Locate an article in a current computer-based journal that deals with micro-computer security. Briefly summarize the article.

3. Develop a short statement for users on micro security policies. This statement would be distributed to all users and posted in computer areas.

4. Prepare a list of various passwords that you would use for files.

5. Describe how passwords are used on your campus for both mainframe and microcomputers. What are some of the strengths and weaknesses of those password schemes?

PROBLEM-SOLVING EXERCISES

1. If you had a notebook or laptop computer, what are some of the security and control measures you would apply to protect the information and the equipment?

2. Ask several of your friends and classmates to list passwords they have used. Are you able to easily guess the identity of the passwords' owners due to physical characteristics, known factors, etc.?

3. Involving more people in the application of micros requires more security as security is a people problem, not just a technical problem. How would you involve people in micro security to ensure that company-wide policies are effectively adopted?

4. The cost-effective benefit analysis of microcomputer security is a paramount concern as companies are very reluctant to spend more on protection than the equipment originally cost. What measures could be adopted to secure micros, yet be justified financially?

5. It is management's responsibility to provide direction in assuring security for PCs. Management should focus on protecting information, not just computers. How is the value of information determined?

CASES

Babcock Poultry Company

Whit Babcock, president of Babcock Poultry Company, located in Broadway, Virginia, is a regional supplier of poultry products covering Virginia, West Virginia, North Carolina, Tennessee, and Maryland.

Recently, Mr. Babcock converted his record-keeping operations from a manual process that was done by production workers in the processing plant to a computerized process completed by bar coding and scanners. Everything was working out very well, and Mr. Babcock was extremely pleased with the up-to-date, thorough reports made available to key managers.

Recently, Mr. Babcock has noticed, however, that one of the managers is making copies of disks and additional printouts on a regular basis. Mr. Babcock was able to determine, after watching this manager for several weeks, that one of the items he was leaving with was a copy of customers, sales volumes, suppliers, etc. This information could be extremely valuable to a competitor.

1. How could this problem have been avoided?

2. How could Mr. Babcock determine what the manager was doing with the information?

3. What course of action should be followed—fire the manager, confront the manager, tighten up the system, etc.?

Green Dynamics, Inc.

Mrs. Alicia Green, owner and founder of Green Dynamics, travels extensively to her clients' sites to conduct public relations reviews and communications strategies. Green Dynamics' primary function is to advise clients on how to project a positive image and foster clear communication regarding their products and services to their customers.

Mrs. Green recently purchased a portable notebook computer, weighing 6 1/2 pounds, to take to her on-site appointments to collect information and give quicker feedback to her clients in brainstorming sessions.

Mrs. Green has been hearing some horror stories, however, about stolen portables and would like to secure the computer and disks as she travels from location to location.

What advice would you give Mrs. Green on the proper use, protection, and control of her notebook computer?

VIRUSES

LEARNING OBJECTIVES

After studying this chapter, you will be able to:

1. Consider the historical role of viruses worldwide and their impact on today's computer systems.
2. Understand the properties of viruses and how they enter a computer system.
3. Identify the various strains of known viruses.
4. Explain how viruses are spread throughout the computer system.
5. Identify the various types of pseudo viruses and how they are used in conjunction with viruses.
6. Understand the motivation of virus program authors.
7. Compare and contrast some of the known viruses that are currently attacking systems.
8. Describe some of the software products that are used to prevent, detect, and eradicate viruses.
9. Relate how viruses are affecting computers on an international scale, both in industry and military applications.
10. Define the following terms.

TERMS

- Virus
- Virus Marker Byte
- Memory-Resident Virus
- Error Simulation Virus
- Data Targeted Virus
- Crasher Virus
- Computer Time Theft Virus
- Call-Me Virus
- Hardware Damaging Virus
- Pure Data

- Overwriting
- Non-Overwriting
- Transient Virus
- Resident Virus
- Signature
- Pseudo-Virus Program
- Trojan Horse
- Logic Bomb
- Time Bomb
- Trapdoor

TERMS (Continued)

- Salami Slice
- Phone Phreaking
- Worm
- Hamburg Virus
- Macintosh Virus
- Pakistani Virus (Brain)
- Hebrew University Virus
- Flu-Shot Virus
- IBM Christmas Virus
- Scores Virus
- Amiga Virus

- Israeli Virus
- Falling Tears Virus
- Michelangelo Virus
- Local Memory Infection
- Local Disk Storage Infection
- Shared File System Infection
- Infection of System-Wide Media
- Prevention Software
- Detection Software
- Identification Software

DEFENDING AGAINST COMPUTER VIRUSES

Even the world's largest computer maker isn't safe from them. On a Monday morning in December 1987, IBM was forced to shut down its worldwide network for two hours as a result of an attack by a virus, a computer program that infects other programs by modifying them to include a copy of itself. In this way it spreads from program to program and machine to machine, corrupting programs and data. In IBM's case, thousands of mainframes were shut down; at the peak of the attack, 500,000 instances of the virus appeared in two hours. Even weeks after the attack, "aftershocks" continued to plague the network.

In 1988 IBM rival Digital Equipment Corporation also suffered an attack, this time by a worm that roved from machine to machine throughout the company's network, again worldwide. A worm is a program that was originally developed to tap unused network resources in order to run another very large program. Worms can tie up the computing resources of a network (and thus an organization) and essentially shut it down. The Digital attack resulted in a day-long shutdown of the network, but no serious long-term damage.

Continued

DEFENDING AGAINST COMPUTER VIRUSES

From page 245

The lesson: if it can happen at the world's major computer manufacturers, it can happen to you.

Because all computers, networked or not, are vulnerable, all computer users are vulnerable as well. Most threatening is the fact that viruses have entered the commercial software field, where more than 10,000 legitimate packages from major software vendors were distributed containing viruses.

What can users do? The top management of organizations must be able to gauge the significance of an organization's risk. Users must be aware of the dangers of sharing software and data diskettes that have not been subjected to a virus detection program, and should keep an eye out for strange behavior on the part of their computers. In general, organizations that depend on computers must implement new policies and procedures to deal with the threat from viruses and worms.

As reported in a *Computer Security Journal* advertisement in *The Wall Street Journal*, April 4, 1989.

In Chapter 1, computer viruses were briefly mentioned. In this chapter, we will discuss viruses in detail so that a complete picture of this latest menace to computers is introduced. We will take an in-depth look at viruses by first studying their history, learning how they enter and damage computer systems, and then offering solutions and remedies against their attack and spread. "Forewarned is forearmed" is a time-honored adage that may well prepare computer systems managers and users to "know thy enemy" in order to defend against it.

Messages similar to the following are posted on university electronic bulletin boards frequently:

> Several nearby universities have had serious cases of computer viruses infecting their IBM-compatible microcomputers during the past month. The infections have been primarily limited to the student population and in at least two cases the viruses were obtained from outside the university via Internet.

> Symptoms of computer viruses vary widely, but generally consist of the machine refusing to boot up or running much slower than normal. In many cases a notice such as "Legalize Marijuana," "Your Computer is Now Infected," or "Hi, I'm Casper the Virus," will appear on the screen.

Protect yourself by maintaining two or more copies of important data on separate diskettes. Don't use any software unless you know where it came from and that it is virus-free. If you use Internet to obtain free software and shareware, use virus scanners.

Viruses were once thought to be the "playthings" of ambitious hackers on college campuses. It is quite apparent from the numerous reports circulating in practitioners' journals and at various professional meetings that the "illness" has spread to industry as well. Understanding the problem is the first step toward protecting computer software from the virus menace.

HISTORY OF VIRUSES

The term virus was first used to describe an unwanted computer code in a 1972 science fiction novel, although the description in the book does not quite meet the accepted definition created by Fred Cohen, then a graduate student at the University of Southern California. His virus took a week to complete and was used in demonstrations in his class. His advisor suggested that he call his creation a computer virus.

It has only been within the past few years that the computer virus problem has gained much attention since the first infection of the Brain virus in 1986, with many of its variations. Viruses now affect every type of personal computer currently in use. Viruses have been allowed to flourish in the personal computer environment as a result of their weak security. The personal computer was originally designed for use by individuals so that securing the system was not of great concern. The only security was an optional key-locked hard drive and a few weak software packages.

As the uses for personal computers have evolved, improvements in their security have not kept pace, thus rendering many of today's PCs susceptible to intrusion by viruses.[12]

Historical Crimes

In the mid-1970s, the network of computers at a Silicon Valley research center was taken over by a program that loaded itself into an idle workstation, disabled the keyboard, drew random pictures on the screen, and monitored the network for other idle workstations to invade. The entire network and all the workstations had to be shut down to restore normal operations.[18]

Gene Spafford of Purdue University, in his article entitled "The Internet Worm Program: An Analysis," credits David Gerrold with being the first person to use the word "virus" as a computer attacker in Gerrold's science fiction stories about the G.O.D. machine.[63] These stories were later combined and expanded into the book, *When Harlie Was One* (Ballantine Books, First Edition, New York, New York, 1972). In later editions of the book, the virus plot was removed. Spafford describes the origins of the virus as:

A subplot in that book described a program named VIRUS created by an unethical scientist. A computer infected with VIRUS would randomly dial the phone until it found another computer. It would then break into that system and infect it until it found another computer. This program would infiltrate the system software and slow the system down so much that it became unusable (except to infect other machines). The inventor had plans to sell a program named VACCINE that could cure VIRUS and prevent infection, but disaster occurred when noise on a phone line caused VIRUS to mutate so VACCINE ceased to be effective.[56]

Kenneth Thompson, in his article entitled "Reflections on Trusting Trust," further explains this growing menace by describing the development of what can be considered the first computer virus (even though he didn't call it that).[71] Thompson wrote a self-reproducing program in C programming language code and modified the program to "learn" new syntax. He then planted a Trojan Horse (to be described in detail later in this chapter) that deliberately miscompiled the UNIX *login* command, thus enabling him to log in to the system as any user. He then added a second Trojan Horse aimed specifically at the C compiler. After compiling the modified source with the normal C compiler to produce a "bugged binary," he installed this binary as the official C and removed the bugs from the source of the compiler. Thereafter, whenever the new source of the compiler was compiled, the new binary reinserted the bugs. The *login* command remained bugged, but with no trace in the source.[18]

A number of people continued the trend in the early 1980s by developing computer viruses on IBM PCs and Apple II computers. Virus activity, even though it existed, was proceeding at a slow, well-hidden pace.

Viruses leapt into the forefront in November 1983 when Fred Cohen, a doctoral student in electrical engineering at the University of Southern California, presented the idea of a computer virus to a computer security class led by his advisor, Len Adelman.[56, 29] Cohen demonstrated five viruses on a VAX 11/750 running UNIX. Each virus obtained full control of the system within an hour.[29] Cohen later demonstrated that similar results could be obtained on a Tops-20 system, a VM/370 system, and a VMS system. According to Cohen, a computer **virus** is:

> . . . a program that can "infect" other programs by modifying them to include a possible evolved copy of itself. With the infection property, a virus can spread throughout a computer system or network using the authorization of every user using it to infect their programs. Every program that gets infected may also act as a virus and thus the infection grows.[7]

A large number of viruses now inhabit computer systems—particularly in uncontrolled PC environments. The definition of viruses, the number of occurrences, and the severity of the damage is open to conjecture, but there is general agreement as to the properties of viruses. (See Figure 9.1.)

John McAfee, founding Chairman of the Computer Virus Association, a group of leading manufacturers of antiviral software products, offers a very extensive history of the growth of the computer industries and the emergence of devious programs in his book, *Computer Viruses, Worms, Data Diddlers, Killer Programs and Other Threats to Your System* (St. Martin's Press, New York, 1989). (See Figure 9.2.)[45]

ANATOMY OF VIRUSES

Computer viruses have become an international problem that plagues computer users of every type. The term "virus" itself is derived from the Latin root word

FIGURE 9.1
Properties of Viruses

> ☑ Are present on the system without the consent of the system owner
>
> ☑ Have the capability of moving from one computer to another
>
> ☑ Potentially have the capability of destroying or altering files
>
> ☑ Have the capability of denying service to legitimate users.

virus, meaning "poison." In biological terms, viruses are known for injecting themselves into a larger body, reproducing themselves, and spreading throughout the entire organism being invaded, causing untold damage.

Similarly, a computer virus enters an unprotected system and carries out a similar mission as the biological virus. In order to adequately protect from viral attacks, it is helpful to know:

- What viruses are
- What viruses are made of
- How viruses spread

Definition of a Computer Virus

A **virus** is a computer program created to infect other programs with copies of itself. The virus has the ability to clone itself in its constant search for a new host environment. The virus may have the single mission of replicating itself or it may be written to:

- Damage other programs
- Alter data
- Self-destruct, leaving no trace of itself behind[22]

Numerous definitions have been offered for the term virus (relating to the computer strain, not the biological one). Among them are:

A **virus** is a segment of self-replicating code that attaches itself to application programs or to other executable system components. These code segments move from program to program and machine to machine. They can replicate an indefinite number of times or as limited by their creator.[45]

A **virus** is a specific type of Trojan Horse that attaches itself to another block of code in order to propagate, has the ability to replicate itself (whether or not it does so), and is damaging or at best neutral (dormant).[29]

A **virus** is a code fragment that copies itself into a larger program, modifying that program. A virus is not an independent program and executes *only* when its host program begins to run.[56]

FIGURE 9.2

Computer Chronology and Virus Growth

. . . from the end of World War II to 1988

1945   ENIAC, the world's first electronic digital computer, is introduced. Originally developed as an aid with ballistics during World War II, it is 100 feet long but has less processing capabilities than a modern laptop computer.

1949   Neumann's *Theory and Organization of Complicated Automata* publishes and presents the first theories about replicating organisms.

1950s   Computer development accelerates rapidly in the United States and Europe.

1953   The Bell Telephone System publishes technical details of frequencies, which enable the early "phone phreaks" to break into the system.

1959   AT&T Bell Laboratory programmers begin playing Core Wars games, developing programs that could consume data. Researchers at the MIT artificial intelligence laboratory and the Xerox Research Center in Palo Alto, California, experiment with core memory killer programs.

1965   An identifiable subculture began to emerge when ardent computer enthusiasts were dubbed "hackers."

1969   ARPANET was developed as the world's first large computing network.

1971   Computer crime begins to escalate, initially with data diddling to alter credit files, change inventory records, and divert funds.

1972   Phone phreaks find new ways to break into telephone systems without paying, thus developing the expertise for the hackers to gain unauthorized access into computing systems.

1973   The first details of the Equity Funding Corporation's scam are uncovered. Over the previous 10 years, programmers had created 64,000 bogus policies.

1974   The first self-replicating code is demonstrated at Xerox Corporation. Administrators at the research establishments subsequently stop the Core Wars games.

1975   Microcomputers are introduced.

1976   The Red Brigade terrorist group begins a series of 10 raids on computer installations in Europe. The U.S. Department of Justice warns a Senate Committee about the potential seriousness of computer crime.

1977   The Data Encryption Standard (DES) is adopted to protect data in the computers of federal government agencies.

1978 $10.2 million is stolen from a Los Angeles bank by unauthorized telephone use of passwords and bank codes.

1979   Arizona is the first state to enact computer crime laws.

1980   Microcomputers become increasingly more powerful, and less expensive worm programs, which can be designed to destroy data, are invented at the Xerox Corporation laboratory.

1981   There is an estimated 300% increase in the number of hackers in a single year.

FIGURE 9.2, continued
Computer Chronology and Virus Growth

1982   A logic bomb is found in the Montgomery County, California, library computing system.

1983   The movie "War Games" is released and hacking gains momentum. The secret of self-replicating mechanisms is revealed in a speech by Ken Thompson, the software engineer who originated the UNIX operating system, to the Association of Computing Machinery.

1984   *Scientific American* publishes details of Core Wars, and information about writing viral programs begins to circulate more widely. Fred Cohen identifies viruses.

1985   American universities experience infections from early virus programs (e.g., the Cookie Monster). The Middle Core faction leads left-wing groups in physical attacks on 20 computer installations that disrupt train systems used by 10 million Japanese commuters. The Pakistani Brain virus is created in Lahore, Pakistan, and begins to circulate internationally on pirated software. Donald Gene Burleson is fired by a Fort Worth, Texas, securities trading firm and subsequently plants a worm program to destroy data.

1986   The first viruses to cause widespread infection appear.

1987   The Lehigh virus is identified and begins to cause damage on a large scale in universities. The Pakistani Brain virus is identified at the Universities of Pennsylvania and Wyoming. The Christmas virus crosses the Atlantic from Germany and seizes up the 350,000-terminal IBM network.

1988   Viral attacks reach epidemic proportions. Hebrew University computers are infected by the Israeli virus programmed to destroy data on the anniversary of the ending of the State of Palestine. Georgetown University experiences a persistent seven-month-long infection of the Pakistani Brain virus. The Brain spreads to 300 computers at the *Providence Journal* in Rhode Island. The Scores virus infects NASA and other government agencies, spreading to Congressional offices and thousands of other systems, including those at the Boeing Aircraft Company and Ford Aerospace. Ford's systems are later infected by the NVir virus. The MacMag virus goes off on March 2, the first anniversary of the Mac II introduction. (Estimates are that it spread to 250,000 systems.) Hamburg's Computer Chaos Club claims to have infected NASA systems. The Club's virus expert is arrested in Paris. The first known case of proprietary commercial software being infected is reported when the MacMag virus gets into Aldus FreeHand programs and is widely disseminated. Later in the year, beta test versions are infected with the NVir virus. The Software Development Council creates a task force to propose legislation and develop defenses against virus attack. The world's largest viral infection to date becomes visible on November 2 when Robert Morris, Jr., plants malicious code on a Cornell University computer, spreading through Internet and ARPANET networks to thousands of systems. Over 30 strains of virus are now identified. Approximately 40 major corporations are now believed to have experienced infections.[45]

Source: Computer Viruses, Worms, Data Diddlers, Killer Programs, and Other Threats to Your System, McAfee, John, and Haynes, St. Martin's Press, Inc., New York, NY. Copyright ©1989.

A **virus** is the newest term for the most serious of all "demon" programs. It takes on most seriously the personalities of both the Trojan Horse and the worm, but goes one step further. The virus, most prevalent on the PC, will hide in memory on an executable program. When executed, a set of instructions will cause it to infect another program, spreading itself like a common biological virus.[10]

A computer **virus** is a segment of machine code (typically 200-4,000 bytes) that will copy its code into one or more larger "host" programs when it is activated. When these infected programs are run, the viral code is executed and the virus spreads further.[18]

These various definitions point out that, even though the exact description of a virus varies somewhat, the experts agree that a virus has the potential for extreme damage to systems and data.

How a Virus Works

In recent years, as viruses have become more prevalent and sophisticated, the term *virus* has come to describe a program written:

- To *attack* existing programs
- To *ensure* its survival
- To *destroy* the integrity of the host program to the extent that it no longer functions in the intended manner

The term *virus* seems, at first glance, a rather innocuous term to apply to a computer program because the very name implies a biological entity. However, a computer virus acts in much the same manner as a virus that attacks a living organism. To illustrate this idea, a parallel can be drawn between what the biological virus does and what the computer virus does.

- The biological virus attacks specific body cells; a computer virus attacks specific programs, such as files with *.com and *.exe extensions.
- The biological virus can modify the genetic makeup of the infected cells; the computer virus can manipulate the infected cell to perform tasks.
- The biological virus grows in the infected cell, as does the computer virus that uses the host program to replicate itself.
- A biological cell cannot be infected more than once by the same virus; the computer virus does not infect the same program over and over.[52]

Properties of Virus Programs

Knowing how to write programs that manipulate or make changes is a small step to understanding exactly how computer viruses work. A virus is a manipulating program because it modifies other programs and reproduces itself in the process. (To see how this is done, refer to Figure 9.3.)

Once a virus program is initiated, the current disk drive is searched for a user program—that is, one the virus program can change. If one is located, the virus

FIGURE 9.3

How a Virus Is Spread

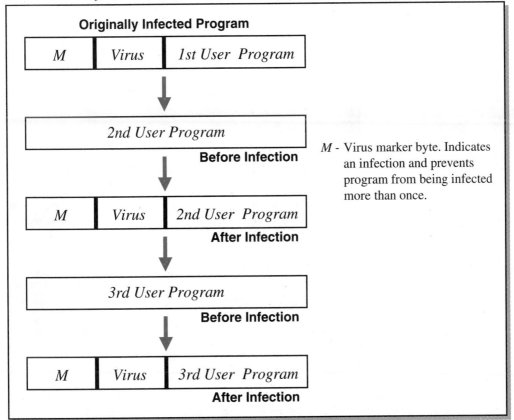

tests the program to see if it has already been infected. To do this, the virus scans the program to check for the presence of a marker byte. A **virus marker byte** resides at the beginning of the program to notify the virus that an infection has already occurred. This marker byte ensures that a virus does not expend its energy infecting a program that it already infected. To illustrate, assume that the virus has found a marker byte; after finding it, the virus moves on to the second user program and repeats the test. If the second user program is not found to contain a virus, then the virus transfers itself into the program by overwriting the start of the program on the disk with a copy of itself. The virus is now spreading; only the trained user may notice a write access to the disk drive. If this second user program is executed, the virus program is executed because it has overwritten the program code of the program. The virus then reproduces itself in the same manner in a third user program. After the virus has successfully copied itself, serious program errors begin to occur in the infected program because part of the program is missing due to the space needed for the virus code.

Dissecting a Virus

A *virus* program performs two basic operations (see Figure 9.4):

- Infection
- Operation

FIGURE 9.4
Operations of a Virus

☑ Infection
    includes reproduction

☑ Operation
    the actual intended viral action

For infections to occur, the virus must be able to gain access to a target computer system. In the multiuser computing world, this access can be from a modem connection that allows the virus writer to gain access; then the virus writer can try random user codes/passwords.[9]

The article "Stalking the Wily Hacker" in *Communications of the ACM* (May 1988), points out an actual case involving a virus when an unauthorized person used the Lawrence Berkeley Laboratories' computers to attempt access to Internet computers to perform espionage activities. Of the more than 450 computers accessed, approximately 5 percent were entered by using common codes such as FIELD or SYSMAN.[9]

In a microcomputer environment, where shared software is used widely, gaining access is very simple. By placing a program (without source code) on a bulletin board and advertising it as a useful utility or game, the virus writer is virtually assured of widespread dissemination. If a time lag between infection and activation of the viral function is long enough, the infection process becomes virtually unstoppable.[9] Known deficiencies or supposedly restricted features in the operating system can provide "loopholes" for virus attacks. The exceptions to this "loophole" activity are operating systems environments where operating details are kept strictly proprietary (e.g., military, government, corporate research and development). Systems personnel are well aware of technical manuals that detail the fact that, during the boot process, the system console is logged in as a privileged user.

Once access is gained, infection is usually introduced using commonly known technical information about the system being infected. By using a book on the machine code and executable file format of an IBM PC, the virus writer can identify a particular instruction that will be used to invade an existing program's operation.

Using a binary debugger, the writer loads the target program containing this instruction and replaces the instruction with a subroutine call to the viral operation. The viral operation's code is then appended to the program saved.

This method presupposes that the writer concocts and converts to machine code the viral function itself. Once a single program has been infected, this process can be automated as a part of the viral function. Each time the virus code is activated, it can search the disk directories available for another program to infect. When one is found, first it is marked as infected (by placing a "flag" in the program), and then the viral code is inserted. A self-replicating virus is now turned loose on the world![9]

See Figures 9.5 and 9.6.

FIGURE 9.5

Program Infection by a Virus

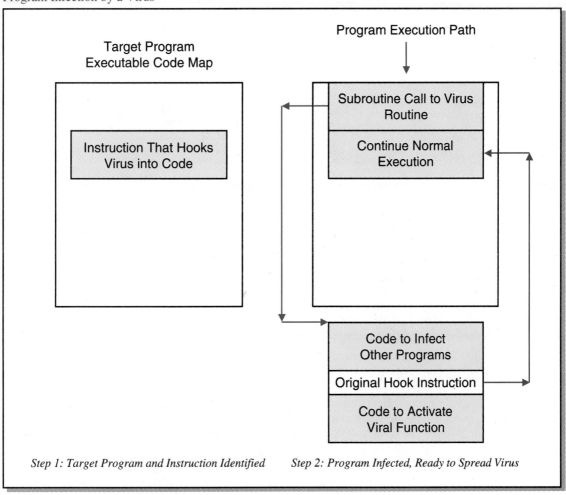

Step 1: Target Program and Instruction Identified   *Step 2: Program Infected, Ready to Spread Virus*

Source: *Computer Viruses* by Mark A. Servello, American Management Systems, Inc., 1777 North Kent Street, Arlington, VA 22209, 1990, p. 3.

Virus programs are often small, composed of just a few lines of programming code that can often be hidden in healthy software, thus rendering it very difficult to detect. Potentially, these infections can attack any type of computer from the smallest laptop model to the largest of mainframes. Experience to date suggests

FIGURE 9.6
Self-Replicating Virus Activates

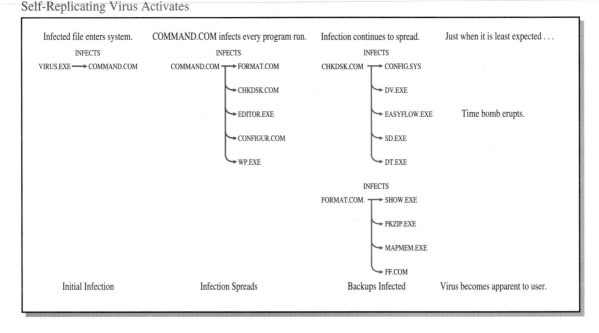

FIGURE 9.6
Self-Replicating Virus Activates

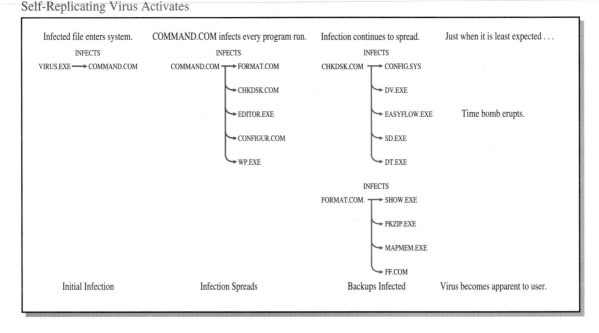

that most of the virus attacks have occurred in a personal computer environment and in some networks, but not directly on large mainframes. A virus can be created that can spread over communication lines to another computer system, where further propagation takes place throughout the network.

CATEGORIES OF VIRUSES AND HOW THEY WORK

Computer viruses come in many strains, variations, sizes, and shapes. Some range in size from the smallest space available, that of "0" or "1" in binary code, to those that contain many thousands of lines of code. On the average, however, most computer viruses contain between 200 and 300 lines of coded information. Some viruses take effect immediately and spread very quickly. Others may take weeks, months, or years to show their effect. Nonetheless, any virus, no matter what shape, size, or form, should be considered a threat to a computer system, and common sense dictates that organizations should take precautions to deal with them.

One way to categorize viruses is by the type of machine they target. Just as there are different operating systems, there are different viruses for different machines. For instance, a virus written for an Apple Macintosh could not infect an IBM or compatible machine. This is similar to a situation in which a floppy disk formatted for an Apple cannot be used on an IBM system, and vice versa.

Due to the extensive efforts of hackers, vigorous attempts to classify certain viruses are shot through with hundreds of different variations or "strains" becoming available. Similarly, this development parallels the generation of new strains of the AIDS or flu viruses.

Actually, the term "strain" refers to a particular virus, targeted at a specific computer type (as determined by the operating system), to perform a specific function. The original virus is called the "strain," while each variation is called a "variety."

Interestingly enough, most computer viruses have affected IBM PCs and their clones. Seventy percent have affected the IBMs, while 24 percent have affected Macintosh and Amiga systems. The remaining 6 percent involves other hardware and operating systems.

Currently on the rise is the emergence of viruses that can affect more than one operating system. The famous Internet virus was the first to do this. A precedent has been set.

For a virus to be most effective, it must be written in a code that can directly talk in machine language (i.e., binary code). Binary code is the lowest level of machine communication. Thus, altering this code alters any other languages that could possibly be running on the system. The most popular codes include the C programming language and Assembler language.

To better explain the concept of binary code, think of a computer as an electronic circuit. The "on" condition can be represented by a "1" and "off" by a "0." Again, all information in a computer system is at some point represented by this code, for it is the foundation of any computer.

Programmers designing viruses strive to disguise their strains but, almost always, viruses target three areas of a system during the initial attempts to infect it. These include:

- The boot segment
- The operating system
- One or more application programs

The boot segment is the startup procedure to install the operating system. The operating system is the software programming that controls all inputs to the system and manages the execution of programs. The application programs execute the useful business functions.[22]

Viruses differ in their duplication, their intended function, or their specific target. Some of the more common virus types are shown in Figure 9.7.

FIGURE 9.7
Types of Viruses

> ☑ Memory-resident viruses
> ☑ Error-simulation viruses
> ☑ Data-targeted viruses
> ☑ Crasher viruses
> ☑ Computer time theft viruses
> ☑ Call-me viruses
> ☑ Hardware-damaging viruses

Memory-Resident Viruses

Memory-resident virus programs make use of a property of computer systems that allows them to be very difficult to detect and very dangerous to the system and other programs. Programs in memory are not overwritten by data or other programs because their memory area is specially managed and is not made available to other programs. After a memory-resident program is loaded, the system behaves as if the memory occupied is not present. In the extreme case, a user can fill the memory completely with memory-resident programs, which under MS-DOS leads to the error message "program too large to fit into memory." (Refer to Figures 9.8, 9.9, 9.10, and 9.11.) Random Access Memory (RAM) consists of three specialized sections. One memory section is reserved for system functions; another is reserved for the operating system itself. All other memory is considered "free memory" that is available for user programs. Once a particular program is loaded, it takes up part of the free memory available to the system. The program in memory can be activated at any time if certain conditions are met. This generally happens through a system interrupt or a call from the user program loaded into memory.[48]

FIGURE 9.8
Normal MS-DOS
Memory

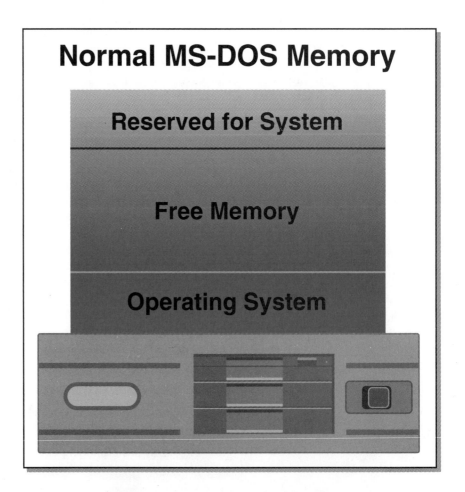

FIGURE 9.9
MS-DOS Memory with
Resident Program

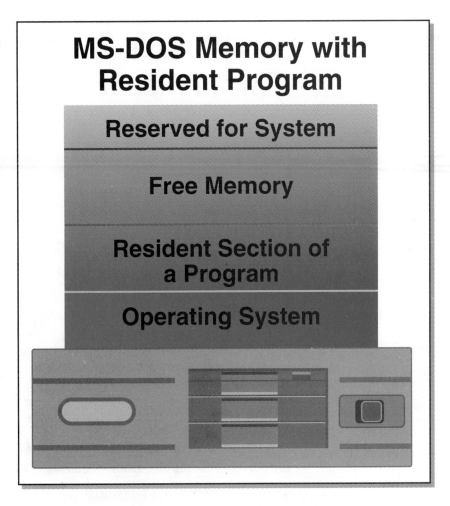

Error-Simulation Viruses

Another type of virus leads the user to believe that there are errors in the system. Such "false errors" have been used for some time by software houses, not in connection with virus programs but to expose pirated copies of software. For example, a message displaying the serial number of the program might be created so that printed copies can be traced to determine the origin of the copy. Viruses can also employ this method. A harmless example of such an **error-simulation virus** is one that simulates a defective keyboard and produces noise over the system speaker each time a key is pressed. It does this after the computer has been turned on for awhile, leading the user to believe that there is something wrong with the keyboard.

Those programs that just display an error message directly on the screen or printer must be distinguished from programs that actually cause errors. It is difficult to draw a line between the simulation viruses and the destructive viruses. A virus which continually identifies more and more sectors on the hard disk as defective

FIGURE 9.10
Normal Memory Calls
from Resident Program

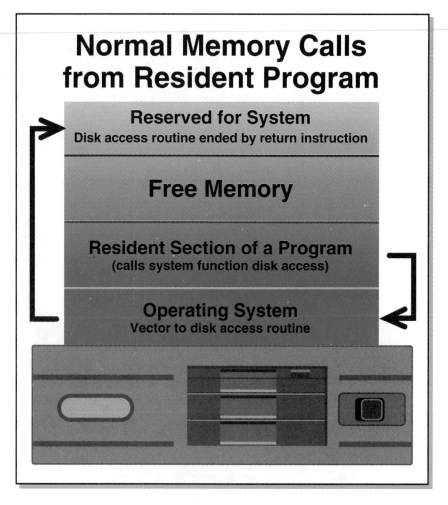

Normal Memory Calls from Resident Program

Reserved for System
Disk access routine ended by return instruction

Free Memory

Resident Section of a Program
(calls system function disk access)

Operating System
Vector to disk access routine

and thus decreases the storage space available cannot be clearly assigned to either one of these groups. An error is simulated for the user, but no actual damage is done to the hardware because the disk can be restored by reformatting. This behavior may, however, cause some users to switch to a different brand of hard disk when they see the number of bad sectors increasing. Basically, there are no limits to the imagination of virus programmers when they are simulating a defective system. All that is needed to make the user begin to doubt his or her computer is to display the error message "Parity check 1" at regular intervals. The hard disk example above shows that such programs can increase hardware sales and leads one to believe in the possibility that some companies use viruses to stimulate demand in the increasingly competitive hardware market.[48]

Data-Targeted Viruses

The worst damage, far more consequential than that caused by other viruses, is caused by data-destroying or altering viruses. A comparatively harmless variant

FIGURE 9.11
Memory Calls with
Resident Virus

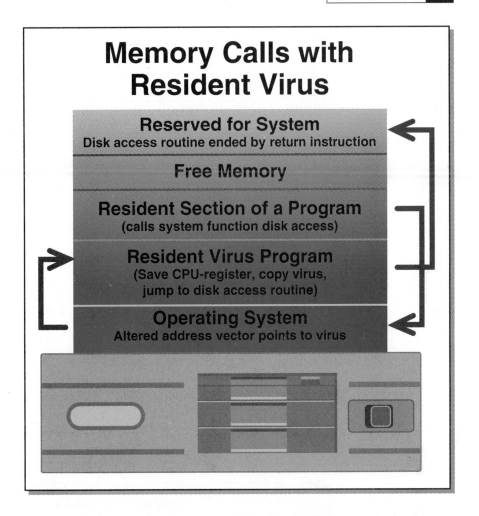

simply erases data; backup copies, however, can be used to repair the damage. Much more destructive are changes to data that are not as easy to detect. Identifying changes often requires detailed knowledge of the data structures, but damage can also be caused without this information. (See Figure 9.12.) If a person wishes to make a search for data and the number 9 is used in the search, the virus located inconspicuously within the program could change the 9 to an 8 if the search is successful and save the changed number to the file. It is easy to imagine the effects of such a change if this search were performed on a payroll file.

Another form of data manipulation involves the inflation of data. If a customer file is filled with imaginary names, the result is more than increased access time. If the file is used for direct mailing, unnecessary postage and advertising costs are incurred. Once such an inflated file is backed up, it is virtually impossible to free it of the unwanted names. Manipulations of this type make it difficult to determine the actual extent of the damage, as the search and print times also increase because of the increased number of file entries. Who can count the additional wear and tear, the delays caused to the users, the wasted storage space, and more, in dollars and cents?[48]

FIGURE 9.12
How Data Can
Be Changed

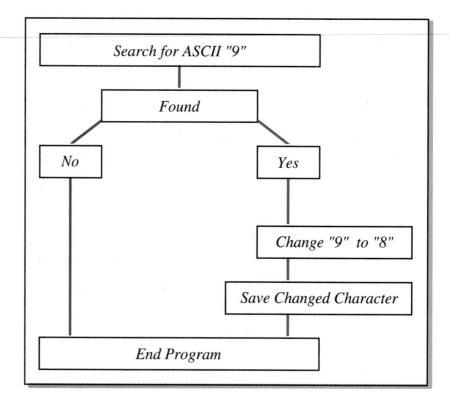

Crasher Viruses

Not all errors caused by viruses are programming errors. Some types of virus programs have no other function than to create fatal system errors—the most common
is the system crash. When this occurs, the system no longer allows any access from
outside. It is impossible to find any clues about the cause of the error. Here, other
operating systems have clear advantages over MS-DOS. A typical example of this
is the SYSLOG (System Log), a file in which all the error messages are noted.
Even with a system crash, the cause can be determined, given sufficient knowledge
of the operating system. This file can be used to find out what happened prior to the
crash of the system and what programs were involved. System crashes caused by
viruses can have various sources:

- A programming error in the virus itself
- Incompatibilities with the system or the software
- The intentional system crash

Computer Time Theft Viruses

The more one examines the manipulations that can be performed by viruses, the
more one realizes the great costs involved with virus programs—the theft of computer time is a good example. If you assume that each program in the computer

takes a certain amount of the computer's time, even if it is just loading time, then one must come to the conclusion that virus programs always hurt the users because they steal system time. The user isn't initially aware of the problem because the time requirements of virus programs are relatively small. But as the computing ability of the system continues to degrade, the user suffers a detriment that theoretically can be expressed in dollars and cents. In practice, it is difficult to calculate the exact extent of the damages.

Call-Me Viruses

On a computer equipped with modems and dial-up lines, a virus program is installed through a disk "forgotten" by the manufacturer; the disk is completely silent during the day and does nothing except replicate itself. When the system time reaches 3:00 a.m., the virus becomes active and calls up the virus programmer and gives him/her access to the system. This not only gives the virus program access to the data, but the owner of the computer must also pay the telephone costs for this access. For some time, hackers have gained access to mainframes in this manner.[48]

Hardware-Damaging Viruses

It is normally assumed that it is not possible to damage or destroy the hardware of a computer through software commands. Certainly the manufacturers make efforts to protect the system from programming errors as completely as possible. In early home computers, it was possible to cause irreparable damage to the computer through a POKE command. The problem has been corrected, but computers did fall victim to this command for some time.

It is more difficult to damage hardware today, but the developers of Killer Programs are quite inventive. The fact that some of the destructive programs have not yet appeared in virus form is probably nothing but sheer luck. For example, there is a routine that instructs the disk controller to place the read/write head of the disk drive on a nonexisting inner track. On some drives, this causes the head to jam against a stop on the inside of a hard disk. It can be freed only by opening the drive and moving the head by hand.

The susceptibility of peripheral devices is another important consideration. Many printers have in their command sets a command to move the paper backward. This is useful in plotter mode or for adjusting the paper. Anyone who has tried to move a large number of pages with the backward-feed command probably ended up with a paper jam in the printer, requiring the printer to be dismantled and cleaned.

Other Virus Types

Another type of program is one that erases a control track from a hard disk in such a manner that the disk cannot even be formatted. So far, no reliable source has examined this program, but confirmation from a number of hackers leads us to believe that it can be done.

A special category of programs includes those that don't cause any direct measurable damage because they do not destroy something directly—they merely wear it out. A small change to the CONFIG.SYS file can cause the number of accesses to the hard disk to increase dramatically. The operating system has to continually move various programs in and out of memory, even when no user work is being done. Such procedures access the hard drive considerably more in a single day than would a week of normal use.[62]

HOW VIRUSES SPREAD

A natural question to ask is: How does a virus spread throughout a computer system? *A virus is spread when infected programs are run.* The viral code is then executed, and the virus spreads. A virus can also spread when a computer is booted from an infected disk or when an infected program is run. The infection can come from any form of writable storage such as a hard disk, floppy disk, tape, or memory. Or "viruses can enter computer systems from an external software source, often hidden in a seemingly innocent program, much like the Greeks within the proverbial Trojan Horse."[45]

Many viruses target the computer operating system to quickly infect other programs when the computer is booted. Experience indicates that operating systems such as MS/DOS, PC/DOS, and UNIX are the most commonly infected. This type of infection can be especially dangerous because the viral code may be executed before the operating system can invoke an antiviral utility to detect the virus.

A virus cannot spread by infecting pure or inactivated data. **Pure data** is data that is not executed; therefore, it cannot be infected. *With viruses, infection must occur via program execution.* This can be done in two ways:

- The first method, called **overwriting**, involves the virus inserting its code so it is executed before the host program, after which the host program no longer functions.

- The second method (**non-overwriting**) involves relocation of the host program, after which it can still function normally.

One important type of virus is the memory-resident virus. This virus remains in memory after the code is executed and the host program is terminated. By doing so, once the infected program is run, the virus can spread to any program in the system during the work session.

Viruses fall into two categories: transient and resident. **Transient viruses** can infect only when the infected portion of the host program is run. **Resident viruses** copy themselves into memory and remain active even after the host program is terminated. For example, a virus infects the standard interrupts used by DOS so that the virus is invoked by other applications when they make service requests. DOS services are invoked by the user in an operation, causing an interrupt. The operating system then calls the interrupt table. The virus has modified this table so that the interrupt causes the viral code to be executed and spread.

However, if a virus keeps spreading itself throughout a system, it becomes easier to detect. As the virus spreads, it uses up more memory and processing time, which may draw the user's attention. To combat this problem, many viruses contain a "signature." The **signature** marks the program as already being infected and thus prevents the virus from infecting the same program over and over again. There are a few different ways to make a signature. A signature can be a character sequence of bytes at a known offset on disk or memory. It may also be a special system call available only when the virus is active in memory. A final way is to use a special feature of the directory entry.

Figures 9.13 and 9.14 illustrate the virus-spreading mechanism for a mainframe and for a network, respectively. Figure 9.15 illustrates the internal logic of a virus.

Triggering a Viral Attack

The goal of most viruses is to spread to as many programs and systems as possible. A virus may replicate itself one program at a time, target a specific set of programs, or infect the entire system. What causes a virus to begin this execution? As covered earlier, the viral code may be executed when the computer is turned on or when a specific program is read. A viral attack may also be triggered by an occurrence of a specific event. Some examples of common triggers are:

- On a certain date
- At a specific time of day
- When a certain job is executed
- After replicating itself a certain number of times
- When a certain combination of keystrokes occurs
- When the computer is rebooted[24, 25]

After the trigger occurs, a number of manipulative or destructive activities may occur. Some of these include unexpected:

- Removal of files
- Reformatting of disks
- Screen displays
- Sound effects
- System reboots[22]

How Viruses Are Being Spread Today

There are many ways in which a virus may spread throughout a system. Raymond Glath, of RG Software Systems, Inc., has found many occurrences of viruses spreading in the workplace. For example, a service technician unknowingly took an infected disk to a customer worksite to diagnose a problem. When the disk was inserted into the customer's system, it quickly infected their network.

Universities are not safe from viruses. One college purchased a mail order computer that was already infected upon arrival and then proceeded to infect about 100 workstations on the local area network.

Many viruses spread because of public ignorance. An example of this is an infection that took place at a major computer retail store. The store had computers for sale which all displayed the message "Your PC is stoned." The salesclerk, unaware that the computers were infected, continued to sell the computers, which later continued to spread the virus at the clients' sites.[24, 25]

FIGURE 9.13

Mainframe Impact

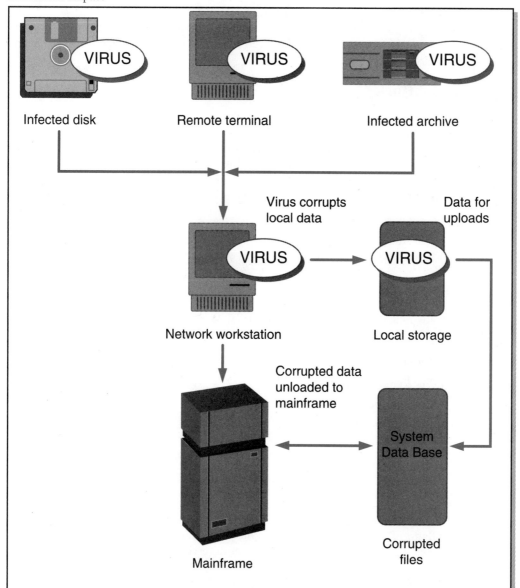

A virus can enter the system either 1) through an infected disk when loaded into a workstation linked to a mainframe, 2) when the workstation is linked to a remote terminal, or 3) when a workstation downloads data stored in an infected archive. The workstation then passes the virus to the mainframe with other software and corrupts files in the system database. The virus can spread to other workstations which access them. The virus may also move from the initially-infected workstation to corrupt data and infect programs stored locally.

FIGURE 9.14
Network Infection Process

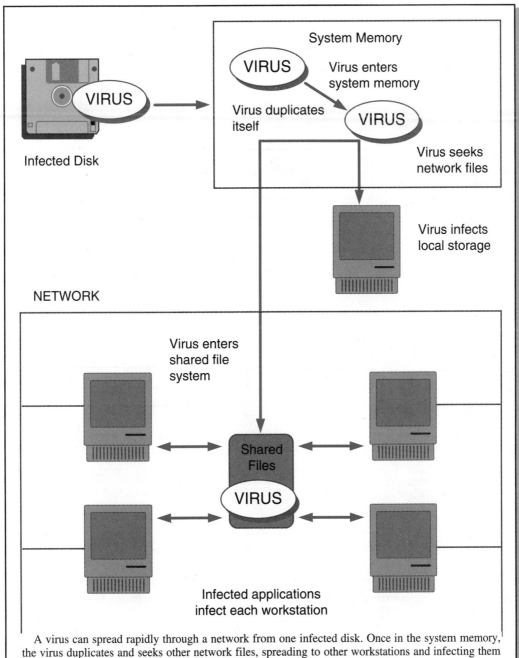

A virus can spread rapidly through a network from one infected disk. Once in the system memory, the virus duplicates and seeks other network files, spreading to other workstations and infecting them when they access shared files.

FIGURE 9.15

Virus Internal Logic

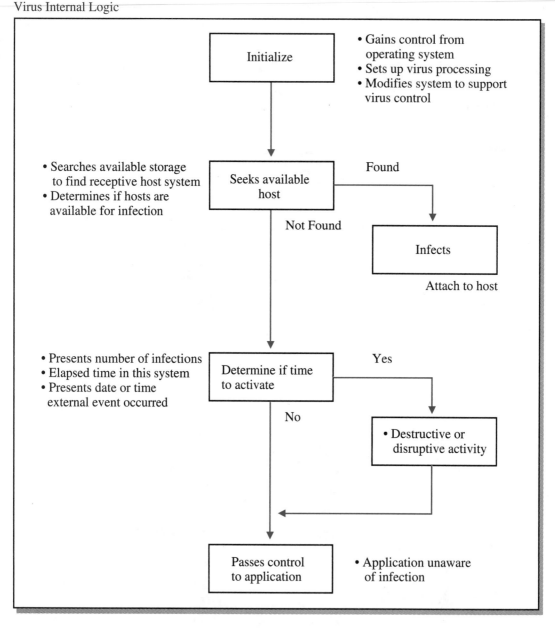

To make matters worse, anyone who wants to infect a system will find plenty of help. Many viral codes are publicized on bulletin boards and in hacker newsletters. In addition, some books are published containing sections of complete viral code. Also, books that are published as programming aids concerning the inner works of DOS, disk operations, and data structures provide plenty of information for those desiring to write viral codes to cripple a system. Programming tools can also help viral authors. These tools, such as disassemblers, can help people who want to take apart software packages to gain knowledge about the inner workings of a package in order to infect it. Thus, it is relatively easy for viral originators to gain the knowledge necessary to infect a program or system.[22]

PSEUDO-VIRUS PROGRAMS

Now that you have been thoroughly versed on how viruses are created and spread through a computer system, a discussion of **pseudo-virus programs** (sometimes called "look-alike viruses") is appropriate (See Figure 9.16.) Often, these pseudo techniques occur independently from viruses. Lately, though, hackers may use these techniques in tandem with viruses, resulting in a very powerful and disastrous outcome.

FIGURE 9.16
Pseudo-Virus
Programs

☑ Trojan Horses
☑ Bombs
☑ Trapdoors
☑ Salami slices
☑ Phone phreaking
☑ Worms

Trojan Horses

A **Trojan Horse** is a code fragment that hides inside a program and performs a disguised function. It is a very popular mechanism for disguising a virus or a worm.

The basic idea of this type of program is as old as the original Trojan Horse. The operation is as simple as it is dangerous. While the user is mesmerized by fantastic graphics displays, perhaps accompanied by music, the program reformats the hard drive—virtually unnoticed. Many viruses are known to hide in Trojan Horses, but Trojan Horses do *not* have the ability to replicate themselves.

A recent example of such a Trojan Horse is the "AIDS" virus. Distributed to major corporations and PC user groups around the world, this virus arrives as a diskette entitled, "AIDS Information—An Introductory Diskette." When installed, the program immediately prints out an invoice for the software which asks that $378.00 be mailed to PC Cyborg Corporation in Panama. When it is rebooted, the system then self-destructs. This virus has clearly perpetrated the largest single targeting of destructive code yet reported.

Logic Bombs

A **logic bomb** is a set of instructions that will execute when certain conditions are met, usually resulting in a system error. The logic bomb is similar to the Trojan

Horse in its programming and ability to damage data but has a built-in timing device so that it will activate at a particular time. (Thus, logic bombs are often referred to as **time bombs**.) Virus programs often include coding similar to that used in logic bombs, but the bombs can be very destructive on their own, even if they lack the ability of the virus to reproduce.[45] Several examples of the time or logic bomb method of attack are the so-called "Columbus Day" that strikes in October and the "Jerusalem-D" and "SYS-B," which will format the infected system's hard disk on any Friday the 13th after 1990.

Trapdoors

A **trapdoor** allows unauthorized access through hidden weakness in the code and thus provides easy access. Since the trapdoor is a secret, undocumented entry point into a module that exposes the system to modification during execution, this type of attack does not always produce a negative effect on the system.

These viruses are commonly placed into authorized programs by the programmer as command insertions that aid in the testing stage. Also, auditors use trapdoors to trace the flow of specific transactions. Unfortunately, they can also evolve from poor error checking of a system's design; this allows unacceptable data to enter the system as acceptable, thus becoming a trapdoor. Trapdoors may be exploited by the original programmer or by individuals who accidentally or deliberately find a vulnerability in the program. When used intentionally, trapdoors must be completely removed or tightly controlled.

Salami Slices

A **salami slice** is a technique that takes off small slices or sums unlikely to be noticed on any particular run. These slices are moved to a secret account being operated by the thief and accumulated until a sizable amount builds up. Salami slices are particularly devastating when applied to payroll accounts or banking records.

Phone Phreaking

Phreaking (or freaking) attacks the telephone network and related services for fraudulent purposes. In 1978, $10.2 million was stolen from a Los Angeles bank by unauthorized telephone use of passwords and bank codes to get into the computer system.[45]

Worms

A **worm** will burrow itself into the computer memory and attempt to replicate itself until the memory is exhausted, possibly triggering a system crash. A worm destroys data, but it does not replicate as a virus does. Worms were innocently invented at the Xerox Corporation laboratory, where early work on self-replicating programs has taken place.[45] Worms and viruses are often confused as they share similar characteristics. Most computer security experts will be quick to point out that the most famous worm known to date is the one planted by Robert Morris, Jr. in 1988.

The trial of Robert T. Morris, Jr., the son of the well-respected computer guru, Robert T. Morris, Sr., has brought to the attention of the American public the seriousness of computer viruses. The trial was given much attention, not just because it was ironic that it was the son of a computer scientist who consulted with many large corporations and the U.S. government, but because the outcome of the trial determined the severity of punishment computer hackers can expect. At age 23 and a graduate student at Cornell University, Robert T. Morris brought the 60,000-computer Advanced Research Projects Agency Network (ARPANET) and another unclassified computer network operated by the Department of Defense, MILNET, to a halt on November 2, 1988. Because of lax programming, the ARPANET system was penetrable. Robert Morris didn't pick the lock; he simply retrieved the key under the mat, according to Eugene Stafford, a Purdue computer science professor.

The system was part of an international grid of telephone lines, buried cables, and satellite hookups, established by the Department of Defense in 1969. Using a terminal at Cornell, Morris sent a highly sophisticated, 47,000-character program into the Massachusetts Institute of Technology's computer system. This program was feasible due to a flaw that Robert Morris had discovered in Berkeley UNIX, the standard operating system of ARPANET's computers. The flaw was actually a type of back door, which the author of Berkeley UNIX had installed for debugging purposes.

Using a feature in UNIX that allowed electronic mail to be sent not only to people but to other computers, Morris first sent a harmless program to a new computer, then sent an electronic mail message containing the virus to the new program. Once in the new system, the worm went into action. It first called up other systems that it could potentially infect, broke into higher security levels, and sent a "birth" notice to another system located at Berkeley to foul an audit trail. The worm was actually successful only in approximately 10% of the computers. The other 90% seemed to ignore the initial message. Still, that meant approximately 6,000 computers were infected. Computer scientists who analyzed the worm after it was stopped were amazed at the simplicity of the code. For example, to crack the passwords of several systems, Morris's program used a subroutine that guessed at the password using only 400 common English words. His subroutine worked like this: If someone wanted to access the computer room in the Pentagon while incognito, that person would throw 400 words at the guard posted at the entrance. The guard would let the person enter if he answered correctly but would not be suspicious of all of the wrong answers.

Computer crime causes a lot of anger and frustration, but how should we determine who gets punished and the severity of the punishments? It can be argued that there are several levels of computer crime, and several levels of viruses and worms for that matter. In the case of Robert T. Morris, Jr., no damage was actually done. The systems were broken into and used to access other systems.

Even though Robert Morris admits to writing the program as a misguided stunt, he was convicted of violating federal computer crime statutes that prohibit unauthorized access to federal computers. This conviction finally cost Morris $10,000 and 400 hours of community service.

To appeal this conviction, Morris debated that because he had legitimate accounts on Cornell University's computer system, he was authorized to use the special mailing programs that enabled him to transfer documents within networks. However, despite this, the United States Court of Appeals for the Second Circuit said that he exceeded his authorization when he used the nationwide computer network.

The *Washington Post* called Robert Morris "the rogue nerd from Cornell" and "the most famous computer renegade" that had ever been caught by federal prosecutors. But whether Morris is a rogue, nerd, or famous computer renegade, this man is very gifted when it comes to computers. Although his friends and teachers consider him to be brilliant, they also agree that he is "slightly strange." He often solved intricate and difficult problems that even his Harvard professors had difficulty completing.

It's obvious to anyone that Robert Morris is indeed a genius. However, it is also very obvious that his motives are misguided. To take his genius and use it in the capacity that he chose leads one to the question: "Can there indeed be too much education?" But whether the answer is an absolute "yes" or a resounding "no," the direction that Mr. Morris chose for his education, knowledge, intelligence, and genius unmistakably led him to a deranged life of hacking and breaking into knowledge that indeed was not his.

MOTIVATION TO CREATE VIRUSES

Although government computers, in many cases, are likely targets for a viral attack, they are not the only possible victim. Other likely victims are businesses, students, computer hobbyists, and often the generators of viruses themselves, although one could not help but take delight in the prospect of a hacker having his own virus turn on him.

Who are the most common creators of viruses? John McAfee, president of Interpath Corporation, states that there are three primary groups of people who originate computer viruses:

- **Legitimate researchers.** These researchers may be studying the replication processes of viruses or working with artificial intelligence. Because these people are well aware of the dangers of an unleashed virus and thus guard against a virus's release, they are seldom a threat to the community at large.

- **Pranksters.** Pranksters usually write only harmless code and thus are more of a nuisance than a threat.

- **Outlaw technologists.** This third group poses the greatest threat for the production of damaging viruses as they write deadly code in an attempt to get back at society.[45]

What Makes Virus Creators Tick?

Why do programmers create viruses? There are clearly a wide array of reasons. Some of the possible motivations include the following:

- Personal vendetta against organizations or individuals
- Practical joke
- Publicity for a cause or a product

- Political terrorism
- Expression of sociopathic behavior
- Attention

KNOWN VIRUSES

To illustrate that viruses are real and do attack systems, following is a partial listing of known viruses that have played havoc with computer systems in the past few years. Even though this list is far from inclusive, as new viruses and strains of existing viruses emerge almost daily, it does point out that, viruses are, indeed, dangerous and deadly.

The Hamburg Virus

In the fall of 1987, the Hamburg Chaos Computer Club in West Germany planted a virus in NASA's Space Physics Analysis Network, which is a semipublic bulletin board, and a second virus in the system of Lehigh University. The results from the NASA attack have still not been made public, but the attack at Lehigh was publicized to warn other universities.

At the university, students checked out computer disks at the library just like they were checking out a book. The **Hamburg virus** attached itself to the COMMAND.COM, which is a system file within the disk. When the virus was eventually triggered, the DOS interrupt 26H (absolute disk write) wrote a series of zeros to the first 12 sectors in DOS, which are the most important, and any information on the remaining 20 sectors. The Hamburg virus made the disk fail to boot and caused the hard disks to crash.

The Macintosh Virus

This virus was on a disk shipped with Freehand, a graphics package manufactured by the Aldus Corporation of Seattle. The benign virus displayed a universal peace message from MacMag, a Canadian Macintosh magazine, on March 2, 1988. After the message appeared on the screen, it attempted to duplicate, but in the process it destroyed itself.

The Pakistani Virus (Brain)

In the fall of 1987, this virus infected several American university computer systems including the University of Delaware, the University of Pittsburgh, the University of Pennsylvania, George Washington University, and Georgetown University. If it could not find a volume label, it created one called (C) **Brain**. The resulting sectors caused damage to students and faculty. One percent of affected disks were determined to be completely unusable, hundreds of other disks were determined to be partially unusable, and a large number of other disks were reported to be affected to some extent. One graduate thesis was completely destroyed. Finding its way into the computer system of a large eastern hospital, the virus destroyed almost 40% of the medical records before it was stopped. When the virus code was broken down into assembler language, a message appeared advertising the creators of the virus and a telephone number that could be called for the virus vaccination.

The Hebrew University Virus

This virus was designed to wipe out all files on Friday, May 13, 1988. The virus would have affected the University's 1,000 microcomputers, as well as thousands of personal disks. It was discovered before it could do serious damage because it replicated itself on Friday the 13th of each possible month for several months prior.

The Flu-Shot 4 Virus

In March 1988, this virus appeared on bulletin boards and allowed users to think that it was an updated version of Flu-Shot 3, which was a legitimate virus detection program. Sign-on screens and information about the program were identical to the original program. It was designed to corrupt the disk parameter table in the zero sector of the present floppy disk.

The IBM Christmas Virus

This virus attacked IBM's worldwide electronic message network during the holidays of 1987. The program first sent holiday greetings to computer users, and then passed the same message to everyone on that user's mailing list. In a matter of hours, the massive electronic chain letter overloaded the network, forcing the system to shut down.

The Scores Virus

This virus invaded the Macintosh network at Miami University and a number of federal agencies. Its purpose was to destroy data through system crashes. Not until the code was analyzed was it known that the author's true objective was to seek and destroy software carrying the signature of Electronic Data Systems, headquartered in Dallas, Texas.

The Amiga Virus

The **Amiga virus** works by attaching itself to a computer's memory from an infected disk that boots the system. It will wait and then infect the next disk that boots the system. After this occurs several times, a message will appear on the screen such as "A wonderful thing has happened. Your amiga is alive! And, even better, some of your disks are infected by a virus!"

The Israeli Virus

Also known as Black Hole, this virus projects a black dot on the screen to indicate its presence.

The Falling Tears Virus

This virus infected the Soviet Union as well as the United States; it caused all of the characters on a computer screen to "cascade" down the screen into a jumbled pile.

Michelangelo Virus

The latest instance of a potential widespread virus is the **Michelangelo**, named after the Italian Renaissance artist. This virus was set to activate on March 6, 1992, the 517th anniversary of Michelangelo's birth. Targeted to affect IBM compatible computers and primarily transmitted by booting from floppy disks, this virus gained widespread attention from the news media as warnings were widely disseminated in an effort to lessen the impact. The March 6 date was set to trigger a "real blow" to systems worldwide.

John McAfee, a renowned expert in computer security, was quoted in the March 7, 1992, issue of the *Washington Post* as estimating that only 1,000 companies were affected on doomsday, a fraction of the nearly 47 million computers worldwide. This was due for the most part to the high publicity that Michelangelo generated. Computer systems for the House of Representatives, the Potomac Power Company, the Central Intelligence Agency, and the Bell Atlantic Company were among those saved from Michelangelo.

DETECTION AND ERADICATION

To minimize the risks of virus infection, safe computing practices and procedures must be followed. When precautions are paired with antiviral products, the risk of infection is decreased substantially. Following these 12 fundamental safe-user practices can help prevent the vulnerability of a system:

1. Never boot from any floppy other than the original write-protected disk.
2. Assign only one boot disk to each and every floppy-based PC.
3. Never boot a fixed disk system from a floppy drive.
4. Treat public domain and shareware with caution.
5. Create meaningful volume labels on all fixed and floppy disks at the time they are formatted.
6. Watch for changes in the pattern of a system's activities.
7. In a multiuser environment, minimize the exchange of executable code whenever feasible.
8. Never place public domain or shareware programs on a file server that could be accessed by all users.
9. Allow no one, other than the system administrator in a network environment, to use the file server node.
10. Separate emulation software from other files.
11. Write-protect all boot floppies.
12. Remove floppies from drive slots and store in filing cases when they are not in use.

To protect against viruses, the following measures should be instituted:

1. **Keep file trading to a minimum.** Don't use any files if you don't know where they came from and aren't sure they are clean because file trading is the primary way the virus is spread.

2. **Use software bought from a manufacturer.** These companies screen the programs for viruses, and the programs will not operate if one is present in your system.

3. **Write-protect all your disks.** If you do this and make your disk read-only, even if it is exposed to another computer with a virus, the virus cannot be passed along to your disks.

4. **Be careful of shareware.** Because many people use the programs, the likelihood of a virus will be greater than with normal disks.

Recovering from a Virus

Once you know your system has been infected with a virus, you can then move to eliminate the unwanted visitor. Removal techniques vary with the type of virus. According to John McAfee, three factors determine the complexity of removing a virus from your system:

1. The *type of virus* to which you have been exposed

2. The *amount of time* the virus has been allowed to develop in your system

3. The *number of potentially infected PCs* and workstations you are working with.[45]

Some viruses take up residence in a specific file or program in your system. These are the easiest to remove—simply remove the program from the diskette. However, use close scrutiny to ensure that no other areas in your system have also been infected. Prevention is the key to dealing with computer viruses!

If a virus has been in your system for a while, then your problem is exponentially magnified. How many diskettes have made a temporary home in the infected machine? Where are they now? Each machine an infected diskette comes in contact with probably has been infected. A virus that has been on a hard disk for some time has probably wriggled its way into most of the files. Every floppy you used is a carrier, thus presenting another problem—reinfection. You can shut down your system, build it back up from the original write-protected software you use, and be totally free from infection. However, remember that every floppy you've used is diseased. Even if you spend the time and effort to cleanse your system, all it takes is for one person to reinsert an infected diskette. Then you're back at square one. Once again, prevention plays a major role in dealing with computer viruses—be careful!

Stages of Infection

Just as the type of virus plays a role in your "exterminating" decisions, the level of infection is also important to consider. A virus goes through four stages in your system. (See Figure 9.17.) The first stage is known as **local memory infection**,

FIGURE 9.17
Stages of
Virus Infection

☑ Local memory infection
☑ Local disk storage infection
☑ Shared file system infection
☑ Complete infection of system-wide removable media

which occurs when you catch the virus at a young age, when it hasn't had time to infect your entire hard disk and it is isolated to one or two files. Recovering from this level of infection is relatively easy. Power down the system, reboot it (from the master write-protected boot disk), check for the virus, and eliminate the infected files.

The second level of infection is called **local disk storage**. At this stage, the virus has had time to copy itself into most of your executable and system files. Basically, you'll have to reformat your disk and copy everything back from the original, write-protected floppies. Reinfection is likely to occur as someone will eventually reinsert an old, infected diskette into the system.

The third stage of an infection involves the server in a local area network. Known as a **shared file system infection**, it indicates that once a file server has been infected, any program executed or file copied carries the virus to a new workstation. Removal involves a complete shutdown. Just as with local disk storage infections, you'll have to start again from the original application software.

The fourth level of viral infection deals with complete **infection of system-wide removable media**. Every piece of software ever in contact with the infected system must be closely examined and repaired.

Recovering from a Virus Attack

Recovery from a virus attack may come only by tearing down your system and starting over from the beginning. Obviously, this is very costly. Is it worth the cost? What kind of virus do we have? Is it harmless or aggressive? You must look at the situation as a whole to determine what you should do. Also, unless you are technically proficient with your machines, don't try to deal with an infection on your own. Don't dig around your computer system without knowing what you are doing. You could do more damage than the virus ever would have. The best thing you can do is set apart all infected systems (don't use them to prevent spreading the infection) and contact an expert. Specialists are available to help you. To get in touch with one of them, simply contact the Computer Virus Industry Association. This organization has also released guidelines for dealing with an infection:

1. Don't panic.

2. Power down your machine.

3. Seek professional assistance if the infection is widespread.

4. If the infection is local, reboot your system from the original system diskette.

5. Back up all your non-executable files (they won't be infected).

6. Format your diskette at the base level.

7. Replace all application files from the original, write-protected diskettes.

8. Restore your data files.

And, as always, prevention is the key to successful computing.

VIRUS PROTECTION PACKAGES

With all of these viruses circulating, products that can combat the threat of viruses are clearly needed. Industry has felt this need and has supplied users with many choices. Almost all of these virus protection programs can be divided into three main categories: **prevention products**, **detection products**, and **identification products**. With so many virus protection programs entering the market, knowing which ones are valid is important.

Virus Prevention Software

Virus prevention software programs prevent the initial infection of the virus. The virus never actually infects the system when these products are used. These packages detect an attempted infection and then notify the user. Programs with checksum or signature-verification abilities should be able to indicate that an attempt is being made to modify a file. Many of these products are referred to as "virus filters."

Virus Detection Software

These products detect the virus after it has already infected the system. These programs usually tell the user that there is a virus on the system, but do not identify the specific virus. Some of these programs tell the user what the file code says. It takes an executable program file (.COM or .EXE), removes all the unintelligible machine language characters, and lists the ASCII character string (in English language, not machine code). This allows the user to quickly see if a program contains any statements such as "now trashing your hard disk" or "your fixed disk is now history."

Virus Identification Software

These more complex programs actually identify the specific strain of virus and then remove that infection. These programs must first identify the specific virus because each strain is removed differently; consequently, the program needs to know which virus it is so it can follow the removal specifications for that particular virus. These programs usually isolate the particular virus by locating and matching extra code found on the end of executable files.

Choosing Virus Software

In deciding which one of the many available products to choose, it is important to first realize that no program is going to be 100 percent effective. Beyond that, is it possible to test whether an antivirus program is both valid and effective? The answer is a guarded yes. To sufficiently test an antivirus program, one must test against every known virus, in many different computer environments. As one can see, this could be a very long and difficult task.

Some programs are not effective because they are outdated. Virus protection programs must be updated continuously because new strains of viruses are appearing all the time.

In judging one of these products, one must look at the program's effectiveness first, then at how it works in that user's personal system with his or her particular needs, and finally how easy it is to use. The most realistic way for a user to judge a program's effectiveness is to review tests done by independent laboratories. To determine how well the program works in a user's personal system, he or she will just have to test it out. If a program is difficult to operate, the user might not use it correctly, making proper detection and removal impossible to guarantee.

Antivirus Software Products

Numerous virus products on the market today can aid computer users in the prevention, detection, and eradication of known viruses. The products share some common features, yet each has individual characteristics. The listing that follows should not be construed as an endorsement of any one product, to the exclusion of others.

Anti-Virus. Anti-Virus by Central Point Software is a comprehensive program designed to protect a system from virus attacks. Published by the developers of PC Tools, this program detects, immunizes, and cleans more than 400 known computer viruses in files and removes them.

Additional utilities include a choice of memory-resident virus protection programs, VSafe and VWatch. VSafe occupies approximately 22K of system memory and includes eight different monitoring options, while VWatch requires only 8K of RAM in its single protection mode. Both display a warning message, giving the operator the opportunity to continue or to run Anti-Virus to eliminate an attacker.

The vaccine component of the program is Bootsafe. A boot sector and partition table security utility guard protects these vital areas against data corruptions.

The program, which is easy to install and operate, is well documented and supported. Support includes a 24-hour Virus Hotline, a quarterly newsletter, quarterly program updates, the Central Point Bulletin Board with program update downloads, and a forum on CompuServe. Also, the manual provides general information on the viruses and prevention strategies useful in the development of a system security plan.

List price of the package is $129; it requires an IBM compatible processor, 512K RAM, DOS 3.0 or higher. Hard drive is optional, but recommended. One free virus software update is included with purchase, with others being offered quarterly.

Flu-Shot +. Marketed as an MS-DOS antivirus product, this program stages "residence" in computer memory and allows other programs to run while it watches the background. Whenever computer memory is modified or a disk is modified or formatted, Flu Shot + is activated and warns the user that something was attempted. The user can allow the action to proceed by pressing Y or G; otherwise, the action is halted. Flu Shot + also does a check of systems files when it is installed. If any modifications are made afterward, the user is alerted. To allow trusted programs to run, the user gives Flu Shot + a list to prevent it from giving false alarms.

Virus Clean. The Virus Clean package is a very powerful combination of virus protection and removal programs for personal computers and networks. Its outstanding detection and containment capabilities are based on the most rigorous virus analysis techniques available today. Virus Clean is based on the findings of the British Computer Virus Research Center since its foundation in 1988. Written and maintained by Joe Hirst, Virus Clean will not only recognize known PC viruses, but can remove them and even prevent infected programs from executing. It will also warn of infected floppy disks. Hirst's programming excellence has enabled him to minimize the TSR (memory resident) size while maximizing the speed and power of his detection and removal modules.

Virus Clean enables:

- automatic disabling of memory-resident viruses
- fast and accurate detection and removal
- simple menu-driven user interface
- accurate prediction of new virus variants
- precise assessment of the effects of the virus
- fewer false alarms

Viruscan. This program scans the memory, the boot sector, the partition table, the files of the PC, and its disks. Over 251 variations of viruses can be detected using Viruscan. It first identifies the virus and then identifies what area of the system is infected. During the execution, Viruscan runs a self-test. If any modifications have been made to Viruscan during the self-test, it will continue checking for viruses but will display a warning that the possibility of infection exists. One strong point of this program is that it allows for updates with the use of an "external virus data file," to look for new viruses.

Netscan. Recently, a network version of Viruscan has been introduced called Netscan. The Netscan program scans the virtual drives of the computer. The program is designed to identify any preexisting PC virus infections in specific files or system areas, and it will also identify the strain of virus that caused the infection. Netscan does not remove the virus automatically, but when used with Viruscan, it can identify 161 major viruses including 12 of the most common ones. These 12 viruses make up over 95 percent of reported infections—a very important fact to consider for network security.

Netscan also checks the common places that viruses infect, including:

- The hard disk partition table
- The DOS boot sector of hard disk or floppies
- One or more executable files within the system

Another feature of the Netscan program is that it enables the user to scan individual directories as well as individual files. The program operates at a speed of three minutes per 1,000 files. It is recommended that the designated drive be inactive while it is being scanned or an error message will appear.

Clean-Up. Clean-Up is a program that will disinfect viruses from various components, including:

- The partition table
- The boot sector
- Files

This program will isolate and remove the virus indicated by the Viruscan program. Generally the Clean-Up program is able to restore the area that was infected to normal operations. When Clean-Up is first activated it will run a self-test to check for and add modifications to the program. If Clean-Up finds any modifications, a message will appear notifying the user that there may have been changes made to the program while continuing to remove the viruses. Version 72 adds the removal of two new viruses, Liberty and Pakistani, which brings the total to 37 common viruses that it will remove. If a lesser known virus is found, the program will ask if the user wishes the infected area to be overwritten or deleted. One important element of Clean-Up is that some of the viruses contained in the .EXE files are not removed. In most cases when Clean-Up does not work it is because the virus is not attached to the end of a file.

VShield. VShield is a program that stops viruses from infecting a computer system. VShield actually checks a program before it is loaded into the computer and, if a virus is detected, VShield will not allow a program to be executed. VShield contains three levels of protection. Level I checks the Cyclic Redundancy Check Validation Code. Cyclic Redundancy Check searches for matching validation codes. Level II checks the files for a virus signature in the previously mentioned memory areas. Level III protection is simply a combination of the first two levels. If a virus is detected in any of the three levels, the program will not be executed. When VShield does detect a virus, the user should use Viruscan to determine the extent of infection.

Validate. Validate is a program that is used to check shareware and other software for any harmful tampering. Validate is easy to use because the user simply runs the program, receives authentication numbers, and then compares the authentication numbers to the ones on file. If the numbers do match, the program has not been tampered with in any manner. McAfee Associates is willing to distribute this program at no charge as long as the proper documentation is given with the program.

VIREX. This software product has both an IBM and Macintosh version and has the ability to search for and destroy a virus at user discretion. It also has the impressive ability to protect a disk or system on the front end by scanning each new disk when it is put into the machine.

The user-driven portion of the package includes the ability to scan individual files, folders, a single disk, or a number of disks. This scan indicates either a clean bill of health or the presence of a known virus. Once a virus is diagnosed, the repair function is easily engaged to eradicate the problem. Another feature in this program allows the user to define a certain virus strain. This feature is found in the expert mode and allows the user to keep protection as current as possible while waiting for updates. The application is driven almost exclusively through the use of the mouse and easily recognized icons on the screen.

The portion of the package that can be installed in the system folder of the hard drive or the boot disk is manipulated in a similar manner. This part of the package can be set to scan any new disk inserted in the machine; it is known as VIREX INIT, named appropriately for its stealthy placement in the system folder. The obvious advantage to the INIT application is its ability to prevent any virus from getting a chance to infect the system or any data files, thus preventing the possibility of damage in the first place.

Once placed in the system folder, the INIT application is manipulated through the control panel found under an icon in the desktop main menu. A password can be set to prevent access to the control features of the program. This is an excellent security measure, preventing a meddlesome user from disabling the application. Among the most important control features is the setting to diagnose all disks inserted in the machine or to diagnose only those designated by the user. Another control feature allows the application to be locked in the system folder, thus preventing its accidental or intentional removal. The program thus not only provides security for the system, but is thoughtful enough to ensure its own safety and integrity.

Cautions on Using Virus Products

Raymond M. Glath has developed a list of 12 special considerations that the user should heed when considering a virus software product:

1. **The impact the package will have on the computer's performance**. If the package is RAM resident, does it slow down the machine's operations? And if it does, which operations does the package interfere with?

2. **The level of dependency on user intervention**. Does the package require users to regularly perform certain tasks? Does it check for viruses only on command? Does it require much time to install and maintain? Does the package need to be used each time you install new software?

3. **The impact on productivity (the annoyance level)**. Does the package periodically stop processing or require users to take some action? If so, does the package have any capability to learn its environment and stop the interference?

4. **False alarms**. How does the package handle situations that appear to be virus attacks but are the legitimate actions of legitimate programs? Are there situations where jobs will have to be rerun or the system rebooted because of the protection package?

5. **The probability that the package will remain in use**. Will interference or usage requirements discourage users? No package will be effective if users de-install it or only pretend they are using it when managers are around.

6. **The level of effectiveness it provides in combatting viruses**. Will it be effective against viruses produced by people with varying levels of experience: the typical end user, power user, applications programmer, or systems engineer? Will the package detect a virus attempting to clone itself? Does it

detect a virus attempting to place itself into a position to be automatically run? If a virus gets into the computer, what kind of damage will it protect against? Massive destruction like formatting a hard disk, partial or selective destruction, random alteration of data, or annoyance? Does the software detect a virus before or after it has infected a program or made its attack? Does the publisher claim total protection from all viruses?

7. **Assistance for post mortem analysis of suspected problems.** If a virus symptom is detected and the computer is halted, is there any supporting information for analyzing the problem other than what the user remembers?

8. **The impact on the machine's resources.** How much RAM is used? Is any special hardware required?

9. **Compatibility with your hardware configuration,** the version of your operating system, your network, or other installed software packages, including TSR programs.

10. **Use by current computing personnel without substantial retraining.** What type of computing experience is required to install the package?

11. **The background of the publisher.** Who uses this or other products from the vendor? Does the company provide ongoing support?

12. **Other useful benefits besides virus protection.** Viruses aren't always meant to be destructive. The MacMag virus flashed a peace message, and a virus on the BitNet network sent Christmas greetings. Even so, they can cause programs to act in quirky ways or slow down the operation of a stand-alone machine or a network. Some viruses are merely annoying, but who wants to be annoyed?[24, 25]

INTERNATIONAL PERSPECTIVE

Despite the potential damage of computer viruses within the United States, some hackers aspire to have influence on an international basis. These individuals have motives that range anywhere from a personal gain to a political statement or an act of terrorism. On a global scale, the potential for destruction is almost entirely unrestricted. It is not difficult to assimilate the horrifying possible ramifications of an individual with ill intentions who breaks into the defense systems of a nation.

Throughout Asia and the European Economic Community, computer viruses are a severe problem. In France, it is estimated that computer security is breached 20,000 times per year. In Japan, one virus struck the users of the NEC shopping network. It instructed systems to disclose the owners' passwords and identification numbers so that the creator of the virus could make purchases using their accounts. The problem has escalated to the point that the government is giving tax incentives to companies that improve their computer security.

Despite the impact of viruses on an international scale, much of their destruction in other countries goes without notice in the United States, largely because viruses have not caused as much damage globally as they have locally. There are not as many computers in other parts of the world as there are in the United States. For example, Europe hosts 5 million personal computers while there are over 30 million PCs in the United States. Additionally, computers tend to be more expensive in other countries and thus less affordable.

Military Applications

The U.S. Army intensified its computer efforts during the Persian Gulf War, which showed how electronic warfare has become an essential aspect of the modern battlefield. However, the idea of using computer viruses as weapons has caused a tremendous amount of controversy. Many experts believe that the United States has the most to lose if these types of weapons are produced. The risks to the Pentagon's own computers may well outweigh any potential benefits.

The U.S. Army first became interested in computer virus weaponry in early 1989. In a controversial article, electronic warfare expert Myron Cramer suggested that the application of computer virus concepts has made possible the advent of a new class of electronic warfare. After the Persian Gulf War, the Army redoubled its efforts to develop electronic weapons and security measures for its own systems. In May 1990, it began accepting bids to examine the potential use of a computer virus as a battlefield electronics weapon to disrupt or destroy data in targeted communications computers. The Army eventually awarded $50,000 contracts to two small software companies to examine the feasibility of such weaponry.

The Pentagon sees computer viruses as inexpensive yet potentially powerful ways to disrupt an enemy's operations. In the Persian Gulf War, intelligence agents intercepted a computer printer that was to be smuggled from Amman, Jordan, to a military facility in Baghdad. The printer was to be used in a computer network that controlled Iraq's air defense artillery batteries. The agents replaced a computer chip in the printer with another designed by the National Security Agency (NSA). The chip contained a computer virus that attacked and disabled the mainframe computer to which the printer was linked. Although it probably was not needed to knock out the Iraqi ADA batteries, this "attack" proved that inexpensive computer viruses could be very effective in disabling enemy computer networks.

THE FUTURE OF VIRUSES

Even though virus attacks can be quite devastating, a proper antivirus system can foil or allow recovery of lost data caused by a virus. The number of virus attacks are proportionally small in comparison to other types of data loss but, potentially, virus attacks on personal computers will probably increase in the future. This fact implies that virus protection must increase to ensure that data remains safe. Viruses are not invincible, and with the proper methods of prevention, the personal computer will be able to continue to perform vital tasks without interruption in the future.

With the steady increase of computer usage, one must also realize the growing potential for problems. The fact that one no longer must walk to a file cabinet to store information also means that an unauthorized person no longer needs a key to open the cabinet. There is now more than one way in. We must become aware of these vulnerabilities and take proper precautions. Passwords, encryption, and controlled access are just a few ways to reduce the chances of a virus entering a system. Each system is unique and must individually be evaluated to determine the needed security to protect itself and its data.

SUMMARY

Computer viruses and other types of "killer coding" are the current wave of misuse and abuse facing computer centers worldwide. Even though malicious code has been in existence for nearly 20 years, its recent, rapid growth indicates that there is a great deal of activity going on. A **virus** is a program that can "infect" other programs by modifying them to include a possibly evolved copy of itself. With the infection property, a virus can spread throughout a computer system or network using the authorization of every user using it to infect their programs. Every program that gets infected may also act as a virus, and thus the infection grows.

Viruses are designed to damage other programs, alter data, and self-destruct, leaving no trace of themselves behind. Numerous viruses (and strains of existing viruses) are being turned loose on systems. Each day new viruses emerge to play havoc with computer files and systems. The different types of viruses currently in existence are **memory-resident viruses**, **error-simulation viruses**, **data-targeted viruses**, **crasher viruses**, **computer time theft viruses**, **call-me viruses**, and **hardware-damaging viruses**. Some of the pseudo-viruses that are used separately or in combination with a virus for a particularly deadly result are **Trojan Horses**, **logic bombs**, **trapdoors**, **salami slices**, **phone phreaking**, and **worms**.

Numerous virus software products are available to detect, eradicate, and eliminate viruses. The user should be cautioned that an individual product may not remove the particular virus infecting an individual system, so care should be used to probe deeply so that all possible infections are detected and eliminated.

Viruses are definitely not confined to any geographic location and have been found in all parts of the globe, thus intensifying their exponential growth and spread.

All users of computer systems should pay particular attention to their systems, assume that the "worst could happen," and take measures to protect against any possible infection. Constant vigilance is necessary to be fully assured that your systems are "clean."

REFERENCES

1. "A Manager's Guide to Computer Viruses . . . Symptoms and Safeguards," Computer Security Institute, 1989.
2. Barnhill, William, "Privacy Invaders," *AARP Bulletin*, Vol. 33, No. 5, May 1992, pp. 1, 10.
3. BloomBecker, J. J. Buck, *Spectacular Computer Crimes* (Homewood, IL: Dow Jones-Irwin, 1990), pp. 1-242.
4. BloomBecker, J. J. Buck, J.D., Editor, "Commitment to Security," National Center for Computer Crime Data, 1989.
5. Bradley, Barbara, " 'Non-Techies' Could Be Hurt by Virus," *The Christian Science Monitor*, November 22, 1988, pp. 7-8.
6. Brothers, M. H., "A 'How-To' Guide for Computer Viruses Protection in MS-DOS," *Proceedings of the 12th National Computer Security Conference*, National Institute of Standards and Technology and National Computer Security Center, Baltimore, Maryland, October 10-13, 1989, pp. 349-358.
7. Cohen, Fred, "Computer Viruses: Theory and Experiments," *Computers and Security*, Vol. 6, No. 1, February 1987, pp. 22-25.

8. "Computer Security . . . Virus Highlights Need for Improved InterNet Management," United States General Accounting Office, Report to the Chairman, Subcommittee on Telecommunications and Finance, Committee on Energy and Commerce, House of Representatives, Washington, D.C., June 1989.

9. "Computer Viruses," American Management Systems, Inc., Arlington, Virginia, 1990, p. 3.

10. *Computer Viruses*, Proceedings of an Invitational Symposium, October 10-11, 1988, Deloitte Haskins and Sells, New York, 1989.

11. "Computer Viruses," *The Wall Street Journal*, April 4, 1989.

12. *Computers at Risk . . . Safe Computing in the Information Age*, Systems Security Study Committee, Computer Science and Telecommunications Board, Commission on Physical Sciences, Mathematics, and Applications, National Research Council, National Academy Press, 1991.

13. Cordani, John, and Douglas Brown, "Virus Ethics: Concerns and Responsibilities of Individuals and Institutions," *13th National Computer Security Conference*, National Institute of Standards and Technology and National Computer Security Center, Washington, D.C., October 1-4, 1990, pp. 647-652.

14. "Could a Virus Affect Military Computers?" *U.S. News and World Report*, November 14, 1988, p. 13.

15. Danca, Richard A., "Michelangelo Virus Eats Hard Disks," *Federal Computer Week*, February 17, 1992.

16. Danca, Richard A., "PC Security Under Sneak Attack," *Federal Computer Week*, December 2, 1991, pp. 35-39.

17. "Data Disk Viruses a Threat," *ISP News*, September/October 1991, p. 6.

18. Denning, Peter J., Editor, *Computers Under Attack . . . Intruders, Worms, and Viruses* (New York: ACM Press, 1990).

19. Dvorak, John C., "New Stealth Viruses: A Menace to Users," *PC Magazine*, April 14, 1992, p. 93.

20. Esters, Stephanie, "Avoid It Like the Plague: Protecting Yourself from Computer Virus Attacks," *Black Enterprise*, February 1989, p. 55.

21. Forcht, Karen A., "Virus Alert," *IBSCUG Quarterly*, Vol. 3, No. 4, Winter 1992, p. 11.

22. Forcht, Karen A., G. Bauer, Mike Donlan, Lisa Byrne, Beth Burlingame, and R. Howard, "Computer Viruses—Anatomy and How They Spread," *Computer Audit News and Developments*, Vol. 9, No. 3, 1991.

23. Gibson, Steve, "At Last, How to Protect Yourself From Polymorphic Viruses," *Infoworld*, Vol. 14, Issue 17, April 27, 1992, p. 36.

24. Glath, Raymond M., Sr., "Computer Viruses: A Rational View," RG Software Systems, Inc., 1988, pp. 1-4.

25. Glath, Raymond M., Sr., "PC Viruses . . . Reports from the Front Lines," RG Software Systems, Inc., May 1, 1990.

26. Green, James L., "The Father Christmas Worm," *Proceedings of the 12th National Computer Security Conference*, National Institute of Standards and Technology and National Computer Security Center, Baltimore, Maryland, October 10-13, 1989, pp. 359-368.

27. Greenberg, Ross M., "Know Thy Viral Enemy," *Byte*, June 1989, pp. 275-281.

28. Hafner, Katie, and John Markoff, *Cyberpunk* (New York: Simon and Schuster, 1991).

29. Hoffman, Lance J., Editor, *Rogue Programs . . . Viruses, Worms, and Trojan Horses*, (New York: Van Nostrand Reinhold, 1990).

30. Honan, Patrick, "Avoiding Virus Hysteria," *Personal Computing*, May 1989, p. 85.

31. Honan, Patrick, "Beware: It's Virus Season," *Personal Computing*, July 1990, p. 36.

32. "Host Busters," *Washington Post*, January 14, 1990, p. B3.

33. "How to Find and Whip Michelangelo," *U.S. News and World Report*, March 3, 1992, p. 19.

34. Hutt, Arthur E., "Information Security—Is There More Than Hackers and Viruses?," *Security and Audit News*, Vol. 89.1, 1989, pp. 25, 26, 29.

35. "Iraqi Virus Hoax," *Security Insider Hoax*, (Nashville, TN: Inter Pact Press, 1992).

36. Israel, Howard, "Computer Viruses: Myth or Reality?" *Proceedings of 10th National Computer Security Conference*, National Bureau of Standards, National Computer Security Center, September 21-24, 1987, pp. 226-230.

37. Joseph, Mark K., "Towards the Elimination of the Effects of Malicious Logic: Fault Tolerance Approaches," *Proceedings of 10th National Computer Security Conference*, National Bureau of Standards, National Computer Security Center, September 21-24, 1987, pp. 238-244.

38. Kenner, Hugh, "Print Queue," *Byte*, November 1990, pp. 466-467.

39. "KGB Hackers Get Off Lightly," *Nature*, Vol. 344, March 1, 1990.

40. Lickson, Charles P., "Those Nasty Little Viruses," *Tech Exec*, March 1989, pp. 46-48.

41. Lockhart, K.E., "Virus Task Force Stresses Preventative Medicine," *Information Technology Bulletin*, Commonwealth of Virginia, Department of Information Technology, Richmond, Virginia, Vol. 2, No. 3, Fall 1991, p. 4.

42. *Mad . . . Computer Virus Edition*, No. 75, Summer 1991.

43. McAfee, John, "The Virus Cure," *Datamation*, February 15, 1989, pp. 29-31.

44. McAfee, John D., "Managing the Virus Threat," *Computerworld*, February 13, 1989, pp. 89-96.

45. McAfee, John, and Colin Haynes, *Computer Viruses, Worms, Data Diddlers, Killer Programs, and Other Threats to Your System* (New York: St. Martin's Press, 1989).

46. McDonald, Chris, "Reflections on an Internet Worm," *ISSA Access*, 2nd Quarter, 1989, p. 12, 19.

47. Melka, Paul, "Wishful Thinking Will Not Make Publicity-Seeking Viruses Go Away," *Infoworld*, Vol. 14, Issue 17, April 27, 1992.

48. Menkus, Belden, "The Computer Virus Danger Grows," *Modern Office Technology*, February 1989, p. 38-39.

49. Morrison, Perry R., "Computer Parasites," *The Futurist*, March-April 1986, pp. 36-38.

50. Nordwall, Bruce D., "Rapid Spread of Virus Confirms Fears About Dangers to Computers," *Aviation Week and Space Technology*, November 4, 1988, p. 44.

51. Page, John, "An Assured Pipeline Integrity Scheme for Virus Protection," *Proceedings of the 12th National Computer Security Conference*, National Institute of Standards and Technology and National Computer Security Center, Baltimore, Maryland, October 10-13, 1989, pp. 378-388.

52. Pournelle, Jerry, "Dr. Pournelle vs. the Virus," *Byte*, July 1988, pp. 197-200.

53. Pozzo, M. M., and T. E. Gray, "An Approach to Containing Computer Viruses," *Computers and Security*, Vol. 6, No. 4, August 1987, pp. 321-331.

54. Riemer, Michael, "How to Manage the Two-Legged Virus," *ISSA Access*, 4th Quarter, 1989, pp. 32-34.

55. Robins, Gary, "When a Virus Strikes," *Stores*, January 1990, pp. 78, 83, 84.

56. Russell, Deborah, and G. T. Gangemi, Sr., *Computer Security Basics*, (Sebastopol, CA: O'Reilly and Associates, Inc., 1991), pp. 1-441.

57. Sampson, Karen L., "Computer Viruses: Not Fads, Not Funny," *The Office*, October 1989, pp. 56, 57, 59, 61.

58. Schlack, Mark, "How to Keep Viruses OFF Your LAN," *Datamation*, October 15, 1991, pp. 87-88.

59. Schneidawind, John, "PC Users Fear Damage from Rogue Virus," *USA Today*, February 13, 1992.

60. Schwering, David A., and H. Gerald McGuire, "I Am Virus, Hear Me Roar!" *ISP News*, September/October 1991, pp. 45-48.

61. Seymour, Jim, and Jonathan Matzkin, "Confronting the Growing Threat of Harmful Computer Software Viruses," *PC Magazine*, June 28, 1988.

62. Smith, Wade, "How Deadly Is the Computer Virus?" *Electrical World*, July 1988, pp. 35-36.

63. Spafford, Gene, "The Internet Worm Program: An Analysis," *Purdue Technical Report* (CSD-TR-823, November 29, 1988).

64. Stoll, Cliff, "An Epidemiology of Viruses and Network Worms," *Proceedings of the 12th National Computer Security Conference*, National Institute of Standards and Technology and National Computer Security Center, Baltimore, Maryland, October 10-13, 1989, pp. 369-377.

65. Stoll, Cliff, *The Cuckoo's Egg . . . Tracking a Spy Through the Maze of Computer Espionage* (New York: Pocket Books, 1989), pp. 1-356.

66. "Stopping a Plague," *Administrative Management*, March 1988, p. 20.

67. Stover, Dawn, "Viruses, Worms, Trojans, and Bombs," *Popular Science*, September 1989, pp. 59-63.

68. Sullivan, Kristina B., "Virus Policies Vary Widely," *PC Week*, December 9, 1991, pp. 33, 36.

69. "The Dirty Dozen—An Uploaded Program Alert List," maintained by Eric Newhouse, originally by Tom Neff, Issue #8, February 5, 1988.

70. Thimbleby, Harold, "Can Viruses Ever Be Useful?," *Computers and Security*, Vol. 10, 1991, pp. 111-114.

71. Thompson, Kenneth, "Reflections on Trusting Trust," *Communications of the ACM*, Vol. 27, No. 8, August 1984.

72. "Viruses Watch," Leprechaun Software International Ltd., Marietta, GA, U.S. Special Edition, 1991.

73. Wack, John P., and Stanley A. Kurzban, "Computer Virus Attacks," *NCSL Bulletin*, United States Department of Commerce, National Institute of Standards and Technology, August 1990, p. 1.

74. "Watchdog PC Data Security . . . Computer Virus Protection for Your PC," Fischer International Systems Corporation.

75. Welter, Theresa R., "Sick Computers," *Industry Week*, August 15, 1988.

76. Wiener, Daniel, "When a Virus Makes Your PC Sneeze," *U.S. News and World Report*, February 26, 1990, pp. 62.

77. Young, Catherine L., "Taxonomy of Computer Virus Defense Mechanisms," *Proceedings of the 10th National Computer Security Conference,* National Bureau of Standards, National Computer Security Center, September 21-24, 1987, pp. 220-225.

REVIEW QUESTIONS

1. Briefly describe the introduction and growth of viruses from the 1970s to the present.

2. Define a virus and explain how it enters a computer system.

3. What are some of the properties of known viruses?

4. What are the different categories of viruses?

5. Describe some of the different types of viruses.

6. Describe some of the pseudo-virus programs known to exist today.

7. What motivates people to write and plant viruses in computers?

8. Describe some of the known viruses that exist today.

9. How is virus software used to detect, eradicate, and destroy viruses?

10. What are some of the precautions that users can employ to lessen the threat and spread of viruses?

DISCUSSION QUESTIONS

1. Some hackers assert that there are "good viruses" that can merely play around with systems without the intent of destroying. Is there such a thing as a "good virus"?

2. When Robert Morris planted his Internet worm in 1988, the security community was aghast at the consumer movement that surrounded Morris and portrayed him as a hero for pointing out flaws in systems. In your opinion, was Robert Morris's action a blessing or a curse?

3. If your corporation has no set policy or action plan in place to detect and eradicate viruses, how would you go about trying to build momentum (and funding) for antivirus support throughout your organization?

4. How would you compare and contrast a virus and a worm? Is one more destructive than the other, or are they just "different streets leading to the same house"?

5. If you were assigned the task of selecting a virus product that would be installed throughout your organization, how would you organize the task, set criteria, and determine outcomes before ordering multiple quantities (or site licenses) of the software?

EXERCISES

1. Interview a computer laboratory assistant or an employee in your university's computer center to ascertain whether your university has experienced any "hits" by computer viruses.

2. Locate several articles in your university library relating to recent virus attacks.

3. Locate some virus software vendors by browsing through computer-related security journals and ask these software vendors to send you a demonstration (or full) copy of antivirus software for review.

4. Choose an individual in the corporate or industrial environment and interview him or her concerning what his or her organization is doing to ward against viruses.

5. Draw a chart, depicting how a virus enters a system and spreads. In contrast/comparison, draw a separate chart depicting how a worm is spread.

PROBLEM-SOLVING EXERCISES

1. Many corporations and organizations today are designing computer systems that will not allow any "unscreened" software to be loaded in any computer system throughout the organization. If an employee has a game, program, or miscellaneous piece of software he or she wishes to run on the computer in his or her office, it must first be run through a virus filter before it will boot into the system. This policy has met with mixed reviews by many employees. If you were to introduce this filtering policy to employees, how would you draw up a statement to explain the corporation's position? How would you distribute this statement throughout the organization?

2. Drawing on the analogy that a computer virus and a biological virus are similar in many ways, draft a diagram that would illustrate the likenesses of these two distinctly different viruses.

3. To show how a virus attacks the .EXE files and COMMAND.COM of computer systems, draft a chart showing the likenesses and differences in a virus attacking these two areas.

4. Draw a flowchart to show how a virus enters a system, moves throughout the system, and then attacks at a given time or activity. Your flowchart should depict a visual representation of your particular virus. Ask a class member to describe what your virus is doing.

5. Compare and contrast how viruses perform in a stand-alone PC environment, a mainframe environment, and a network (or shared file) environment. What are the likenesses in all three systems and the unique characteristics that each of the three systems have?

CASES

University of Wollongong

Dr. Graham Winley, professor in the Business Information Systems Department at the University of Wollongong, New South Wales, Australia, would like to introduce the basic elements of computer viruses to his Management Information Systems classes. His main objective is to train the students in good computer practices to thwart spreading viruses and to make them fully aware of the dangers that viruses pose. Dr. Winley feels that he may be dealing with a two-edged sword as:

1. Too much knowledge can be a dangerous thing as students may become intrigued and experiment with planting viruses on university or corporate computers.

2. Too little knowledge produces ignorance and is certain to result in havoc. Students should not be kept in the dark and find out about viruses the hard way when a "hit" occurs.

Determine the best method of introducing the virus knowledge to the students so that they are aware of the possibility of these malicious programs, yet are not tempted to experiment with viruses.

Roderick Design Systems

Allison Roderick, president of Roderick Design Systems, located in New Braunfels, Texas, has been reading in newspapers and magazines and hearing on TV about the spread of viruses throughout computer systems. Her firm is an organization that distributes unique office layout designs for corporate offices in the San Antonio and Austin, Texas, area. In most cases, the designs are custom-fit to each client and take a great deal of time and effort to create. The client usually sets strict completion deadlines, so time is of the essence as the project nears completion. (Otherwise, RDS must assume a substantial completion penalty.) Ms. Roderick does not want the "messiness" and inconvenience of a virus entering the organization's computers and designs, shutting down the system or altering data in any way. She has decided to call in a well-known computer security expert as a consultant to assess the organization's current risk and to set up a workable virus protection program.

What criteria should Ms. Roderick adopt in working with the consultant so that the situation can be assessed quickly, efficiently, and at a reasonable cost?

PART

IV

LEGAL AND
ETHICAL ISSUES

LEGAL ISSUES AND CURRENT LEGISLATION

LEARNING OBJECTIVES

After studying this chapter, you will be able to:

1. Define computer crime and compare it to other types of crime.
2. Explain the various methods that are used by computer criminals.
3. Describe the types of computer crimes that are committed.
4. Develop a profile of the typical computer criminal.
5. Explain why computer crimes are committed by internal employees and external criminals.
6. List the various management controls that can be used to deter and detect computer crimes.
7. Explain the various software violations that occur.
8. Describe software piracy and its far-reaching impact.
9. List the types of crimes against computer systems.
10. Summarize some of the current computer crime legislation.
11. Consider the issue of privacy and how it relates to computer crime.
12. Define the following terms.

TERMS

- Computer Crime
- Computer Abuse
- Computer Fraud
- Data Diddling
- Trojan Horse
- Salami Technique
- Superzapping
- Trapdoors
- Logic Bombs
- Asynchronous Attacks
- Scavenging

- Data Leakage
- Piggybacking
- Impersonation
- Wiretapping
- Simulation
- Modeling
- Telecommunications Fraud
- Embezzlement
- Hacking
- ATM Fraud
- Records Tampering

TERMS (Continued)

- Disgruntled Employees
- Child Pornography
- Drug Crimes
- Organized Crime
- Boiler Room Operation
- Patent
- Copyright
- Trade Secrets

- Unfair Competition
- Theft
- Fraud
- Sabotage
- Espionage
- Wiretapping/Bugging
- Terrorism
- Privacy

SECURITY BREACHES: WHO IS LIABLE?

Computer systems security is a crucial factor in today's business and governmental organizations. But the situation is heating up to the point that failure to adequately protect corporate computer systems against hacker attacks, viruses, and other security breaches could lead information systems managers into the courtroom—as defendants.

Sanford Sherizen, president of Data Security Systems, Inc., a computer crime prevention firm in Natick, Massachusetts, likens today's computer security situation to a famous late 1980s environmental disaster, saying, "Corporate and governmental computer systems are like the *Exxon Valdez*—unprotected, under inadequate leadership control, operating through dangerous channels, and loaded with valuable and messy stuff." Like the captain of the *Valdez*, information systems managers are responsible for the proper securing and handling of corporate data and networks, and when security is violated, they must answer to that responsibility.

"Corporate exposure and vulnerability have outstripped the rate of acceptance and proliferation of computer systems and networks," says Kenneth Weiss, chief technical officer of Security Dynamics, Inc., and chairman of the computer security division of the American Defense

Continued

SECURITY BREACHES: WHO IS LIABLE?

From page 295

Preparedness Association. "If [senior management] really understood the potential risks to corporate assets and to their reputations, they might shut down all networks and computer centers."

What has some computer experts nervous is that there are relatively few IS security standards in place today, even though experts agree that the prospect of a serious computer security breach—one that causes a company to go bankrupt or that otherwise harms its employees or customers—is likely to increase. The situation is complicated by the proliferation of distributed information systems and the networks that interconnect them. Networks have made corporations more competitive while simultaneously making them more vulnerable to attack by disgruntled employees, hackers, and criminals. The distribution of computing power to end users also brings with it a distribution of responsibility for computer security, a fact that is not understood or readily accepted by end users.

As reported in Alexander, Michael, "Lax Security Invites Liability Nightmare," *Computerworld*, March 26, 1990, pp. 1+.

Now that you have studied the security issues of databases, networks and telecommunications, microcomputers, and viruses, Part IV will introduce two very relevant and important topics having a great impact on security and control—legal issues and ethical use of computers. These two issues represent the "people side" of security that, coupled with the technical perspective, will give a complete picture of the complexity of security.

DEFINING COMPUTER CRIME

Crime usually does its ingenious best to keep pace with technology, so it is probably inevitable that some of the more interesting specialized manifestations of criminality to emerge recently have had to do with deliberate misuses of the computer...computers have driven some people who work with them to crimes that have taken on the quality of ideological acts.[57]

In recent years, computers have become an integral part of our society. Computers speed up operations, simplify tasks, organize information, calculate numbers, and generally aid us in processing the large volumes of data that are essential to the operation of our organizations. Conversely, computers can also confuse, intimidate, and ultimately be abused. The integration of computer systems into our business world and personal lives has occurred so quickly that many people do not fully appreciate the benefits of automation. Most people do not fully appreciate the potential for harm that computers can generate. As is usually the case with many great inventions, the computer has the potential for good, coupled with its potential for harm.

Basically, **computer crime** can be defined as the violation of a computer whereby a victim suffers from the wrongful intentions of a perpetrator. To understand the term fully, certain aspects must be examined and explained. Computer crime, being a relatively new concept, has emerged as the result of increased use of computer systems for storing data. As organizations increase their use of and dependence on computers, the potential for harm or misuse increases also. The National Center for Computer Crime Data, in its publication entitled "Computer Crime, Computer Security, and Computer Ethics," states that by increasing "the number of people with access to computers, you increase the number of potential criminals. The portion of our labor force involved with computers has been growing for quite some time. It gives no indication of starting to stop."[12] When an organization stores all its important data in computer data banks and does not properly safeguard this vital company asset, then the potential for vulnerability is paramount. All types of organizations that use computers are potential victims of computer crime. Statistically, though, banks appear to be the hardest hit—possibly due to the immediate cash advantages of "ripping off" a financial institution. Experts believe that the large amounts of money and transactions in the banking industry make it the most likely target for computer crime—a modern day Bonnie-and-Clyde scenario.

When the typical computer crime is compared to to other crimes, the financial losses can be staggering. (See Figure 10.1.)

FIGURE 10.1
Comparing Computer
Crime to Other
Major Crimes

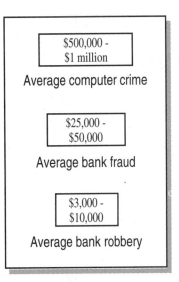

$500,000 -
$1 million

Average computer crime

$25,000 -
$50,000

Average bank fraud

$3,000 -
$10,000

Average bank robbery

Computer crime usually takes on three forms:

1. Crime *supported by* the computer, where the computer is the *tool or instrument* used to commit the crime.

2. Crime *because* of the computer, where the computer is the *object* of the crime (e.g., theft of equipment or information).

3. Crime *around* the computer, where the computer is the *target* of the crime (e.g., physical attack on computer, sabotage, espionage)

The Federal Bureau of Investigation (FBI) defines computer crime as "prohibited acts which violate criminal laws under existing statutes where we have jurisdiction and where the computer and/or its peripheral equipment are the instruments of the victim of the crime."[22]

The FBI says computer crime may involve illegal acts directed against a computer and its peripherals, or the use of a computer in illegal activity. "We've found that most so-called computer crimes are nothing more than traditional crimes using the computer as the instrument—much like a gun, a knife, or a forger's pen," states Floyd Clarke, inspector-deputy assistant director of the FBI's Criminal Investigation Division in Washington, D.C.[22]

Two other closely connected terms that are often used to refer to computer crime are computer abuse and computer fraud. As defined by the National Institute of Justice, Washington, D.C., **computer abuse** encompasses a broad range of intentional acts that may or may not be specifically prohibited by criminal statutes. Any intentional act involving knowledge of computer use or technology is computer abuse if one or more perpetrators made or could have made gain and/or one or more victims suffered or could have suffered loss.

Computer fraud is any crime in which a person "may use the computer either directly or as a vehicle for deliberate misrepresentation or deception, usually to cover up the embezzlement or theft of money, goods, services, or information."

METHODS OF COMPUTER CRIME

Determining just who will become the criminal is not always possible, but experts do agree on what methods are used. The most common methods are the following:

- Data diddling
- Trojan Horse
- Salami technique
- Superzapping
- Trapdoors
- Logic bombs
- Asynchronous attacks
- Scavenging
- Data leakage
- Piggybacking and impersonation
- Wiretapping
- Simulation and modeling

Data Diddling

Data diddling can be defined as the unauthorized changing of data *before* or *during* input into a computer system. Changes can be done by *anyone* who has access to the data. Diddling is the safest, simplest, and most common method of computer crime used today. An example could be a timekeeping clerk who completes data forms of hours worked by employees. The computer, internally, processes only numbers; externally, all records are kept by name. The clerk fills out overtime cards on employees who frequently work overtime hours, yet inputs the hours to his own employee number when entering the data into the computer, thereby increasing his income several thousand dollars each year.

Trojan Horse

The **Trojan Horse** method utilizes computer instructions secretly inserted in a computer program to perform unauthorized acts when the program is executed. This method tends to be the most common used for intentional computer-based fraud and sabotage. The method is fairly "clean" because instructions can be inserted and removed so that no evidence remains.

Salami Technique

The **salami** can be defined as the unauthorized, covert process of taking small amounts (slices) of money from many sources with the aid of the computer—one of the most popular being the "round-down" fraud. For example, when figuring interest rate calculations, the fraction of one cent that is normally rounded up or down to the nearest cent is put into the criminal's own account. Rather than distributing the rounded down remainders to the other accounts as they build up, the amount is accumulated in a separate account. The accounts will still balance, but over a period of time a sizable amount can be accrued.

Superzapping

Superzapping can be defined as the unauthorized use of utility computer programs that violate computer access controls to modify, destroy, or expose data in a computer. The term is derived from superzap, which is a macro-utility program used in most IBM computer centers as a systems tool. A universal access code is needed in a computer for various reasons and can be compared to a master key, which is used when other keys are lost. A good example of superzapping occurred in New Jersey when a computer operations manager was using a superzap procedure to make changes to account balances to correct errors as dictated by management. He soon learned that he could use this superzap process to his personal benefit and transferred funds to three friends' accounts. He was eventually caught only because a customer's complaint led to investigation.

Trapdoors

Trapdoors can be defined as functions, capabilities, or errors in a computer program that facilitate compromise or unauthorized acts in a computer system. In developing computer programs, it is necessary to insert debugging aids that provide breaks in the program for insertion of additional code and intermediate output capabilities. Trapdoors then make modification of the program easier; they are usually eliminated at the end of the program. When these trapdoors are used for unauthorized purposes, they are referred to as negative specifications. Recently, this method was discovered in a commercial time-sharing service. It was able to use large amounts of computer time free of charge and obtain data and programs of other time-sharing users.

Logic Bombs

Logic bombs are usually placed in the computer using the Trojan Horse technique. A hypothetical example could be a payroll systems programmer who puts a logic bomb in a personnel system so that if his name were ever removed from the personnel file, indicating termination of employment, a secret code would cause the entire personnel file to be erased.

Asynchronous Attacks

Asynchronous attacks can be defined as those that take advantage of the asynchronous structure of computer operating systems to perpetrate an unauthorized act. A computer normally performs requests asynchronously, based on resources available rather than in the order they are received. Asynchronous attacks are highly sophisticated methods of confusing the operating system so that it allows the isolation of one job from another to be violated.

Scavenging

Similar to a scavenger hunt where random, loosely defined items are collected, **scavenging** is a covert, unauthorized method of obtaining information that may be left in or around a computer system after the execution of a job. This includes both a physical search and a search for residual data within the computer storage areas.

Data Leakage

Data (or information) **leakage** can be defined as the unauthorized, covert removal or copying data from a computer system (i.e., sensitive data that may be hidden in otherwise innocuous-looking reports). There are other ways to leak data such as interspersing data, encoding data to look like something different than they are, or controlling and observing the movement of equipment parts. All of these acts are deliberate and are not unintentionally or haphazardly performed.

Piggybacking and Impersonation

Physical **piggybacking** is a method of gaining unauthorized physical access to guarded areas when control is accomplished by electronically or mechanically locked doors. Electronic piggybacking occurs when a computer or terminal covertly shares the same communication line as an authorized user. **Impersonation** occurs when one person falsely assumes the identity of another. An example of impersonation has been reported in which an individual stole electronic funds transfer (EFT) cards and then called the owners, posing as a bank official, requesting their personal identification numbers (PINs) associated with each card. The card owners were led to believe that the bank was attempting to prevent the thief from gaining access to their accounts.

Wiretapping

Interception of data communication signals with the intent to gain access to data transmitted over communications circuits is known as **wiretapping**. At the present time, there are few instances of actual, verified experimentation of this method, but as more and more computer systems are tying into networks, wiretapping's potential could become the newest "nightmare" of computer crime. Wiretapping requires expensive equipment, which tends to limit its potential. With the prices of modems and communications equipment becoming ever lower, this method of theft could become within nearly every criminal's reach all too soon. (Wiretapping and bugging will be covered later in this chapter in more detail.)

Simulation and Modeling

Both of these techniques are somewhat sophisticated by design, but they are well worth the efforts of the ardent, dedicated criminal. **Simulation** attempts to create (or simulate) as closely as possible a real-life situation (e.g., a company's operation). **Modeling**, somewhat different in design, attempts to recreate (usually in somewhat smaller form) a real entity (e.g., robots, model airplanes). In both of these methods, the computer is used as a tool for planning, controlling, and monitoring a crime. There have been several documented cases in which the computer was used in insurance companies to "model" their operations and determine the effects of the sale of large numbers of insurance policies—a kind of "what if" game. By playing around with the numbers and creating hundreds and thousands of fake insurance policies, companies were virtually wiped out when benefits were paid on nonexistent policies or premiums were paid only "on paper" on supposedly valid policies.

TYPES OF CRIMES COMMITTED

Using some of the methods just stated, computer crime takes on many forms. According to the National Institute of Justice, some of these crimes include telecommunications fraud, embezzlement, hacking, ATM fraud, records tampering, acts of disgruntled employees, child pornography and abuse, drug crimes, and organized crime.

Telecommunications Fraud

During the 1970s, some individuals defrauded phone companies by using stolen account numbers or reversing charges to pay phones. **Telecommunications fraud** is defined as avoiding paying telephone charges by misrepresentation as a legitimate user. A Philadelphia company's voice-mail system was infiltrated when callers from Arizona used stolen Sprint codes to tap into the voice-mail system and reprogram it to exchange information regarding their own illegal business. The callers made hundreds of dollars' worth of illegal Sprint calls and blocked company employees from using the voice-mail system.

Embezzlement

Embezzlement involves using the computer to steal or divert funds illegally. Embezzlement, one of the oldest white-collar crimes, is now considered easier to accomplish with the speed and impersonalization of the computer as financial transactions can occur entirely within the computer system. An embezzlement in Denver, Colorado, involved theft from a brokerage firm by one of the firm's agents. The crime involved changing a cash account into a margin account, altering the symbol that represented shares owned, and moving a decimal point, resulting in $178,000 being stolen. In another case, a Philadelphia department store employee used the store's cash register, which was linked to the computer system, to clear her own charge account and those of some of her friends.

Hacking

Hacking denotes a compulsive programmer or user who explores, tests, and pushes computers and communications systems to their limits—often for illegal activities.

Hackers often misuse telephones, become involved in stolen credit card numbers, arrange for mail-order purchases for delivery to friends or vacant homes, access on-line subscription service and avoid paying the charges, snoop through files, destroy or alter files, and generally cause havoc to legitimate business activities.

Automatic Teller Machine Fraud

Automatic teller machine fraud involves using an ATM machine for a fraudulent activity—faking deposits, erasing withdrawals, diverting funds from another person's account through stolen PIN numbers. ATM fraud is rapidly replacing robberies and burglaries as a method of gaining some "quick cash."

Records Tampering

Records tampering involves the alteration, loss, or destruction of computerized records. Although information has always been regarded as valuable, computers have made tampering much easier than a physical break-in to a file cabinet or person's desk. Because of the abundance and accessibility of computer records, everyone is vulnerable to having their credibility and finances destroyed by individuals using computers.

In 1985, a Golden, Colorado, district attorney was prosecuted for having another public official change his driving record to remove speeding tickets.

Acts of Disgruntled Employees

Disgruntled employees often use a computer for revenge against their employer. Small businesses are particularly prone to these acts as their computer systems do not have adequate safeguards and many employees have the knowledge to commit the crimes. In one small business, an employee changed all the employee passwords, potentially affecting the production of W-2 tax statements. Later, the former employee demanded a ransom for the altered passwords. In another case, a former employee, in retaliation for his dismissal, channeled calls to shut down the phone system in all of the company's big accounts.

Child Pornography and Abuse

Illegal or inappropriate arts of a sexual nature committed with a minor or child, such as photographing or videotaping, can be considered **pornography** or abuse. Police officers in Philadelphia began monitoring publicly available bulletin boards after receiving information that the boards were being used to transmit information on child pornography. Computers may also be used more directly in crimes related to child sexual abuse. Often, the adult offender uses a home computer and electronic bulletin board to develop relationships with minor children. At times, in an effort to silence difficult victims, offenders will use computer networks to transmit threats of physical harm.

Drug Crimes

Drug offenders are now using computers to further their business deals. Drug dealers use computers to communicate anonymously with each other and to keep records of drug deals. In a recent case in Florida, federal agents raided a motel room obtaining evidence of a multimillion dollar drug smuggling ring. One piece of confiscated information, a slip of paper listing two names, led to the discovery of two individuals who were IBM employees hired by the ring to help computerize the operation.

Organized Crime

Of growing concern is the role of computers in furthering the ventures of **organized crime**. Reported cases include the use of computers in connection with organized prostitution, pornography, fencing goods, money laundering, and loan sharking. In a typical scheme known as a **boiler room operation**, criminals organize a fake company. They use a computer to uncover long-distance phone codes and make calls to companies and individuals (generally out of state), offering unusual discounts on products of interest. Using credit cards, the recipients of the calls place orders for the fictitious products. After the boiler room operators process the payments, they close down operations and move on—never sending the merchandise.

Typical Computer Criminals

Computer criminals fall into distinct categories, each with differing characteristics, motives, and methods for carrying out their crimes. Employees represent the most significant type of computer criminal. Because they have the greatest opportunity to harm the system and gain secret information, employees will continue to be the greatest threat. The number of "whiz kids," or young people committing computer crimes, has been increasing as well. These young people show potential to become great computer criminals in the future. Other computer lawbreakers include "traditional criminals" who have expanded into computer crime. Irrational individuals who perform destructive acts can be called deranged computer criminals. Still, some criminals perform wrongdoings for a specific cause or ideal in which they believe. Looking more closely at the different types of computer criminals, one can gain a better understanding of the problem and the steps that can be taken to combat it.

Novices. Most computer crimes are performed by novices. Usually these beginners in crime are computer professionals that discover, in the course of their job, a way to gain entrance to the system and choose to take advantage of this unauthorized access. Such "rookies" of computer crime are not psychotic criminals, but simple trespassers. These amateurs do not rely on crime as a means for living, but often have problems, such as gambling debts, that they feel can be solved by using their job to gain access to cash or other valuables.

Hackers. The next category of computer criminals is made up of hackers, crackers, and "whiz kids." These "unauthorized system explorers" have emerged due to advances in computer technology. A hacker is typically male, is between the ages of 12 and 19, and interacts with computers rather than other people. A hacker is often addicted to computer antics and seeks recognition for achievements by accessing a computer system.

In some cases, parents of these "whiz kids" are themselves computer professionals. These parents are aware of the power of the equipment their children have, yet they fail to censor or monitor the use of the computer.

Traditional Criminals. An increasing number of computer violators are "traditional criminals" who have expanded into the field of computer crime. An example is a purse-snatcher who uses stolen access cards to steal from automated teller machines. Another example is a burglar who takes a boxful of computer components.

Career Criminals. Career criminals make up another category of computer criminals. They are usually repeat offenders who began as computer professionals. They found a flaw in the system and realized the tremendous payoffs of the crime. These offenders usually live off the rewards from their crimes for substantial periods. Often, their friends and families admire these criminals for their violations.

Deranged Computer Criminals. Deranged computer criminals are psychotic or irrational individuals who perform destructive acts against computers. These unstable people are hard to predict because of their state of mind. An example of such a criminal is the deranged computer operator in the Southwest who entered a computer center one evening armed with a shotgun, pistol, and rifle and wounded the receptionist and shot the computer.

Extreme Advocates. Finally, there are computer criminals who are considered extreme advocates. These criminals are dedicated to causes or ideals for which they commit violations of the law against others or society as a whole. Among the causes these extreme advocates commit computer crimes "for" are political, economic, religious, and human rights issues.

The "typical" computer criminal is between the ages of 18 and 46, intelligent, highly motivated, and often overqualified for his job. Three-quarters of computer criminals are employees of the organization they abuse. Most of these criminals gain enough knowledge from their on-the-job training to commit the computer crimes. Overall, it is the overqualified, bored, or vengeful employee who is the organization's greatest vulnerability. (See Figure 10.2.)

FIGURE 10.2
Profile of a
Computer Criminal

- ☑ Functional end user, not the technical type and not a hacker
- ☑ Holds a nonsupervisory position
- ☑ No previous criminal record
- ☑ Bright, motivated, desirable employee
- ☑ Works long hours; may take few vacations
- ☑ Not sophisticated in computer use
- ☑ The last person you would suspect
- ☑ Just the person you would want to hire

Why Computer Criminals Commit Crimes

Criminals have different perceptions of the computer before they decide to commit the wrongdoing. (See Figure 10.3.) The computer can be perceived as a playpen, land of opportunity, cookie jar, war zone, soap box, fairyland, or tool box. A criminal commits a crime for a reason related to his or her perception of the computer. For example, usually those who perceive the computer as a "playpen" commit a crime because it is satisfying and challenging. On the other

FIGURE 10.3
Why People Commit
Computer Crime

- ☑ Personal or financial gain
- ☑ Entertainment
- ☑ Revenge
- ☑ Personal favor
- ☑ Beat the system, challenge
- ☑ Accident
- ☑ Vandalism

hand, a criminal perceiving the computer as a "cookie jar" commits a crime because of his need for money. These different perceptions also lead to different types of computer crimes. A criminal who sees the computer as a "soap box" will physically damage the computer, whereas a criminal who perceives the computer as a "war zone" will use the computer to get back at the company (by deleting files, for example). Because these varied perceptions lead to different types of computer crimes, security measures against such perpetrators differ.

Generally speaking, computer criminals, the "bloodless crime perpetrators," are ordinary people with a problem to solve who are in a position of trust. Many of the same characteristics that make them highly desirable as data processors also make them fit the profile of a computer criminal. Employers must be aware of this "matching" and use rigid pre-employment testing, as well as periodic reviews about job satisfaction. Job dissatisfaction has been shown to be a strong motive for computer crime, either out of boredom or out of revenge. The criminals will often rationalize that they aren't hurting any individual, only a large, impersonal company.

The employees who seem to have the greatest ability, due to their positions, in covering up their crimes are:

- Programmers
- Managers
- Bank clerks

Attempting to determine who will become the next criminal is not an easy task. It does not always hold true that only those with easy access or computer knowledge will be most intent on breaking into the system. There seems to be no single answer, so other methods of prevention will need to be employed.

Playpen. A person who sees the computer as a "playpen" feels it is a place to play computer games for as long as he or she likes. To those people playing the games, computers can be quite satisfying and challenging. However, this need for satisfaction can lead to violation of others' rights. Examples of such a case occur when a person obtains computer time without paying for it or penetrates a computer's security system. Computer crime results when the "playpen" perception gets out of hand. The satisfaction one gets from working with the computer comes into conflict with others' rights to use the system, thus resulting in a computer crime. Often, it is the hacker or "whiz kid" who perceives the computer as being a "playpen."

Land of Opportunity. Some criminals feel the computer is a "land of opportunity." He or she feels there is nothing wrong with exploring a computer system's flaw. These people are motivated by the lack of challenge. He or she has come across a vulnerability in the system through the course of a job. An example of a criminal finding the computer a "land of opportunity" is the operator of a check-printing device. He presses the repeat button on the computer as his check is being produced and ends up cashing twelve paychecks. This criminal is not a computer genius, just a person who takes advantage of the opportunity the computer offers him.

Cookie Jar. The criminal who feels the computer can solve his/her problems of gambling debts, drug habits, and the like sees the computer as "a cookie jar." The criminal takes what he needs from this "cookie jar." These criminals are motivated by the pressures of needing money, rather than taking advantage of a flaw in the computer system. An example of a criminal who saw the computer as a "cookie jar" was a head teller at a New York City bank. His gambling debts led him to steal $1.4 million from the bank to support his $30,000-a-day gambling habit.

War Zone. Some computer criminals see the computer as a "war zone" or a symbol of their problems. The computer serves as the field of a battle between employees and their employer. An example of such a criminal is a disgruntled employee who feels management is out to get him or her and uses the computer to attack the company. Employees may attack by erasing or misfiling tapes and attacking the computer with guns. A specific example of such a case occurred in Sacramento, California. An argument over pay raises triggered three employees of the State Department of Justice to delete some arrest records from California's system of criminal justice records. Another case in which the computer symbolized problems for the criminal involved an employee of Houston's Western Geophysical Company of America. He erased nearly 1,000 computer files just before leaving the building. By deleting those files, which were used daily by the company, the employee was able to get back at the company through the computer system.

Soap Box. Some criminals feel the computer itself is their enemy or "soap box" and thus release their anger toward the machine. The war is not between just the employer and the employee, but a number of forces are involved. These crimes are usually described as direct attacks on the computer itself. Examples include attacks on the machine with guns, bombs, and metal objects. One such case involved a programmer who had an urge to shut the computer down, so he short-circuited its memory. Another case involved a newspaper reporter from the *Baltimore Sun* whose stories got lost when the computer center went down. The reporter smashed the computer circuit box with a steel tool. After opening the circuit box, he ripped out the wires. In this crime, the reporter physically attacked part of the computer system to act out his rage. These crimes are also performed by non-employees. The best means of prevention in these cases relate to the physical security of the computer center.

Fairyland. Some criminals act as though the computer environment is a "fairyland" and not real. These cases usually involve employees who do not realize the computer system's power. They act as though computer crime is impossible; these cases often involve crimes of the heart. One such case involved a woman whose boyfriend convinced her to transfer $2 million to his bank account. The boyfriend told her that he wanted to play a joke on a computer operator at the bank. The money and the friend vanished after the transfer of funds was complete. In another case, Patricia Ferguson, an associate of Stanley Rifkin, pled guilty to the attempted theft of $50 million. She did it out of her affection for Rifkin, not because of any motivation to commit a computer crime.

Tool Box. Some criminals use the computer as the "tool" of a crime, not the target. This criminal can be called the "technological crook." He or she sees the computer as a powerful machine that can be used to commit a crime. The people who view the computer as a "tool box" are more likely to be criminals than those mentioned above. One such case involved employees of Equity Funding who used computers to create thousands of false insurance policies. Another involved a Los Angeles brothel owner who used a minicomputer to keep track of his customers.

Preventing Computer Crime

Computer crime cannot be totally eliminated, but it can be substantially reduced if correct preventive measures are intact in an organization. Because nearly 75 percent of all crimes are committed by insiders, organizations should concentrate their efforts primarily on this group by raising the security awareness of the employees and the organization as a whole. Some of the clues that computer crime may be taking place are shown in Figure 10.4.

FIGURE 10.4
Signs of
Computer Crime

☑ Unauthorized use of computer time

☑ Unauthorized use of or attempts to access
   data files

☑ Theft of computer supplies

☑ Theft of computer software

☑ Theft of computer hardware

☑ Physical damage to hardware

☑ Data or software destruction

☑ Unauthorized possession of computer disks,
   tapes, or printouts

Raising employees' awareness of computer security is the most important step; employees must be convinced that they need to be security conscious. Workers must come to realize that since security is good for the company, it is good for them. A good security awareness program will call for a statement of the company's confidentiality policy to be included in the employee handbook. Often, employees will be required to sign such a statement upon being hired. The company must clearly state in the policy that it will discipline violators through prosecution, termination, or some other means. This statement will reiterate the company's stringent policy concerning computer security.

In order to convince employees that it is best to be security conscious, the organizations must make computer security a part of everyday life. This can be accomplished by making the employees interested in security through posters, publications, exhibits, and speakers. Programs that teach employees about safety in their daily lives (for example, self-defense classes) are a good way to spark their interest in security. These programs will improve employee involvement in security and security awareness, which will result in better security in the long run by recognizing and rewarding those employees who have contributed to improved security.

If a crime does occur, the company must take a stand and do something about it. This is a surefire way to convince employees that computer security is a serious issue. However, when cases involve employees, many companies will not prosecute. Instead, the criminal is fired and publicity is avoided. In such cases, it is imperative that the employee be escorted off company grounds as soon as possible. Letting an employee stay around will only give him or her the extra time and incentive needed to harm the computer system.

An organization can take other preventive measures to reduce the problem of computer crime. As an initial step, the personnel department must carefully screen all applicants before they are hired. Recruiters should check references and previous employers to make sure the candidate is who he or she says he or she is and has not broken the law while employed elsewhere. Another step an organization can take is to rotate employees who are involved in sensitive areas—an effective way to reduce temptation. Because one employee will not be the only one in charge of this sensitive material, a control of sorts is placed on the employee. A company can use passwords and user logs as a measure against computer crime. Both of these must be used correctly to be effective. Employees must keep their passwords secret and change them periodically. User logs can track who accessed the system and when. These logs are used primarily to check for any irregularities in an employee's behavior that may warrant suspicion. Also, access to the system can be limited so an employee can retrieve only what is related to his or her job. Therefore, an employee sees only what he or she needs to see to work. This will minimize an employee's temptation to browse through the system. Log-on audit trails are another means of tracking the users' actions in the system. A company should include these basic precautions as a part of the computer security policy in their fight to combat computer crime in the workplace. (See Figure 10.5.)

FIGURE 10.5
Commitment
to Security

> Controls to be implemented should include:
> - ☑ Clear procedures and rules in regard to who will be permitted to use information and what information they will have access to
> - ☑ Written sanctions that would be followed if these procedures or rules are broken
> - ☑ Better internal controls, audit trails, separation and rotation of duties
> - ☑ A close eye on computer employees and education with respect to the security of information

SOFTWARE VIOLATIONS

Computer programs represent a considerable investment on the part of developers who want to reap rewards from their creative efforts. The economic value of programs and databases arises from the ability to exploit their uniqueness either by being the exclusive user or by marketing them and securing payment from others for their use. Various systems of legal protection for computer programs have been used, with varying degrees of success, to ensure these rewards for the developer of software.[37]

Figure 10.6 summarizes the legal protections available to software developers.

FIGURE 10.6
Legal Protection
of Software

☑ Patents
☑ Copyrights
☑ Trade secrets
☑ Unfair competition

Patent Protection

A **patent** protects inventions, not works of the mind. Early procedures involved obtaining patents for new, useful, and unobvious inventions. The United States Constitution expressly provides for patent rights for this purpose in Article I, Section 8, which states, "The Congress shall have power . . . to promote the progress of . . . [the] useful arts, by securing for limited times to . . . inventors the exclusive right to their . . . discoveries."[37] As a consequence of this constitutional mandate, patents have been issued in the United States since 1790 and Congress has, in its latest statutory implementation, enacted the Patent Statute of 1952, found in the United States Code as Title 35. Patents are obtained from the U.S. Patent and Trademark Office. Patents are usually issued for a 17-year life, thus enabling the patentee to prevent others from competing in the manufacture, use, or sale of the novel contribution provided. A patent is designed to protect the *device or process* for carrying out the idea, not the *idea* itself.

Patent infringement usually involves three major legal points:

- The infringer will claim that the two inventions are satisfactorily different.
- If a prior infringement was not opposed, the patent rights may no longer be valid.
- The infringer may try to persuade the judge that the Patent Office acted incorrectly in granting a patent and that the invention is not worthy of patent.[42]

Figure 10.7 summarizes patent infringements.

FIGURE 10.7
Patent Infringement

☑ Products are not the same
☑ Invalid patent
☑ Invention not novel

Some of the computer software that are patented are unique algorithms used in software. Because of the time and effort involved, however, in obtaining and maintaining a patent, this form of protection may be unacceptable for a small-scale software writer.

Copyright Infringement

Copyrights protect the expression of ideas and apply to creative work, such as stories, photographs, songs, and computer programs. The Copyright Law foundation has the same Constitutional foundation as does the Patent Law (Article I, Section 8, Clause 8) of the U.S. Constitution. The Copyright Law states that:

> The Congress shall have the power to promote the Progress of Science and useful acts, by securing for Limited Times to Authors and Inventors the exclusive Right to their respective Writings and Discoveries.[37]

Even though the source for copyright and patent protection is the same, the protection that Congress has sought to grant to each is considerably different. Copyright law was originally developed in response to the desire to protect property interests in books, works of art, musical compositions, etc. The new Copyright Act, passed in 1976 and now in full effect, clearly implies that programs are copyrightable as "literary works."[37]

The two major questions to be addressed relating to the copyright of computer programs are:

1. What precisely does the copyright protect if a program is copyrighted?
2. Is copyright an appropriate and practical form of protection for software even if it is available?

The basic dilemma is reiterated by the legal community as *Copyrights* protect the *expression* of the idea, not the idea itself.

> If one views programs as more similar in nature to inventions than to other forms of literary works, copyright protection is inadequate since protection against unauthorized use is what the owner seeks, thereby protecting not only the form of expression but the idea.[37]

Notice of copyright is generally required on all published, visually perceptible copies of a work. The Copyright Office's regulations specify the placement of notice on machine-readable copies such as magnetic tapes or disks or on the inside flap or first page of a book or other published document. A copyright application must be officially filed with the U.S. Copyright Office, along with a nominal fee and a copy of the work. The filing must be done within three months of the first distribution of the work. A copyright now lasts for 50 years beyond the death of its author or last living coauthor, or a total of 75 years if it is considered a work done for hire.[42]

Trade Secrets

A **trade secret** is information that gives one person or company a competitive edge over others. Though lacking the statutory formalities required for patent and copyright protection, trade secret coverage is easier to obtain and trade secrets are recognized as a property right much like copyrights and patents.

One of the major difficulties is that trade secrets are governed by state law rather than federal law, and definitions are not uniform from state to state. "Liability for the taking of a trade secret arises from a breach of confidentiality . . . A trade secret primarily protects its owner against wrongful disclosure by persons with a legal relationship to the owner."[37] Legal relationships can be defined as employer-employee or contractor-contractee.

Unfair Competition

The law of **unfair competition** is based on the idea of preventing dishonesty in business dealings and aims to promote business integrity and fair competition. Unfair competition is governed by state law, and the rules are not as well structured as those for federal patent and copyright protection. "There is, however, a substantial body of law which has been used to prevent one from duplicating and selling the intellectual property belonging to another person."[37] The types of conduct prohibited are:

1. **Misappropriation of commercial values** (e.g., misappropriating computer software, "pirated" software).

2. **Protection of goodwill.** A developer has built up goodwill and unsatisfactory alterations to the product may result in loss of this goodwill (e.g., inferior quality software—"look-a-likes").

3. **Misrepresentation.** Disguising of a product by means of false statements, inference, etc. (e.g., quality of goods, properties possessed by goods, false pricing, origin of goods).[37]

SOFTWARE PIRACY

Computer software piracy is and has been a hot and cold topic in the computer industry for some time. Computer software piracy refers to the unauthorized duplication of commercially purchased software.[3] Some of the points that have been raised concerning this issue include:

1. Why is there so much hype over duplicating software?

2. Is there any legislation concerning software duplication?

3. What has been done, in addition to legislation, to slow the illegal copiers?

4. Who's responsible to see to it that pirating is curtailed within a company?

Illegal Duplication

One of the most important reasons for the concern over illegal software duplication is the revenues lost each year to software capers. The statistics are astounding. Between 1981 and 1984, an estimated $1.3 billion was lost to pirating. The Association of Data Processing Services Organization (ADAPSO) reported that while $3.2 billion worth of software was sold in 1985, approximately $800 million of software was copied. A study conducted by a market research firm out of Dallas,

Futuring Computing, reinforced these findings after a study it conducted revealed that $750 million in revenue was lost to pirates. Another study conducted by the same company revealed that for every copy of purchased software a pirated copy has been made. Joe Diodati, vice-president of Softguard Systems, Inc., explains that in addition to the revenues and profits that are lost, there is also an increase in overhead cost associated with pirated software. Because a customer service representative answering questions regarding software cannot easily distinguish customers who have actually purchased the software from those who have acquired the software by illegal means, they may be supporting unauthorized users. Geoffrey Berkin, an associate general counsel at Ashton-Tate Inc., commented that as of 1986, the company's total losses from illegal software copying was in the tens of millions of dollars. A senior vice-president of Apple Computer Corp., Albert Eisenstat, believes that foreign and domestic piracy has cost the United States in excess of $20 billion between 1981 and 1986, in addition to a loss of approximately 750,000 American jobs. Software piracy could very well wipe out a small-scale business whose primary thrust is selling its software product.

Legislation Concerning Software Duplication

Up until 1980, there was much confusion concerning what could and could not be copyrighted as far as computer programs were concerned. In 1978, the National Commission on New Technological Uses of Copyrighted Works (CONTU) recommended that the Copyright Act of 1976 be amended to include provisions for computer software in the following ways:

1. To provide explicit copyright protection of computer programs
2. To ensure that the copyright protected all computer uses of the programs
3. To allow authorized users of the copies of the programs to adapt the copies for their use

These recommendations were reviewed and eventually led to the establishment of the Computer Software Copyright Act. The law was signed in December 1978, by President Carter. In 1980 Congress passed the Computer Software Copyright Act of 1980. There have been some amendments to the Act, however. One amendment was added to clarify the meaning of a computer program. Section 101 declares that a computer program is "a set of statements . . . to be used directly or indirectly in a computer in order to bring about a certain result." In addition, the amendment states that a copy is "any material object from which, either the naked eye or other senses, or with the aid of a machine or other device, the work can be perceived, reproduced, or communicated." Another amendment to the act allowed for copies of programs to be made for archival purposes as well as copying essential for the execution of the program. In defense of copyright infringement charges, companies argue four points of fair use that are covered by the Copyright Act:

1. The purpose and character of the use, including its commercial nature
2. The nature of the copyrighted work
3. The proportion that was "taken"
4. The economic impact of the "taking"

In addition to the Copyright Act of 1980, most, if not all, commercially purchased software packages come with a license agreement. The agreement is to be read prior to opening the package because in some instances, when the seal on the package has been broken, that binds a company to the terms of the contract. Some of the terms that could be specified on such an agreement include:

■ Permission to use the application with title still residing with the software company

■ Use of the program on computers manufactured only by that particular vendor

■ Prohibited copying of application except for backup purposes

■ Restricted use for telecommunication purposes; no time-sharing

Litigation for copyright infringement can be very costly. Awards for damages and profits can be left up to the plaintiff and range anywhere from $10 to $50,000. In addition to the damages, court costs and attorney fees can be awarded. Copyright infringement is considered a federal misdemeanor with fines up to $10,000 and a maximum sentence of one year in prison.

Lotus Development Corporation and Ashton-Tate, Inc., along with several other corporations, have established a fierce reputation as "hard balls" when it comes to litigation for pirating. Lotus Development Corporation sued Rixon, Inc., for pirating software and distributing copies of its Lotus 1-2-3 software package to branch offices. The case was not only settled out of court but Lotus acquired more than $10 million and all copies of the software were returned by Rixon, Inc. Eric Deutsch, a lawyer representing Lotus, believes that if software producers do not protect their products then they are, in effect, closing down the market.

In 1986, Geoffrey Berkin of Ashton-Tate, Inc., sued three companies for pirating and was assisting the Federal Bureau of Investigations (FBI) in three more criminal cases of pirating software. In another case, after a two-month undercover FBI investigation, Ashton-Tate, Inc., brought suit against Horn Computer International Inc. for unlawfully duplicating software onto hard disks in personal computers . . . [and] advertising in a popular computer magazine that [they were] selling disks bundled with "tons of software." These piracy cases are not typically what one would consider such a big deal that the FBI would get involved but, due to Berkin's persistence, the FBI took an interest in the cases. Berkin has a two-prong approach to nabbing the copiers. First, attack all suspected duplicators, then set an example. Berkin's attitude is that no matter how small the violation, companies will pay—be it in court or out of court.

Albert Eisenstat estimated that between 1981 and 1986 Apple Computer has spent approximately $5 million on lawsuits for copyright and patent infringement. He advises his colleagues to take a fast and hard approach to counterfeiters to show them that piracy will not be tolerated.

In 1988, INSLAW, a small software firm in Washington, D.C., was involved in legal proceedings against the U.S. Department of Justice. In 1982, a $10 million contract was signed between INSLAW and the Department of Justice for a legal case-tracking software known as PROMIS—Prosecutor's Management Information System. Contract disputes ensued when the Justice Department started withholding payments to the tune of $2 million. In 1985, INSLAW's founder, William Hamilton, realized that they were in the midst of financial difficulties which resulted in filing for bankruptcy. After reviewing some of the company's documents, Hamilton's wife, Nancy, found inconsistencies that led her to believe that she and her husband had a case against the Justice Department. After three years of litigation, Judge George Francis Bason ruled that the U.S. Department of Justice purposely tried to force INSLAW out of business to acquire their software product PROMIS. The Justice Department was ordered to pay INSLAW $6.8 million in licensing fees and $1 million in legal costs. This case received a fair amount of coverage because it was the first time anyone had successfully won a case for software piracy against the federal government. This serves as a reminder that no one is above the law.

To add some clout to laws governing software piracy, a software antipiracy bill was passed in 1992 by the U.S. House of Representatives and the U.S. Senate. This law would make it a felony—rather than a misdemeanor—to copy software illegally. The bill calls for a fine of up to $250,000 and imprisonment up to five years for unauthorized reproduction or distribution of 50 or more copies of one or more computer programs during a 180-day period. For piracy involving between 10 and 50 copies, the penalties are up to $250,000 and imprisonment up to two years. This bill, according to the Software Publishers Association, is vitally needed as the U.S. software industry lost more than $2.4 billion in 1990 because of piracy.

Curbing Piracy

To combat the problems of piracy, a three-pronged approach has been recommended: education, technical support, and litigation. Litigation (laws) can be strengthened by educating users and supporting legitimate software owners and writers.

Education. ADAPSO publishes a brochure entitled, "Thou Shalt Not Dupe," which outlines the issues concerning software piracy and offers some suggestions on different policies that companies can adopt to control the problem. Companies need to be made aware of the legal, as well as ethical, ramifications involved in computer software piracy. Not only do they need to be aware, but companies should let their employees know that such practices will not be tolerated. Some companies post warning labels on the terminals and disks to serve as reminders that copying software is illegal. In addition, the organizations can draw up an employee/employer contract that outlines the policies and procedures regarding unauthorized duplication of programs. Because piracy could lead to substantial monetary loss to the company, some organizations make illegal duplication of any software package or computer programs grounds for dismissal.

Many colleges and universities are also recognizing the "hot bed" of piracy they face as students are introduced to computers. Most colleges of business recognize the problem; software copying is cited as one operational issue of concern. "It is still unclear whether universities are acting as quickly as possible to address the problem before they face legal sanctions or have their ethics called into question."[2] (The issue of ethical use of computers will be thoroughly discussed in Chapter 11).

Technical Support. The most prevalent method of curbing piracy is to use copy protection products that serve to physically prevent duplication of software. These products come in two different formats: hardware-based and software-based. It is important that the information systems personnel overseeing the security of the system understand the benefits and drawbacks of both categories. In hardware-based products, the computer hardware is altered to protect software duplication; in software-based applications, the protection schemes are on disks. In 1986, ADAPSO set out to devise a standard protection scheme to be used by all computer users. Their proposal was hardware-based and involved inserting a "key" into the computer when the software was used. Some copy protection schemes require that the disk remain in the computer for the duration of the execution or running of the program, which allows for only one program to be run at a time. This type of scheme requires more technical knowledge to copy programs, and it causes a great inconvenience to end users. Software-based applications use encryption techniques to prevent duplication of programs. Some things to consider when purchasing software-based schemes are:

- Is the product purchased compatible with the hardware available?
- Can a backup copy be made of the program?
- Can you run the program from the hard disk without having to re-insert the floppy every time you access the program?
- Does the program use the standard media for convenience?

Lindsey Kiang, general counsel for Lotus Development Corporation, uses this three-pronged approach to combat piracy. In cooperation with ADAPSO, Lotus educates its users on copyright laws. Legal action is taken in cases of computer piracy, and it uses protection codes to impede user access to copying applications and programs.

The Institute of Internal Auditors offers some useful recommendations for the education of users and the deterrence of software misuse:

- Have a centralized software library for monitoring of software use on the proper machines.
- Have a periodic training session on the new laws of copyright and misuse of computers.
- Monitor computer sites for illegal use of applications.
- Have management review and enforce policies and procedures.
- Distribute policies and procedures to employees.
- Devise tracking logs to monitor use of software.

CONSULTANTS AND OUTSIDE CONTRACTORS

Today, more and more organizations are using outside system integrators for development projects. However, despite the large use of outside services, organizations are often unskilled in managing and devising agreements with the outside vendor, which often results in legal disputes costing millions of dollars. Often, the users hire the consultants to relieve them of the headaches and problems associated with implementing a new project. The ultimate responsibility for the project is still that of the organization, and users must monitor and manage the system and contractors at all times just like any other project of the organization.

Another primary area where problems between consultant and organization occur is the issue of consequential damages. Most consultants design contracts that exempt them from paying any consequential damages. However, some prior court cases have held the consultants liable for damages despite what the contracts say.

Many organizations hire outside consultants to help with the requirements analysis phase and then eventually award this same consultant the actual project management contract. Using a single consultant helps to establish a close working relationship between integrator and user. This allows for easy, open discussion and detailed specifics to be set.

A checklist for users concerning a consulting contract follows:

- As many performance specifications as possible should be included in the contract.
- The contract's liability limitations should be reviewed carefully.
- Ways to expedite dispute resolutions should be considered.
- Plans for what will happen if the consultant goes out of business should be discussed.
- The consultant should be prohibited from raiding the organization for information system talent.

CRIMES AGAINST COMPUTER SYSTEMS

To more fully define the types of computer crimes covered earlier in this chapter, this section will discuss the six types of crime that have now been updated with a newer, faster "weapon"—the computer:

- Theft
- Fraud
- Sabotage
- Espionage
- Wiretapping or electronic bugging
- Terrorism

Theft

Theft is defined as taking unauthorized possession of property belonging to another person. The computer can be used to pilfer or divert valuable merchandise, transfer funds, or steal information. One of the paramount problems with the prosecution of theft via computer is that, all too often, the original is still in place so the owner is not necessarily denied the use of the item (e.g., software copies, copies of records, bogus bank accounts, etc.).

Fraud

Fraud is defined as the obtaining of assets of an organization or person through intentional misrepresentation or misapplication of information. The potential for "inadvertent fraud"—the unplanned misapplication of information due to carelessness, incompetence, or poor controls—exists.

Sabotage

Sabotage is an intentional act with the purpose of damaging the organization as a means of revenge or personal motivation.

Espionage

Espionage can be motivated by political or economic reasons and is defined as the removal (stealing) of valuable information for profit or political gain. "Corporate espionage" is becoming an all-too-prevalent computer-aided crime worldwide.

Wiretapping/Bugging

The unauthorized tapping into another person's telephone or computer lines is known as **wiretapping**. By simply connecting the "tap" directly to a system's telephone or teleprinter lines, a felon can intercept and record messages. **Bugging** involves planting a microphone in a computer facility for the purpose of listening to or recording conversations or transactions.

Electronic eavesdropping devices are available that enable a person to park outside a building and view a computer screen through a window by reading emanations. Criminals can easily capture information from up to a mile away.

Computers emit electromagnetic waves that are caused by small changes in current and voltage. These waves can be intercepted and interpreted to allow eavesdroppers to view screen or printer activity. Tapping into corporate computers can have great benefits. Competitors can obtain client lists, access codes, passwords, personal information, and other beneficial data with little effort. The law is still vague on dealing with this type of crime, so individuals and companies must protect themselves. Techniques such as covering computers in plastic boxes will limit the computer emissions. This is a real security threat and should be taken seriously by today's businesses. (See Figure 10.8.)

Terrorism

Terrorism is the act of holding a system hostage or causing extreme damage to a system; it is usually motivated by ideological fervor. Electronic funds transfer systems (EFTS) are a popular target for international terrorists. The terrorist poses a two-fold challenge to a cashless society—a threat to the security of its economic institutions and a threat to its political well-being.[8]

COMPUTER CRIME LEGISLATION

In investigating and proving computer crimes, there are many obstacles to clear. In our present justice system, the "ball is in the perpetrator's court" from the very

FIGURE 10.8
Capturing Electronic
Emissions

beginning as the laws usually lag behind the rapid growth of technology. The following problems are those that must be overcome if the criminal is to be prosecuted:

- **Lack of uniform reporting**. Many companies do not report computer crimes, because they do not want the negative publicity.

- **Lack of interest by prosecutors**. In many large U.S. cities, prosecutors are overloaded with more cases than they can handle. When prosecuting white-collar crime cases, they tend to pick the larger and more prominent ones. Many security professionals become frustrated because some DAs will not take a case involving less than a million dollars.

- **Charging the suspect**. The charge depends on the statutes in the state in which the crime was committed. This creates problems of jurisdiction. Rule 18 of the Federal Rules of Criminal Procedure states that the prosecution shall be held in a district in which the offense was committed. Since telephone access is a very common means of computer crime, where was the offense committed? Are you subject to the laws of the jurisdiction where the computer information resides or the laws where the phone call was made? This is a complex issue to tackle, and the courts will have difficulty doing so.

- **Evidence**. Gathering admissible evidence is critical, and in computer-related crimes it can be easily altered, leaving the prosecution with nothing to use. This will be discussed in more detail later.

- **Documentation**. Computer disks and tapes need to be distinguishable from one another when presented to the jury and need to be accurately marked to clearly establish a chain of custody.

- **Technical jargon**. Computer cases involve computer terms and buzzwords, which need to be simplified so that the judge and jury can understand.

- **Informants**. In computer cases, individuals from the computer community are usually involved. When it is time to prepare the case, informants often do not cooperate and sometimes even change their story. Therefore, the investigator should not base a case on one informant's testimony.

- **Audit trail**. In many computer crimes this does not exist. The perpetrator almost always makes sure that his actions are covered up and erased, leaving no track. In a best-case scenario, the prosecution is dealing with a computer system with audits built in, thus making the process of convicting the criminal much easier.

The Data Processing Management Association (DPMA) has published a list of guidelines for use by the legal community that aids legislators and the legal community in defining computer crime and delineates points to be addressed in the prosecution of criminals. (See Figure 10.9.)

Within the past few years, nearly every state in the United States has enacted some form of computer-related crime legislation. The enactments, however, are not uniform and the laws vary considerably as to the types of conduct proscribed, especially in regard to the kinds of property protected by such statutes.

The primary laws that govern computer-related crime are discussed below.

Accounting and Auditing Act (31 USC 65)

The Accounting and Auditing Act establishes requirements for accounting systems and internal audits. The act also assigns to the Government Accounting Office the responsibility for developing standards for accounting systems and audits.

Anti-Deficiency Act (31 USC 665)

Under provisions of the Anti-Deficiency Act, no officer or employee of the government may create or authorize an obligation in excess of the funds available, or in advance of appropriations unless otherwise authorized by law. Before executing any contract, the contracting officer shall (a) obtain written assurance from responsible fiscal authority that adequate funds are available or (b) expressly condition the contract upon availability of funds.

Brooks Act (Public Law 89-306)

The Brooks Act establishes the basic policy for the management of data processing equipment in the federal government. The act assigns specific responsibility to the General Services Administration (GSA) for the purchase of data processing equipment; to the Office of Management and Budget for the development of administrative and management policy; and to National Institute for Standards and Technology for the development of standards to maximize the ability to share computer programs and data. The act was extended to computer-related services such as computer service bureaus and contract programming. OMB Circular A-130 (replacing Circular A-71) implements provisions of the Brooks Act.

FIGURE 10.9
Data Processing Management Association's (DPMA) Computer Crime Legislation Parameter Guidelines

1. The unauthorized use or access of computer resources, including the computer and the information stored in it.
2. The unauthorized release of computerized information.
3. The unauthorized copying or use of computer software by anyone other than the owner.
4. The unauthorized modification of computer resources, including the computer, computer software, and information. Modification would include any unauthorized changes, appending, replacement, or contamination of the resources.
5. The unauthorized destruction of computer resources, including the computer, computer software, and information. The destruction can be contamination or any act that would make the resource unsuitable for its intended purpose.
6. The unauthorized and malicious denial of access to computer resources, including the computer, computer software, or information.
7. The use of computer resources in the commission of a felony. (This is the use of a computer or computer information to aid in the commission of a felony.)
8. The aiding, abetting, or conspiring to commit or solicit computer crime.
9. Civil Relief and Damages—This makes the criminal liable and protects the secrecy and security of the computer resources of the victim.
10. Forfeiture of Property—This allows the materials used in the commission of the crime to be forfeited to the state or federal authorities.
11. Rule of Evidence—This establishes the guidelines for what evidence will be considered "best evidence."
12. The intent of devising or executing any scheme or artifice to defraud or to obtain any tangible property.
13. The intent to illegally obtain information.
14. The intent to establish contact for the purpose of unauthorized experimentation without the intent to defraud or commit any other crime after such contact is established.

Source: Data Processing Management Association, Park Ridge, IL.

Computer Fraud and Abuse Act of 1984 (Public Law 98-473)

This legislation makes it a crime to gain unauthorized access to federal computers. While other legislation and common law had been interpreted to cover unauthorized access, this law eases prosecution by clearly defining unauthorized access alone as a crime. Under the provisions of this law, "whoever knowingly accesses a computer without authorization, or having accessed a computer with authorization, uses the opportunity such access provides for purposes to which such authorization does not extend" and obtains access to sensitive information, uses, modifies, destroys, or discloses information in, or prevents use of such computer (operated for or in behalf of the government) may be subject to prosecution. The Computer Fraud and Abuse Act provides for fines of more than $100,000 and/or 20 years imprisonment for multiple offenses.

Computer Fraud and Abuse Act of 1986 (Public Law 99-474)

This law revises PL 98-473 and provides increased penalties for fraud and related activities in connection with computers and computer access devices. New offenses include intentionally accessing a Federal Interest Computer without authorization and (1) obtaining anything of value (including data), or (2) preventing authorized use, or (3) altering information. Also stated as a new offense is knowingly trafficking in any password or similar information through which a computer may be accessed without authorization if interstate or foreign commerce is affected or if the computer is used by the government.

Computer Security Act of 1987 (Public Law 100-235)

The Computer Security Act of 1987 amends several laws (including the Brooks Act and Federal Property and Administrative Services Act of 1949) to add provisions for the protection of computer-related assets (hardware, software, and data). Among the provisions of the act are:

- Assignment of responsibility for the development of computer security guidelines and standards to the NIST.
- Requirement that within six months after enactment of the Act, federal agencies shall have identified existing systems and systems under development that contain sensitive information.
- Requirement for development of a security plan for each identified sensitive computer system within one year of enactment of the Act.
- Requirements for mandatory, periodic training in computer security awareness and accepted computer security practice of all employees who are involved with the management, use, or operation of federal computer systems.

Copyright Act of 1980 (17 USC)

The Copyright Act applies to sensitive applications because such systems may use commercially available software that is protected by copyright. In such cases, application management must ensure that reasonable procedures are in place to prevent unauthorized use or copying of copyrighted programs. These procedures should include provisions for disciplinary action to be taken against individuals who duplicate copyrighted software for use on home computer systems.

The copyright laws were amended in 1980 to recognize the realities of modern data processing systems. Section 117 was added to permit copying of copyrighted software for back-up or archival purposes if a copy is required to install the software on the computer.

Electronic Communications Privacy Act of 1986 (Public Law 99-508)

This legislation establishes specific protections that update wiretap and privacy statutes to make them more effective relative to advancements in technology. The legislation protects remote computer services, electronic mail, cellular telephone conversations, satellite transmissions, and other communications technologies. The law also establishes clear guidelines that must be met by government officials prior to requiring a remote processing services company to provide information from a customer file.

Federal Managers' Financial Integrity Act (Public Law 97-225)

The Federal Managers' Financial Integrity Act requires that financial systems be evaluated for compliance with internal control procedures and standards prescribed by the Comptroller General. The basic objective of the act is to reduce or eliminate the incidence of waste, fraud, and abuse in government financial systems. The act requires that the internal controls of financial and accounting systems be reviewed annually and an accreditation statement be prepared that states that the internal controls are adequate to provide reasonable assurance that assets are properly allocated, tracked, and managed. In this context, applications such as inventory control and accounting systems are assets that require protection and management.

OMB Circular A-123 implements the provisions of the Federal Managers' Financial Integrity Act. OMB Circular A-130 provides for acceptance of a Sensitive Application Certification in lieu of a separate internal control review.

Freedom of Information Act (Public Law 93-502)

The FOIA is generally perceived as a law that opens agency records to the public. The FOIA does, however, require agencies to review sensitive material and ensure that data such as trade secrets, proprietary financial data, and personal data are not improperly released. For application designers and operators, this means that policies and procedures must be in place to review requests for information and deny release of protected and sensitive data.

Paperwork Reduction Act (Public Law 93-511)

The Congress, recognizing the growing importance of data processing in the federal government, added specific language to the Paperwork Reduction Act that materially affects computer security and internal control. Specifically, the legislation defines information as a resource, for which the responsibility for management and control must be at the highest level within an organization. In many agencies, this has led to the creation of an Office of Information Resources Management. OMB Circular A-130 implements provisions of the Paperwork Reduction Act.

Privacy Act of 1974 (Public Law 93-579)

The Privacy Act of 1974 was enacted to provide for protection of information related to individuals that is maintained in federal information systems. The act establishes specific criteria for maintaining the confidentiality of sensitive data and guidelines for determining which data are covered by the act. According to the act, federal agencies and employees are responsible for:

- Maintaining the confidentiality of data covered by the act.
- Taking those actions necessary to reasonably ensure that data (concerning individuals) maintained in federal information systems are accurate. Failure to comply with the provisions of the act could make the agency and individuals within the agency liable for criminal and civil penalties.

OMB Circular A-130 (replacing Circular A-108) implements provisions of the Privacy Act.

Trade Secrets Act (18 USC 1905)

The Trade Secrets Act establishes specific penalties for the improper disclosure of trade secrets entrusted to government agencies.

Patent and Trademark Laws (35 USC)

Some computer software may be patented. In cases where an application contains or uses patented software, users are responsible for protecting the rights of the patent holder. Specifically, the user must ensure that the patented software is not improperly disclosed, used, or copied.

PRIVACY CONSIDERATIONS

Privacy can be defined as a security principle that protects individuals from the collection, storage, and dissemination of information about themselves and the possible compromises resulting from unauthorized release of that information. Invasion of privacy is an issue today, given the ease and speed of information transfer.

Improvements in computer technology have made access to personal records faster and easier. These improvements, however, have often been used to violate an employee's rights to privacy. An invasion of privacy can occur at any point in an organization where the user, computer, or information can be falsified. The quality of information is of main concern to the manager, who is responsible for the enforcement and design of controls.

Unfortunately, computers hold a lot of data on each employee such as FICA tax, federal withholding, state tax, and Social Security number. Many companies willingly provide employee information to creditors, telemarketers, and charitable organizations. Instead of using this information for a specific purpose—for example, to help customers establish a credit record—some companies sell it without approval. While many large organizations, such as IBM, allow employees to review and correct their own files, information still gets into the wrong hands. Organizations could be held legally responsible if erroneous information harms an employee. Therefore, it is the responsibility of the systems analyst to design a system that will avoid a possible lawsuit.

Proper collection, correct use, and accurate protection from unauthorized access is critical to the integrity and reliability of employee data. Although there are countless ways for an unauthorized person to access employee data, there are many laws, principles, and protective measures to increase privacy. The major laws on privacy include the Fair Credit Reporting Act, Privacy Act, Right to Financial Privacy Act, Video Privacy Protection Act, Computer Matching & Privacy Protection Act, and Freedom of Information Act.

A company needs to be aware of the potential privacy risks computer data holds for the employee and should avoid disclosing inaccurate and unreliable information to other companies. Standard policies and roles need to be set so the employee will understand where they fit into a data control system and will responsibly adhere to security measures for protecting personal, as well as corporate, data.

Computer technologies have put into the hands of governments, businesses, and private individuals powers enabling them to invade the "personal space" of others. One of the problems associated with the issue of privacy is "Who really owns information—those who create the file or the person about whom the information is gathered?"

Technological advances have made it easier to store, analyze, and trade personal information. Therefore, privacy experts have recently been concerned about information collected without the employee's or citizen's knowledge. The Privacy Protection Study Commission conducted a study on the abuses of private information and made recommendations that included use of employment records. Although employees were made aware of what personal information was kept, what it was being used for, and how that information could be verified or modified, a majority of companies released private information to credit bureaus, some of which was erroneous or outdated. Furthermore, medical records, which relied on computer information that was often inaccurate, were used to make hiring and promotion decisions. Most people aren't aware of the private information they release in applying for mortgages, credit cards, and driver's licenses. The problems of privacy don't just concern sufficient security against unauthorized access, but also adequate control of authorized use.

The employee data a company holds such as FICA tax, federal withholding, state tax, and Social Security numbers have to be released quarterly for the Social Security Administration or at the end of the year in a W-2 form to the Internal Revenue Service. The data for FICA and federal withholding contains information about wages earned and withheld, identifying the employee with a Social Security number that travels through banks before it goes to the government. Credit bureaus get most of their information automatically once a month when banks and retailers release tapes and files containing purchase and payment records such as mortgages, credit card payments and balances, income, family makeup, employment histories, driving records, bank balances, descriptions of legal tangles, and Social Security numbers. This information can be provided to charitable organizations or applied for marketing strategies without having a specific purpose for use.

Obviously, since a company cannot stop this flow of information, it needs to be sure that it properly collects and uses data and maintains accurate protection from unauthorized access. The biggest privacy concern should be the integrity of data so any information published about an employee is accurate. A firm should be aware that there are problems associated with government intrusion, communication surveillance, First Amendment rights, and privileged communication. PCs, especially when publicly networked or connected by modem, are vulnerable to security problems and are easily used to access passwords, copy or modify information, or create viruses, Trojan Horses, or bombs. Problems of wiretapping and eavesdropping can also occur that are primarily due to centralized communication and multiple user access to large amounts of data. There is hope, however, that computer technology in the future may come to a point where an organization will use information to meet the government's needs without compromising employee privacy. Areas of concern encompass all citizens, institutions, and business—anyone creating information.

CONCLUSION

After analyzing who the computer criminal is and what their motives are, we can make some basic conclusions. First of all, it is obvious that the problem of computer crime is a reality and has been steadily increasing over the past years. In order to determine how to approach this dilemma and decrease the number of crimes committed, an organization must identify the culprit. According to most reports, a large percentage of computer crimes are committed by insiders or employees of the organization they exploit. Good management and internal controls are the key to solving the problem of the insider criminals, as a majority of the employees who commit such crimes do so out of revenge or boredom. To rid the employee of the feeling of need for revenge, management can make certain that communication channels between the employer and the employee exist and are open. Secondly, management could develop a means for employees to air their grievances which, in turn, would combat the vengeful nature of some of the company's employees. In order to deal with the problem of boredom that some employees feel, management could periodically review the job description and necessary job skills and compare them with the employee in the position. If they no longer match, some steps could be taken to expand the job duties or promote the employee. Another alternative to deal with employees who complain about routine work would be to alter the work environment. The change of scenery may be just what is necessary to make the employees content and no longer bored. Or the company could offer incentives for those employees who engage in good security practices. This will show that security is something that the organization really cares about and that the actions of the employees are important to the computer security of the company.

If the problem of the insider criminal is not solved by solid management principles, other controls must be put in place. These include the physical security of the computer center, the use of passwords and user logs, access limitations, and possibly encryption. In the event of an insider criminal, management must take a firm view and discipline the perpetrator, either through termination, prosecution, or some other means. The company's firm stand regarding security gives a clear message to all employees, those that are security conscious and even those that are not, that computer security is a serious issue, one that is of grave concern to all organizations utilizing them.

SUMMARY

This chapter discusses the legal issues and current legislation relating to computer crimes.

Computer crime is the violation of a computer whereby a victim suffers from the wrongful intentions of a perpetrator. Computer crimes can be *supported by* the computer (tool), because of the computer (object), or around the computer (targets).

Some of the methods computer criminals use are: **data diddling, Trojan Horse, salami technique, superzapping, trapdoors, logic bombs, asynchronous attacks, scavenging, data leakage, piggybacking** and **impersonation, wiretapping**, and **simulation** and **modeling**.

The types of computer crimes that are committed are **telecommunications fraud, embezzlement, hacking, ATM fraud, records tampering, child pornography** and **abuse, drug crimes**, and **organized crime**.

Typical computer criminals can include novices, hackers, traditional criminals, career criminals, deranged computer criminals, and extreme advocates. Computer criminals commit crime for personal and financial gain, for entertainment, for revenge, as a personal favor, to beat the system, by accident, or to commit vandalism.

Computer crimes can be prevented, deterred, or controlled by solid management controls and sound security practices.

Software violations include infringements on **patents, copyrights, trade secrets**, and **unfair competition**.

Crimes against computer systems take the form of **theft, fraud, sabotage, espionage, wiretapping/bugging**, and **terrorism**.

Computer crime legislation is being enacted and enforced to curb crime and prosecute criminals.

Privacy issues are of paramount concern, and laws are now firmly in place to protect individuals' rights to privacy.

REFERENCES

1. Alexander, Michael, "Lax Security Invites Liability Nightmare," *Computerworld*, February 20, 1989, pp. 1, 127.

2. Athey, Susan, "A Comparison of the Fortune 500 and AACSB—Accredited Universities' Software Copying Policies," *CIS Educator Forum*, Vol. 2, No. 4, pp. 2-10.

3. Baker, Donald I., and Roland E. Brandel, *The Law of Electronic Fund Transfer Systems,* Cumulative Supplement, 1988 Edition (Boston, MA: Warren, Gorham and Lamont, 1990).

4. Baker, Richard H., *The Computer Security Handbook* (Blue Ridge Summit, PA: Tab Books, 1985), pp. 1-281.

5. "Bar Chart," *Washington Post*, February 18, 1991.

6. Barnhill, William, "Privacy Invaders," *AARP Bulletin*, Vol. 33, May 1992, No. 5, pp. 1, 10.

7. Becker, Regis, and Thomas P. Costello, "The Reporting and Collection of Management and Employee Fraud Data: Current Practices and Developing Trends in American Business," *Security Journal*, Vol. 2, No. 3, 1991, pp. 152-171.

8. Bequai, August, *How to Prevent Computer Crime . . . A Guide for Managers* (New York: John Wiley and Sons, 1983), pp. 1-308.

9. Betts, Mitch, "A $46 Tape Becomes a $10K Print Job," *Computerworld*, April 20, 1991, pp. 120-121.

10. Billingsley, K.L., "Welfare Fraud: It's a Matter of Loopholes," *The Freeman*, April 1992.

11. BloomBecker, J. J. Buck, *Spectacular Computer Crimes* (Homewood, IL: Dow Jones-Irwin, 1990), pp. 1-242.

12. BloomBecker, J. J. Buck, Editor, *Commitment to Security*, National Center for Computer Crime Data, Santa Cruz, CA, 1989.

13. Burnham, David, "Flaws Are Cited in Treasury Computers," *The New York Times*, February 13, 1986, p. D5.

14. Butler, Alison, "The Trade-Related Aspects of Intellectual Property Rights: What Is At Stake?" Federal Reserve Bank of St. Louis, November/December 1990, pp. 34-45.

15. Clavenger, Thomas P., "Audit Opportunities to Control Software Piracy," *Internal Auditor*, December 1988, pp. 43-47.

16. *Compendium of Federal Justice Statistics, 1985*, U.S. Department of Justice, Office of Justice Programs, Bureau of Justice Statistics, Washington, D.C.

17. "Computer Crime Methods Update," Datapro Research Corporation, April 1988.

18. *Computer Crime . . . Prevention, Detection, and Investigation*, Volumes I & II, FTP Technical Library, Port Jefferson Station, NY.

19. *Computer Security Act of 1987* (H.R. 145), Senate of the United States, June 27, 1987.

20. *Computers: Crimes, Clues and Controls . . . A Management Guide*, Prevention Committee, President's Council on Integrity and Efficiency, March 1986.

21. Conly, Catherine H., *Organizing for Computer Crime Investigation and Prosecution*, U.S. Department of Justice, Office of Justice Program, National Institute of Justice, Washington, D.C., 1989.

22. "Defining Computer Crime," *COMP-U-FAX*, DPMA, Vol. 11, No. 6, November/December 1984, p. 2.

23. "Does the FBI Have a File on You?" *Forbes*, October 1988, p. 184.

24. Earley, Pete, "Watching Me, Watching You," *Washington Post*, Summer 1986, pp. 2-8.

25. Fitzgerald, Kevin, "Computer Crime Detection," *Computer Control Quarterly*, Vol. 9, No. 3, 1991, pp. 41-49.

26. Forcht, Karen, "International Trade and Electronic Data Interchange: Legal Issues," *IBSCUG Quarterly*, Vol. 2, No. 4, Winter 1991, pp. 5, 6.

27. Forcht, Karen A., Daphyne Thomas, and Karen Wigginton, "Computer Crime: Assessing the Lawyer's Perspective," *Journal of Business Ethics*, Vol. 8, No. 4, April 1987, pp. 243-251.

28. Guide, Richard, "The Costs of Free Information," *Public Interest*, Fall 1989, pp. 87-95.

29. Hains, David, "Better a Locksmith Than a Lock Picker," *Computer Control Quarterly*, Vol. 9, No. 3, 1991, pp. 39-40.

30. "How to Detect Software Theft," *Computer Control Quarterly*, Vol. 9, No. 4, 1991, pp. 6-7.

31. Hughes, Gordon, "The Copyright Infringement Penalty in Australia," *Computer Control Quarterly*, Vol. 10, No. 1, 1992, pp. 30-32.

32. Hutt, Arthur E., Seymour Bosworth, and Douglas B. Hoyt, *Computer Security Handbook* (New York: Macmillan, 1988), pp. 1-399.

33. Knight, Peter, and James Fitzsimmons, *The Legal Environment of Computing* (Sydney, Australia: Addison-Wesley Publishing Company, 1990).

34. Lawren, Bill, "Breaking and Entering," *Omni*, December 1990, p. 17.

35. Loch, Karen, Houston H. Carr, and Merrill Warkentin, "Why Won't Organizations Tell You About Computer Crime?" *Information Management Bulletin*, February 1991, pp. 5-6.

36. McEwen, J. Thomas, *Dedicated Computer Crime Units*, U.S. Department of Justice, National Institute of Justice, Office of Communication and Research Utilization and Office of Justice Programs, Washington, D.C., 1989.

37. Mellen, Ruth E., "The Legal Protection of Software," *CompuTopics*, Association for Computing Machinery, Washington, D.C. Chapter, Vol. XXVI, No. 6, February 1983, pp. 12-27.

38. Meyerowitz, Steven, "A Legal Feel to 'Look and Feel' Suits," *Business Marketing*, December 1988, pp. 49-51.

39. Nimmer, Raymond T., *The Law of Computer Technology*, 1990 Cumulative Supplement, No. 2, 1990 and 1985 edition (Boston, MA: Warren, Gorham and Lamont).

40. Parker, Donn B., *Computer Security Management* (Reston, VA: Reston Publishing Company, Inc., 1981), pp. 1-308.

41. Pell, Eve, "FOIAbles of the New Drug Law," *The Nation*, December 13, 1986, pp. 666-668.

42. Pfleeger, Charles P., *Security in Computing* (Englewood Cliffs, NJ: Prentice-Hall, 1989), pp. 1-538.

43. *Privacy Journal*, Washington, D.C.

44. Rocca, Jim, "Panel Weighs Copyright for Federal Software," *LC Information Bulletin*, July 2, 1990, p. 248.

45. Russell, Deborah, and G.T. Gangemi, Sr., *Computer Security Basics* (Sebastopol, CA: O'Reilly and Associates, Inc., 1991), pp. 1-441.

46. Schweitzer, James A., *Computer Crime and Business Information . . . A Practical Guide for Managers* (New York: Elsevier, 1986), pp. 1-195.

47. Shenkin, Peter, "Computer Crime," *Security Systems*, September 1987, pp. 28-29.

48. Sieber, Ulrich, *The International Handbook of Computer Crime* (New York: John Wiley and Sons, 1986), pp.1-276.

49. Sievers, Ruth, "A Hard Look at Software Legislation," *LC Information Bulletin*, September 24, 1990, pp. 329-330.

50. Stone, Bradford, *Uniform Commercial Code in a Nutshell* (St. Paul, MN: West Publishing Company, 1975).

51. Sturtevant, David B., "In Search of a Rational Approach to Software Property Rights," *Journal of Systems Management*, October 1988, pp. 31-35.

52. "The Battle Raging Over 'Intellectual Property'," *Business Week*, May 22, 1989, p. 78.

53. *Using Software*, Educom Software Initiative and ADAPSO, 1987.

54. Van Duyn, J., *Automated Crime Information Systems* (Blue Ridge Summit, PA: Tab Books, 1991).

55. Van Duyn, Julia, *The Human Factor in Computer Crime* (Princeton, New Jersey: Petrocelli Books, 1985).

56. Waller, Douglas, "The Open Barn Door," *Newsweek*, May 4, 1992, pp. 58-60.

57. Whiteside, Thomas, *Computer Capers* (New York: Mentor Books, 1978).

58. Wofsey, Marvin M., *Advances in Computer Security Management*, Vol. 2 (New York: John Wiley and Sons, 1983), pp. 1-268.

REVIEW QUESTIONS

1. Briefly describe how the computer is used to commit crime.

2. What are some of the methods used by computer criminals?

3. List some of the types of crimes committed by computer criminals.

4. Why are disgruntled employees a threat to computer systems?

5. Write a brief profile of a "typical" computer criminal.

6. What are some of the reasons individuals commit computer crimes?

7. How can computer crime be lessened or prevented?

8. Compare patents, copyrights, trade secrets, and unfair competition and how they relate to the computer industry.

9. What is software piracy?

10. Briefly describe some of the legislation that relates to computer crimes.

DISCUSSION QUESTIONS

1. Jay BloomBecker, Director, National Center for Computer Crime Data, stated that, "like any other type of crime, computer crime will ultimately reflect the culture that surrounds our computers. Looking at the 'culture' of computing is not an encouraging exercise." Discuss this statement.

2. John Naisbitt in his best-selling book, *Megatrends*, states that "the new source of power is not money in the hands of a few but information in the hands of many." Discuss this statement as it relates to computer crime.

3. Mailing list compilers know a lot about individuals—current magazine subscriptions, car ownership, movie rental habits, buying habits, pet ownership, etc. What are the invasion of privacy issues involved in this massive profile of each citizen?

4. In September, 1986, Intel Corporation's copyright on the microcode of its 8086 machine and 8088 microprocessors was ruled "good and valid" in the

Northern District of California District Court. This action extended the copyright laws to the internal design of computer chips. Semiconductor manufacturers feel that this affirms their right to develop proprietary products without fear of being copied by their competition.

Could this ruling be construed as restraint of trade, curbing unfair competition, patent law enforcement, intellectual property protection, trade secret, etc.? What other legal issues are involved?

5. Intellectual property is considered to be new products, new services, and new manufacturing processes. Artistic works and scientific advances containing an idea as their origin are also considered to be intellectual property. At the present time, the major forms of intellectual property protection are patents, registered designs, trademarks, and copyrights.

What are some of the computer-related products that could be considered intellectual property?

EXERCISES

1. Find out if your university/school has any legal issues workshops for employees. Are computer-related issues covered?

2. Find an article in a computer-related journal that covers legal issues in computing.

3. Research in legal journals/books to find out the legislation in your state covering computer use.

4. Interview a local attorney to ascertain his or her understanding of legal issues in computing.

5. Prepare a brief questionnaire and distribute it to students at your college to ascertain the level of hacking and software piracy taking place on your campus.

PROBLEM-SOLVING EXERCISES

1. The misuse of software packages can be categorized into three classes:

   a. An unauthorized individual or group that copies and uses or distributes another's product.

   b. A dealer's or distributor's unauthorized copying to include with hardware or textbook to close a sale.

   c. An individual or group who uses an illegal copy to develop a computing derivative product.

   What are the legal implications in each of these situations?

2. Copy protection sometimes inconveniences the law-abiding user, as often installing copy-protected software on a hard disk complicates things like backing up the hard disk. Can copy protection schemes sometimes present a huge disservice to the person who purchases or buys the license to the software—inconveniencing the legitimate user? How could this be changed?

3. For a fraction of the cost of buying a computer program, a user can rent a program for a limited time and return it. But unlike the situation with video-cassettes, copying of the program by the renter is easy, inexpensive, and likely. The software industry is pushing legislation to prohibit renting software because it says that rental businesses are displacing what would otherwise be sales. How would you suggest improving this dilemma so that users can rent software inexpensively, yet the vendor can profit?

4. At Wells Fargo Bank, 1979-1981, L. Ben Lewis, an operations officer for this, the 11th largest U.S. bank, allegedly produced bogus deposits in an account at one branch belonging to a boxing promotion outfit. He did this by using the bank's computerized interbranch account settlement process to withdraw funds from a different branch. To keep the computer from flagging the imbalance, Lewis created new fraudulent credits to cover the withdrawal and allegedly kept the rollover going for two years—$21.3 million was diverted. How could this situation have been better controlled?

5. Using a classroom terminal, teenage students at Manhattan's private Dalton School allegedly dialed into a Canadian network of corporate and institutional data systems. No funds were diverted, but damage was done to the data files. Is the school responsible for this action? Are the parents of these students liable? How could this situation have been avoided?

CASES

Hackers Indicted

In an article in *Investor's Business Daily* (July 9, 1992), the following was reported:

A group of computer hackers has been indicted for breaking into computer systems operated by major telephone companies and credit reporting agencies in what prosecutors said yesterday were "crimes of the future."

The charges mark the first time court-authorized wiretaps were used to obtain conversations and data transmissions of computer hackers, the government said.

The indictment alleges the defendants broke into computer switching systems operated by Southwestern Bell Corp., New York Telephone Co., Pacific Bell, US West Inc., and Martin Marietta Electronics Information and Missile Group.

The defendants also allegedly tampered with systems owned by the nation's largest credit reporting companies, including TRW Inc., Trans Union, and Information America. They allegedly obtained 176 credit reports on various individuals.

The defendants allegedly infiltrated computer systems to obtain telephone, credit, information, and other services without paying for them and to obtain passwords, account numbers, and other information they could sell to others.

The defendants in the case were five males, ages 18-22. All were from New York.

The indictment contains 11 counts of computer tampering, computer and wire fraud, illegal wiretapping, and conspiracy. If convicted, the defendants face maximum prison terms of more than 50 years and fines of more than $2.5 million.

If you were a computer center manager, what would you do after reading this article? Would you alter your organization's operation? In what way?

Heintz Realtors, Inc.

Carolyn Heintz, president of Heintz Realtors in Oak Harbor, Ohio, has recently hired a consultant who specializes in real estate software applications to aid in setting up her multiple-listing service (MLS). Mrs. Heintz learned from a close friend who is also a realtor that the consultant was copying Mrs. Heintz's files and offering them for sale to several competing realtors in the Oak Harbor area. The real estate business, being highly competitive, is vulnerable to dishonest, unethical persons wanting to "get the sale." Mrs. Heintz is now faced with the dilemma of litigation against the consultant for disclosure of proprietary information.

1. How could this situation have been handled to avoid dishonest copying of files?
2. What legal steps would you suggest Mrs. Heintz take?

ETHICAL USE OF COMPUTERS

LEARNING OBJECTIVES

After studying this chapter, you will be able to:

1. Define ethics and its relationship to use of computers.
2. Describe teleological and deontological ethical systems.
3. Compare legal vs. ethical vs. moral standards.
4. List and describe the stages of moral development.
5. Explain the information age's growth and its effect on ethical standards.
6. Describe the purpose of a Professional Code of Ethics.
7. Describe the various computer-related Codes of Ethics currently adopted by various professional groups.
8. Describe several corporate policies on ethics.
9. Explain the Codes of Ethics educational institutions are currently using to instill ethical use of computers in students.
10. Define the following terms.

TERMS

- Normative Ethical System
- Teleological Ethical System
- Deontological Ethical System
- Ethics
- Laws
- Morals

COMPUTER ETHICS BEGIN AT HOME

Parents used to worry about their kids staying out too late. Now they worry about what their kids are doing on their computers.

Just as parents should infuse their children with a sense of morality about other types of behavior, today's parents need to teach children about computer ethics. As Robert Grierson, a teacher at the Latin School of Chicago, points out, "You would not be happy if your child were stealing comic books from the local grocery store, scribbling graffiti on a neighbor's fence, copying homework from a friend, or digging through big sister's closet to find and read her diary. Computers make it easier for children to fall victim to these kinds of temptations through violation of copyright laws, tampering with other people's data files, plagiarism, or hacking protected information. Many kids do not know they are doing anything wrong" when they participate in these activities.

A child's computer ethics education should begin the day the parents plug in the computer, according to Dr. Karen Forcht, professor of information and decision sciences at Virginia's James Madison University. "Parents should take an active role in a visible manner to instill the proper use of computers in their children," she says. "Prevention on the front end is preferable to punishment on the back end."

Dr. Forcht advises parents to teach their kids these concepts:

- The computer should not be used to harass others.
- Use only bulletin board systems that are legitimate and appropriate to the child's age.
- Copying software illegally is stealing.
- Computers should never be used to invade the privacy and confidentiality of others.
- Don't plant viruses or other malicious code into software or computers.

"Most students have a good sense of what is right and wrong," Forcht continues. "They know it is wrong to lie, cheat, and steal." However, she admits that many college students don't understand that misuse of computers is wrong because no one has told them. "Once students understand that respecting someone else's property is part of the responsibility of using a computer, they are usually willing to adopt these ideals," she says.

Continued

COMPUTER ETHICS BEGIN AT HOME

From page 335

One father of three believes kids learn computer ethics by watching their parents' behavior. "Ethics are more caught than taught," he says. "The respect we share with our children and individual rights to privacy and ideas are all things that carry over into technology. In my view, the ethical person is not baffled by the challenges that technology presents. Our children will be confused about right and wrong only when we are."

As reported in "Sowing the Seeds of Ethical Computing," *Compuserve Magazine*, July, 1993.

In the previous chapter, we focused on legal issues and current legislation as a deterrent to computer crime. This chapter will go one step further and discuss ethical issues that affect computer systems. The subjects of computer security and computer-based crime have been the focus of substantial debate during the past decade; however, the issues involved are far from resolved. A variety of measures have been instituted, enforced, and monitored to ensure that computer centers are not vulnerable to human intervention—whether accidental or intentional. Unfortunately, this physical interpretation of security represents only one facet of a complex problem. The misuse of computer software and stored data and information may ultimately prove to be the more significant concern. In short, it is not yet clear to all parties involved in computer use just what acts should be considered as computer crime.

In the past few years, interest in the issue of ethics has been heightened as we now focus on the "people side" of computer security. The copying of a software program for a friend, while in direct violation of copyright laws and, therefore, technically a crime, may not be considered as serious to the user as stealing a physical system component or sabotaging a system for profit or revenge. The paramount question then becomes one of, "What are the definitive responsibilities of computer center employees or persons having access to software and information to the public they serve—the ultimate user or owner of information—in creating an 'environment of security' and in practicing solid ethical standards in regard to the valuable data they use when performing their jobs?"[10] The issue of computer security has tended to fall into the "gray" area that educators and industry alike have avoided for fear that too *little* knowledge can be *hazardous* and too *much* could be *dangerous*. The feeling now is that industry and educational institutions should address the issue of data security as an *attitude*, rather than as a *technology*.

DEFINING ETHICS

The development of an ethic is not an easy task: what and who will be covered, and, more significantly, how they will be addressed, are the obvious problems. Traditionally, ethical problems are examined using the knowledge that one gains from ethical theory. The theories developed over the centuries have been debated, discussed, and revised by philosophers everywhere. Is it possible to find a theory that fits the code of ethics you are seeking? That is the very problem, though, that theories present. Do you base your code of ethics on a teleological or deontological viewpoint?

Classical Systems

A **normative ethical system** is an ordered set of moral standards and rules of conduct by reference to which, with the addition of factual knowledge, one can determine in any situation of choice what a person ought or ought not to do. The two normative ethical systems most widely discussed and defended in contemporary moral philosophy are those that are teleological and deontological in nature. A **teleological ethical system**, such as utilitarianism as proposed by Jeremy Bentham (1748-1832) and John Stuart Mill (1806-1873), holds that an action is morally right either if a person's doing it brings about good consequences or if the action is of a kind that, if everyone did it, would have good consequences. The principle followed in this system is the goodness or badness of the consequences of an action.

A **deontological ethical system**, such as that proposed by Immanuel Kant (1724-1804), holds that an action is right if it accords with a moral rule, wrong if it violates such a rule. Moral rules are based on an ultimate principle of duty that does not specify an end or purpose whose furtherance makes actions right.

Applying these two ethical systems to a practical computer abuse situation shows the difficulty in developing a code of ethics. Suppose an employee of a large city government discovers that a copyrighted software program can be put to good use in conducting the city's business. It would increase the efficiency of a certain operation and result in monetary savings for the taxpayers. Less tax money spent would mean less burden on everyone or more money to spend on more pressing concerns, such as public housing. The employee's budget does not allow for the purchase of the software from the owner. But, the opportunity exists to copy it illegally with no one finding out. Should the employee copy and use the software?

Under a teleological point of view, the copying would probably be allowed because the small monetary loss of the software owner is offset by the significant benefits to the public. Using a deontological point of view, the copying would be wrong because it violates a basic moral rule against stealing.

Philosophers have attempted to deal with this problem by developing variations or combinations of the two basic systems; the result has been rather chaotic. They have expended large amounts of energy to develop a theory that will survive repeated attempts to refute it. This dedication to reducing refutation produces a theory with a life of its own and with very little consideration of practical moral decisions.

The gap between the theoretical and the practical in the computer ethics field has created a body of discussion that has yet to be resolved. Many ethicists believe that a knowledge of ethical theory is absolutely necessary to develop applied ethics and acknowledge that the trend to discard the theoretical foundations of ethics in favor of applied ethics is growing. This movement may be attributed to the fact that we know where we have been, but do not know where we are going—we are attracted to moral problems but skeptical about dealing with them in a rational manner.

The fault in neglecting ethical theory does not lie with laypeople attempting to deal with the practical problem, as they often recognize a problem exists and attempt to do something about it. The theorist is at fault for not translating an idea into a practical or applied ethic that would guide the laypeople in dealing with the problem.

Laypeople are confused because their idea of a theory is that of an end in itself, not a means to an end. The theorists may have caused this confusion by their elaborate sets and subsets of theories, each claiming to be the one that is universal to all situations. The major fault of the theorist is the limitation of the theory in question to the individual making the decision to act. In the practical problem examined earlier, the public employee has to make a decision that conflicts with two theories. The solution is simple—involve the two parties who will benefit or lose in the situation.

The reality of the business world is that we interact with people and organizations everyday. Few of our decisions are simple and singular in nature. We must consider the effect our actions will have on others. To rely only on a theory or theories to guide us may be inadequate. We must use not only both the teleological and deontological ethical systems to show us the possibilities, but we must also rely on our interpersonal skills to help us make the correct judgment.

Legal vs. Ethical vs. Moral Standards

Ethics can be defined as the science of social survival; what is ethical promotes the survival of individuals and civilizations, cultures, and societies. Every culture, no matter how civilized or primitive, has an ethical code. Some codes tend to be rather formal and are entered into, unknowingly, at birth as they are a definite part of the social culture. Other ethical codes develop as we grow, becoming a vital part of our personal and professional lives. Our ethical code is formed and shaped by experience (especially historical experience), experiment, and measurement. Throughout our lives, we are constantly faced with the dichotomous dilemma of right vs. wrong, good vs. bad. When asked to identify the parameters that govern our lives, three key areas generally emerge: laws, ethics, and morals. The delineation between the three becomes cloudy as they tend to overlap, depending on the particular situation and the individual's interpretation. (See Figure 11.1.)

Many laws enforce ethical standards in our civilized society and prohibit citizens from acting out of their own volition, thus ensuring and promulgating a sense of social structure. On the other hand, many acts that are considered to be immoral or unethical are not necessarily illegal. **Laws** are enacted to ensure that a society has definite rules of behavior. Those persons violating acceptable rules are to be punished or dealt with according to our established sanctions. Even though laws vary from one locality to the next and over periods of time, they do serve as an

FIGURE 11.1
Parameters Governing
Our Lives

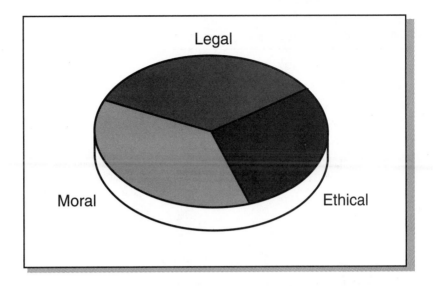

absolute—a known. The real dilemma is in defining or clarifying morality and ethics. **Morals**, generally speaking, are the attempt to control humanity's most powerful social forces—sex, money, and power—while **ethics** is the philosophic or objective study of abstractions called "the good or right conduct."[35] Our religious convictions tend to provide our moral fiber and separate the good from the bad. Ethics, conversely, is a cultural conditioning that builds on our religious convictions and is influenced by our societal beliefs, our family background, our geographic bearing, and other factors. Ethics then becomes the "melting pot" of those gray areas not covered with formal sanctions, such as laws or religious convictions.

If we were to develop a quadrant theory of acceptable behavior (see Figure 11.2), ethics would be a factor in all quadrants as a certain amount of interpretation and assimilation takes place in any situational analysis.

The most simplistic, easy-to-accept quadrant would be the legal and moral (B). If a person donates to a charity, it would be considered both legal and moral. The issue of ethics may not be a valid consideration in this example as the situation is somewhat clear-cut. To further explain the theory, a person might view an act such as robbing a bank as falling into quadrant C—without question. The act is both immoral and illegal as we adhere to the religious doctrine of "thou shalt not steal" and known legal sanctions concerning theft of someone else's property.

On the other hand, not fully divulging information for income tax purposes may fall between quadrants C and D—illegal, but moral—at least in some people's minds, as they tend to rationalize that "everyone does it," thus absolving themselves of personal guilt.

A situation that may fall into the A quadrant—legal, but immoral—could be a country's policy on restricting immigration quotas. The laws currently in effect may enforce limitation, but morally, we face the dilemma of right vs. wrong.

Realistically, people are much more complex than this quadrant theory implies. Our ethics and moral "space" become a matter of personal conjecture. We accept a code we can live with comfortably.

FIGURE 11.2

Quadrants of Acceptable Behavior

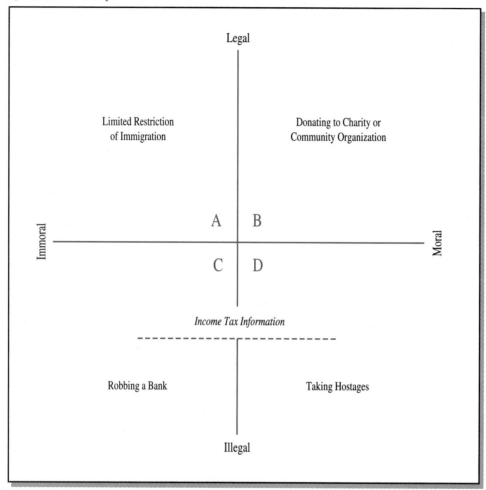

The constant dilemma of deciding between what is legal and ethical is a "cloudy" one that pervades all facets of our personal and professional lives. This "cloudiness" is further compounded by an ever-increasing complexity of our environment.

Moral Theory

Kohlberg hypothesizes that moral development progresses through six stages of increasing social orientation:

1. **Heteronomous morality.** Decision-making behavior is governed by avoiding behavior that leads to punishment.

2. **Individualism, instrumental, purpose, and exchange.** Decision-making behavior is governed by one's immediate interest.

3. **Mutual interpersonal expectations, relationships, and interpersonal conformity.** Decision-making behavior is governed by exhibiting behavior expected by others. Concern for others becomes evident at this stage.

4. **Social system and conscience.** Decision-making behavior is governed by exhibiting behavior that fulfills agreed-upon duties. An attitude that laws are to be upheld except in extreme cases is exhibited at this stage.

5. **Social contract or utility and individual rights.** Decision-making behavior is governed by recognition that people hold a variety of values and opinions and that many values are relative to one's social group.

6. **Universal ethical principles.** Decision-making behavior is governed by following self-chosen ethical principles.[34]

Computer-Related Legislation

Many states have enacted laws dealing with computer crime; however, there is a substantial difference between what the individual states define as illegal, the significance of the act (misdemeanor or felony), and the punishments prescribed. The Computer Security Act of 1987, which was passed by the United States Senate and the House of Representatives, should add impetus to the problem of coping with computer security violations. The general intent as stated in the opening paragraph is that:

> The Congress declares that improving the security and privacy of sensitive information in federal computer systems is in the public interest, and hereby creates a means for establishing minimum acceptable security practices for such systems, without limiting the scope of security measures already planned or in use.

Jay BloomBecker, the founder and director of the National Center for Computer Crime Data in Santa Cruz, California, addresses the need for the development of computer ethics being a solution to the computer crime problem. BloomBecker also points out that there is virtually no way of knowing how much computer crime goes unreported as companies are often reluctant to report a violation, thereby alerting the general public to the vulnerability of their system.

Crimes of Technology and Intellect

The paramount problem with relying on legal sanctions to protect information and punish violators is that our legal system responds to human activity/violations, rather than serving as a predictor—a reactive rather than a proactive approach.

The crimes of technology and intellect, warned Edwin H. Sutherland in the 1940s, are potentially far more serious than the crimes generally associated with the poor and the undereducated. How prophetic this statement was to become as the legal community and general public had no way of knowing that within the next 40 years, misuse of information, hacking, theft of vital data, and diversion of funds would become a reality. As the general public becomes increasingly computer literate, the gap between technology vs. intellect shrinks considerably. Computers are no longer high-powered machines for the knowledgeable few, but tools for the enlightened many. The danger is real: the computer abuse problem will possibly reach epidemic proportions by the year 2000 if not curtailed by legal sanctions, coupled with the adoption of accepted standards of ethical conduct.

The lines dividing ethical, unethical, and illegal practices in using computer hardware, software, and data files is not always easy to distinguish. J. Thomas McEwen, Principal Associate, Institute for Law and Justice, Alexandria, Virginia offers these situations to ponder:

- A parent offers to copy a computer program for a school that cannot afford to buy the program.

- An employee maintains a small database on his employer's computer as part of a sideline business.

- An individual uses someone else's computer account number and password to view the contents of a database.

- A customer gives his telephone number as part of a sales transaction at a store. The store enters the number into a computerized database and later sells the database to a telemarketing firm.

- A school system's computer programmer develops a program to schedule students and classes. The programmer accepts a job with another school system and leaves with a copy of the program.

- An inadvertent error in entering data into a police department's computer leads to the temporary detainment of an innocent person.[38]

McEwen evaluates each according to legal/illegal, ethical/unethical, acceptable/unacceptable. He states that:

> In most states, the first three are illegal actions. Copying a proprietary program is software piracy. Maintaining personal information on a business computer is misuse of a computer. And viewing files with another person's account number and password is illegal access.

> The last three scenarios do not deal with actions that are illegal, but they raise other questions regarding privacy and the need to deal accurately and responsibly with information affecting another's welfare. They illustrate the kinds of situations businesses, schools, and public agencies must address in developing standards for ethical, responsible computer use.[38]

Ethical Standards

Throughout our lives, we are constantly learning right from wrong. As each culture develops specific behaviors and customs, a common law develops—"the greatest good for the greatest number" theory. Common law is an unwritten law that is known simply through living in a specific environment.

Since the information systems field is viewed as a "new" culture, its ethical standards are still being formed. This field, however, has a uniqueness that makes it necessary to develop an ethical code of behavior.

Donn B. Parker, in *Ethical Conflicts in Computer Science and Technology*, suggests that:

- Computer system weaknesses should not be exploited for personal satisfaction or unauthorized gain. A person having the power to subvert a computer system or use it for unfair advantage does not have the right to do so.

- Analysts and programmers should be responsible for a program's performance according to specifications. Users should be responsible for adequate specifications.

- Implicit in any contract are the ethical principles generally accepted in the community.

- Programs are no different from wallets or any other items of value. If you get one by mistake, you have a duty to return it to the rightful owner and to pay for its use.

- A consultant has an ethical obligation to offer services in the best interests of his clients.

- Software products cannot be expected to be entirely free from error or deviation from specifications.[43]

Some of the activities that should be addressed when considering ethical use of the computer are the following:

- The actions of hackers—a term used to describe people who break into computer systems without authorization

- Program piracy—the copying of programs by individuals or organizations rather than paying a fee or license charge to the supplier of the program

- Unauthorized use of systems for personal use, or in some cases, for profit

- Individuals browsing through databases looking for information they are not authorized to have

- Computer bulletin boards that publish information enabling unauthorized users to break into other systems

- People who deliberately set out to deny system use to other users

Who Is Responsible for Ethics?

Our society believes that maturity and responsibility come with age. This is evidenced by many of our state and federal laws which establish legal age for certain rights and privileges. Some examples are:

- Driving
- Voting
- Drinking alcohol
- Enlisting in the military
- Marrying
- Serving as President of the United States

Parents also establish age limit restrictions on their children. They may require them to be at least 16 before they can go out on a date in a car. They may specify in their Last Will and Testament that the children must reach age 25 before they receive any money held in trust for them. The bottom line is that parents and society feel a high degree of confidence that the individuals we've entrusted with certain privileges or resources will behave in a mature and responsible manner when they understand and appreciate the value of a resource and the possible ramifications if a privilege is abused or misused.

The advent of computers has created a paradox for our society. We view computers as an incredible learning tool and something our children must master to be successful in the future. To give our children a head start, we are introducing them to computers *as early as possible*. We give them access and the knowledge to operate a very powerful tool that can be potentially damaging if used irresponsibly or in an unacceptable manner. The paradox for our society is that we have not applied an age restriction to the operation of computers. While we would consider it unconscionable to sell a handgun to a 12-year-old or put a second grader behind the wheel of a motor vehicle, we do, in fact, sell computers and modems to 12-year-olds and put second graders behind the keyboards of computers.

Thus, we have a dilemma. Do we give our youth a head start toward the future, or do we establish an age limit for the purchase and use of computers to promote responsible and ethical use? What a choice—either to enhance the learning abilities of our young or to create a nightmare for law enforcement personnel?

These novice users of computers will eventually become the information workers of tomorrow. Thus it may be somewhat safe to assure that the ethical attitudes of computing developed at a young age will carry over into the workplace. Everyone who uses a computer shares a key responsibility for safeguarding information within the system, and no one who uses a computer has the right to violate the integrity, security, and privacy of anyone else's information.

The Information Age

Peter G. Neumann, in his article "Computers, Ethics, and Values," in the *Communications of the ACM,* offers some insights into the ethical risks computers represent:

> Certainly there is a need for better teaching and greater observance of ethics, to discourage computer misuse . . . We must not assume that everyone involved will be perfectly behaved, wholly without malevolence and errors . . . People seem naturally predisposed to depersonalize complex systems . . . Computers permit radically new opportunities.[40]

As our society moves into an Age of Information, more and more people are involved in collecting, handling, distributing, and using information than previously thought imaginable. The tremendous impact of information is also noted by Richard Mason in his statement, "information forms the intellectual capital from which human beings craft their lives and secure dignity."[37] Mason further postulates that the four ethical issues of the Information Age are as follows:

- **Privacy.** What information about one's self or one's associations must a person reveal to others, under what conditions, and with what safeguards? What things can people keep to themselves and not be forced to reveal to others?

- **Accuracy.** Who is responsible for the authenticity, fidelity, and accuracy of information? Similarly, who is to be accountable for errors in information, and how is the injured party to be made whole?

- **Property.** Who owns information? What are the just and fair prices for its exchange? Who owns the channels, especially the airways, through which information is transmitted? How should access to this scarce resource be allocated?

- **Accessibility.** What information does a person or an organization have a right or a privilege to obtain, under what conditions, and with what safeguards?[37]

Mason's four issues form a composite nucleus of the major issues facing our information society—not an easy solution or a "quick fix." He states that:

> Our moral imperative is clear. We must insure that information handled is used to enhance the dignity of mankind. To achieve these goals, we must formulate a new social contract, one that insures everyone the right to fulfill his or her own human potential (and) . . . create the kind of world in which we wish to live.[37]

Mason's beliefs may well serve as a guide to establishing ethical behavior patterns that are not only acceptable, but plausible.

Ulrich Sieber, in *The International Handbook on Computer Crime*, states that "simple control of employees does not constitute sufficient personnel politics. The application of adequate safety measures and the honest behavior of employees will also be influenced by their *motivation*."[47] Sieber further postulates that the following measures should be instituted:

- Employee counseling and security briefings to develop awareness of risks
- Development of a good business atmosphere, fair practices, and an aura of a privileged elite
- Positive example of moral leadership by directors[47]

One of the major ethical dilemmas managers face is the fact that "lack of interest in information systems ethics appears to stem from the nature of IS people and their tasks, fear of losing their jobs, and a widespread belief that ethics policy is best handled by general management."[46] Michael Simmons of the Bank of Boston Corporation summed up this feeling very poignantly when he stated that, "If the topic of ethics were discussed, it would be a very short meeting."[46]

Basic Codes of Ethics

Figure 11.3 shows a perspective on ethics—a professional code of conduct that could be used to guide all known professions.

Figure 11.4 encapsulates the basic confusion that results from morals, ethics, and legalities—they are not always as clear-cut as they first appear.

PROFESSIONAL CODES OF ETHICS

Many professional groups are attempting to formulate some definitive guidelines in this computer "sea of uncertainty" by proposing formal Codes of Ethics. The concept today in evaluating a computer security program is preventing and planning up front, instead of punishing and reorganizing afterwards.[20] This "preventive maintenance" concept should be practiced by **all** members of the organization—users included—to be truly effective.

FIGURE 11.3
Professional Code
of Ethics

> ☑ Maintain the highest standard of professional behavior.
>
> ☑ Avoid situations that involve a conflict of interest.
>
> ☑ Do not violate the confidentiality of your employer or those that you serve.
>
> ☑ Continue to learn so that your knowledge level keeps pace with the technology.
>
> ☑ Use information judiciously and maintain system integrity at all times.
>
> ☑ Do not violate the privacy of others.
>
> ☑ Accomplish to the best of your ability.
>
> ☑ Do not break the law.

FIGURE 11.4
Trio of Principles

> ☑ Morals—right vs. wrong
>
> ☑ Ethics—code of conduct
>
> ☑ Laws—accepted standard

Objectives of Ethics Codes

In *Ethical Issues in the Use of Computers,* the authors state that ethics codes should emphasize five basic objectives:

- **Inspirational**—to inspire members to be more "ethical" in their conduct
- **Sensitivity**—to alert professionals to the moral aspects of their work
- **Disciplinary**—to enforce certain rules of the profession on its members and defend integrity
- **Advising**—to offer guidance in cases of moral perplexity about what to do
- **Awareness**—to alert prospective clients and employees to what they may and may not expect by way of service from a member of the profession concerned[32]

If a company or organization decides to develop an ethics awareness program or adopt a formal code, several factors should be considered, including the following:

- **Certification**—an affirmation by a governmental or private organization that an individual has met certain qualifications
- **Licensing**—the administrative lifting of a legislative prohibition

- Accreditation—an affirmation by a governmental or private organization that an educational institution meets certain standards
- Ethics—a standard of conduct drawn up by an organization to protect the consumer and competition against "unfair" practices[43]

Accepted Codes of Ethics

At the present time, there are several widely accepted Codes of Ethics in the computer profession:

- The British Computer Society (BCS) *Code of Conduct*
- The Data Processing Management Association (DPMA) *Code of Ethics, Standards of Conduct and Enforcement Procedures*
- The Association for Computing Machinery (SCM) *Professional Conduct and Procedures for the Enforcement of the ACM Code of Professional Conduct*
- The Institute of Electrical and Electronics Engineers (IEEE) *Code of Ethics*
- The Institute for Certification of Computer Professionals (ICCP) *Codes of Ethics and Good Practices*

While there are many similarities and major points of emphasis in the individual codes, there are some fundamental differences based primarily on the nature of the computer-related projects being conducted by the members of each organization.

Several other groups (both industry and educational) have recently offered additional Codes of Ethics, including Computer Learning Foundation (CLF), Information Systems Security Association (ISSA), and EDUCOM. An abbreviated version of each of these rather comprehensive, detailed codes follows.

British Computer Society (BCS) Code of Conduct

Level One

In the practice of your profession you will, to the extent that you are responsible:

1.0 Personal Requirements

1.1 Keep yourself, and subordinates, informed of such new technologies, practices, legal requirements and standards as are relevant to your duties.

1.2 Ensure subordinates are trained in order to be effective in their duties and to qualify for increased responsibilities.

1.3 Accept only such work as you believe you are competent to perform and not hesitate to obtain additional expertise from appropriately qualified individuals where advisable.

1.4 Actively seek opportunities for increasing efficiency and effectiveness to the benefit of the user and of the ultimate recipient.

Level Two

1.1 Others will expect you to provide skills and advice and, in order to do so, you must keep yourself up-to-date.

1.2 Take action to ensure that your hard-won knowledge and experience are passed on in such a way that those who receive them not only improve their own effectiveness in their present positions but also become keen to advance their careers and take on additional responsibilities.

1.3 You should always be aware of your own limitations and not knowingly imply that you have competence you do not possess.

1.4 Whatever the precise terms of your brief, you should always be aware of the environment surrounding it and not work solely towards completion of the defined task and no more.

Principles

As a Professional Member of the BCS, you:

1. Will behave at all times with INTEGRITY.

2. Will act with complete DISCRETION when entrusted with confidential information.

3. Will act with IMPARTIALITY when purporting to give independent advice and must disclose any relevant interest.

4. Will accept full RESPONSIBILITY for any work which you undertake and will construct and deliver that which you purport to deliver.

5. Will not seek personal advantage to the detriment of the Society.

Data Processing Management Association (DPMA)
Code of Ethics and Standards of Conduct

I acknowledge:

That I have an obligation to management, therefore, I shall promote the understanding of information processing methods and procedures to management using every resource at my command.

That I have an obligation to my fellow members, therefore, I shall uphold the high ideals of DPMA as outlined in its Association Bylaws. Further, I shall cooperate with my fellow members and shall treat them with honesty and respect at all times.

That I have an obligation to society and will participate to the best of my ability in the dissemination of knowledge pertaining to the general development and understanding of information processing. Further, I shall not use knowledge of a confidential nature to further my personal interest, nor shall I violate the privacy and confidentiality of information entrusted to me or to which I may gain access.

That I have an obligation to my employer whose trust I hold, therefore, I shall endeavor to discharge this obligation to the best of my ability, to guard my employer's interests, and to advise him or her wisely and honestly.

That I have an obligation to my country. I shall uphold my nation and shall honor the chosen way of life of my fellow citizens.

I accept these obligations as a personal responsibility and as a member of this Association. I shall actively discharge these obligations and I dedicate myself to that end.

The DPMA (Data Processing Management Association) has developed its Code of Ethics and Standards of Conduct to deal with ethical issues in the information systems industry. DPMA's membership represents a cross-section of industry professionals with a foundation of experience that dates back to the mid-50s. DPMA's focus on ethics is reflected in its membership, legislative, and educational endeavors. At DPMA, ethics will always be a major focus of the Association's activities.

Association of Computing Machinery (ACM) Code of Ethics and Professional Conduct

1.   *General Moral Imperatives*
As an ACM member I will . . .

1.1 Contribute to society and human well-being,

1.2 Avoid harm to others,

1.3 Be honest and trustworthy,

1.4 Be fair and take action not to discriminate,

1.5 Honor property rights including copyrights and patents,

1.6 Give proper credit for intellectual property,

1.7 Access computing and communication resources only when authorized to do so,

1.8 Respect the privacy of others,

1.9 Honor confidentiality.

2.   *More Specific Professional Responsibilities*
As an ACM computing professional I will . . .

2.1 Strive to achieve the highest quality in both the process and products of professional work,

2.2 Acquire and maintain professional competence,

2.3 Know and respect existing laws pertaining to professional work,

2.4 Accept and provide appropriate professional work,

2.5 Give comprehensive and thorough evaluations of computer systems and their impacts, with special emphasis on possible risks,

2.6 Honor contracts, agreements, and assigned responsibilities,

2.7 Improve public understanding of computing and its consequences.

3.   *Organizational Leadership Imperatives*
As an ACM member and an organizational leader, I will . . .

3.1 Articulate social responsibilities of members of an organizational unit and encourage full acceptance of those responsibilities,

3.2 Manage personnel and resources to design and build information systems that enhance the quality of working life,

3.3 Acknowledge and support proper and authorized uses of an organization's computing and communication resources,

3.4 Ensure that users and those who will be affected by a system have their needs clearly articulated during the assessment and design of requirements; later the system must be validated to meet requirements,

3.5 Articulate and support policies that protect the dignity of users and others affected by a computing system,

3.6 Create opportunities for members of the organization to learn the principles and limitations of computer systems.

4. Compliance with the Code
As an ACM member, I will . . .

4.1 Uphold and promote the principles of this Code,

4.2 Agree to take appropriate action leading to a remedy if the Code is violated,

4.3 Treat violations of this code as inconsistent with membership in the ACM.

Institute of Electrical and Electronic Engineers (IEEE) Code of Ethics

Article I
Members shall maintain high standards of diligence, creativity, and productivity and shall:

1. Accept responsibility for their actions.

2. Be honest and realistic in stating claims or estimates from available data.

3. Undertake technological tasks and accept responsibility only if qualified by training or experience, or after full disclosure to their employers or clients of pertinent qualifications.

4. Maintain their professional skills at the level of the state of the art, and recognize the importance of current events in their work.

5. Advance the integrity and prestige of the profession by practicing in a dignified manner and for adequate compensation.

Article II
Members shall, in their work:

1. Treat fairly all colleagues and co-workers, regardless of race, religion, sex, age, or national origin.

2. Report, publish, and disseminate freely information to others, subject to legal and proprietary restraints.

3. Encourage colleagues and co-workers to act in accord with this Code and support them when they do so.

4. Seek, accept, and offer honest criticism of work, and properly credit the contributions of others.

5. Support and participate in the activities of their professional societies.

6. Assist colleagues and co-workers in their professional development.

Article III

Members shall, in their relations with employers and clients:

1. Act as faithful agents or trustees for their employers or clients.

2. Keep information on the business affairs or technical processes of an employer or client in confidence.

3. Inform their employers, clients, professional societies, or public agencies or private agencies of which they are members of any circumstance that could lead to a conflict of interest.

4. Neither give nor accept, directly or indirectly, any gift, payment, or service of more than nominal value.

5. Assist and advise their employers or clients in anticipating the possible consequences, direct and indirect, immediate or remote, of the projects, work, or plans of which they have knowledge.

Article IV

Members shall, in fulfilling their responsibilities to the community:

1. Protect the safety, health, and welfare of the public and speak out against abuses in these areas affecting the public interest.

2. Contribute professional advice, as appropriate, to civic, charitable, or other nonprofit organizations.

3. Seek to extend public knowledge and appreciation of the profession and its achievements.

Institute for Certification of Computer Professionals (ICCP) Code of Ethics

Certified Computer Professionals, consistent with their obligation to the public at large, should promote the understanding of data processing methods and procedures using every resource at their command.

Certified Computer Professionals have an obligation to their profession to uphold the high ideals and the level of personal knowledge certified by the Certificate held. They should also encourage the dissemination of knowledge pertaining to the development of the computer profession.

Certified Computer Professionals have an obligation to serve the interests of their employers and clients loyally, diligently, and honestly.

Certified Computer Professionals must not engage in any conduct or commit any act which is discreditable to the reputation or integrity of the computer profession.

Certified Computer Professionals must not imply that the Certificates which they hold are their sole claim to professional competence.

Codes of Conduct and Good Practice For Certified Computer Professionals (CCP)

The essential elements relating to conduct that identify a professional activity are:

- A high standard of skill and knowledge
- A confidential relationship with people served
- Public reliance on the standards of conduct and established practice
- The observance of an ethical code

Computer Learning Foundation (CLF) Code of Responsible Computing

Respect for Privacy. I will respect others' right to privacy. I will only access, look in, or use other individuals', organizations', or companies' information on computer or through telecommunications if I have the permission of the individual, organization, or company who owns the information.

Respect for Property. I will respect others' property. I will only make changes to or delete computer programs, files, or information that belong to others if I have been given permission to do so by the person, organization, or company who owns the program, file, or information.

Respect for Ownership. I will respect others' rights to ownership and to earn a living for their work. I will use only computer software, files, or information which I own or which I have been given permission to borrow. I will use only software programs which have been paid for or are in the public domain. I will make only a backup copy of computer programs I have purchased or written and will use it only if my original program is damaged. I will make copies only of computer files and information that I own or have written. I will sell only computer programs which I have written or have been authorized to sell by the author. I will pay the developer or publisher for any shareware programs I decide to use.

Respect for Others and the Law. I will only use computers, software, and related technologies for purposes that are beneficial to others, that are not harmful (physically, financially, or otherwise) to others or others' property, and that are within the law.

Information Systems Security Association, Inc. (ISSA) Code of Ethics

Membership Requirements. (i) Professionals who have as their primary responsibility information systems security in the private or public sector, or professionals supplying information systems security consulting services to the private or public sector; or (ii) educators, students, attorneys, and law enforcement officers having a vested interest in information/data security; or (iii) professionals with primary responsibility for marketing or supplying information security products or services.

Code of Ethics. The primary goal of the Information Systems Security Association, Inc. (ISSA) is to promote management practices that will ensure the confidentiality, integrity, and availability of organizational information resources. To achieve this goal, members of the Association must reflect the highest standards of ethical conduct and technical competence. Therefore, ISSA has established the following Code of Ethics and requires its observance as a prerequisite and continuation of membership and affiliation with the Association.

As an applicant for membership and as a member of ISSA, I have in the past and will in the future:

- Perform all professional activities and duties in accordance with the law and the highest ethical principles;

- Promote good information security concepts and practices;

- Maintain the confidentiality of all proprietary or otherwise sensitive information encountered in the course of professional activities;

- Discharge professional responsibilities with diligence, honesty;

- Refrain from any activities which might constitute a conflict of interest or otherwise damage the reputation of employers, the information security profession, or the Association; and

- Not intentionally injure or impugn the professional reputation or practice of colleagues, clients, or employers.

EDUCOM and ADAPSO Guide to the Ethical and Legal Use of Software for Members of the Academic Community

EDUCOM (Software Initiative, P.O. Box 364, Princeton, New Jersey 08540) is a nonprofit consortium of over 450 colleges and universities committed to the use and management of information technology in higher education.

ADAPSO (1300 North 17th Street, Suite 300, Arlington, Virginia 22209) is the computer software and services industry association.

Software enables us to accomplish many different tasks with computers. Unfortunately, to get their work done quickly and conveniently, some people justify making and using unauthorized copies of software. They may not understand the implications of their actions or the restrictions of the U.S. copyright law.

Here are some relevant facts:

1. Unauthorized copying of software is illegal. Copyright law protects software authors and publishers, just as patent law protects inventors.

2. Unauthorized copying of software by individuals can harm the entire academic community. If unauthorized copying proliferates on a campus, the institution may find it more difficult to negotiate agreements that would make software more widely and less expensively available to members of the academic community.

3. Unauthorized copying of software can deprive developers of a fair return for their work, increase prices, reduce the level of future support and enhancement, and inhibit the development of new software products.

Respect for the intellectual work and property of others has traditionally been essential to the mission of colleges and universities. As members of the academic community, we value the free exchange of ideas. Just as we do not tolerate plagiarism, we do not condone the unauthorized copying of software, including programs, applications, databases, and code.

Therefore, we offer the following statement of principle about intellectual property and the legal and ethical use of software. This "code"—intended for adaption and use by individual colleges and universities—was developed by the EDUCOM Software Initiative.

Software and Intellectual Rights. Respect for intellectual labor and creativity is vital to academic discourse and enterprise. This principle applies to works of all authors and publishers in all media. It encompasses respect for the right to determine the form, manner, and terms of publication and distribution.

Because electronic information is volatile and easily reproduced, respect for the work and personal expression of others is especially critical in computer environments. Violations of authorial integrity, including plagiarism, invasion of privacy, unauthorized access, and trade secret and copyright violations, may be grounds for sanctions against members of the academic community.

Restrictions on the use of software are far from uniform. You should check carefully each piece of software and the accompanying documentation yourself. In general, you do not have the right to:

1. Receive and use unauthorized copies of software
2. Make unauthorized copies of software for others

Objective of Codes

The primary objective of most established Codes of Ethics is to give some "common ground" to an acceptable standard. Donn Parker, in *Ethical Conflicts in Computer Science and Technology,* states that:

> Unlike the computer field, other sciences and professions have had hundreds of years in which to develop ethical concepts that form the basis for dealing with new issues . . . Computer science and technology have been in existence for only 30 years.[43]

Figure 11.5 and 11.6 point out some of the secondary objectives of codes and some of the possible side-effects that should be addressed.

CORPORATE POLICIES ON ETHICS

Donn B. Parker in his article, "Ethics for Information Systems Personnel," stated that:

> As more corporate resources become directly accessible through the computer, opportunities multiply for information systems professionals to misappropriate those resources. As a result, MIS managers increasingly find themselves forced to make difficult decisions involving ethics and standards of conduct—decisions

FIGURE 11.5
Secondary Objectives
of Codes of Ethics

☑ To enhance the image of the profession in the public eye

☑ To protect the monopoly of the profession in question

☑ To have the code serve as a status symbol—one of the credentials for the occupation

FIGURE 11.6
Side-Effects of
Codes of Ethics

☑ A false sense of complacency to professionals about their conduct

☑ A sense of self-congratulation—codes can be used as a cover-up for "unethical" or "irresponsible" conduct

☑ A tendency to divert attention from the macro-ethical problems to the micro-ethical problems—do we scrutinize as a collective body?

☑ To discourage, if not suppress, the dissenter, the innovator, the critic

that are further complicated by the absence of specific ethical guidelines. This article discusses the role of ethics in information systems and offers practical suggestions for promoting ethical behavior and establishing a workable code of conduct.[44]

Those who argue that information systems professionals are more ethical than those in other occupations can usually offer several reasons. They state that IS professionals:

- Have fewer opportunities for unethical conduct because they do not have direct access to assets that could easily be converted to personal gain

- Closely identify with their organizations' goals and activities

- Are unwilling to jeopardize careers that usually have been built on substantial educational investments

- May sense that, because they work in a high-technology field that pays well and offers an abundance of job opportunities, society expects them to follow a high standard of conduct

- Feel a special loyalty to their colleagues and specialized profession, which encourages a high standard of conduct[44]

Plan of Action

With more and more business activities going on-line and the general public becoming increasingly computer literate, unethical computer use is taking place at an alarming pace. Donn Parker suggests the following plan of action for codifying conduct and enforcing ethical standards:

Step 1.   Formulate a code of conduct.

Step 2.   Establish rules of procedure.

Step 3.   Establish sanctions.

Step 4.   Recognize ethical behavior in job performance.

Step 5.   Focus attention on ethics.

Step 6.   Promote computer crime laws.

Step 7.   Impose accountability and minimize temptations.

Step 8.   Encourage the use of rehabilitation programs.

Step 9.   Encourage participation in professional societies.

Step 10. Set an example.[44]

Codes Must Be Realistic

Are we asking too much of computer users? The following statements are generally accepted by most computer practitioners.

- Ethical behavior toward information and information resources does not come naturally to most people.
- The rules of ethical behavior are not intuitively obvious when it comes to information.

People may not always apply their norms of ethical behavior about tangible assets to information and computers for the following reasons:

- Ethics focuses on our relations with others and their property. Information technology can alter existing relationships and create new and unfamiliar relationships.
- Intangible property *is different,* and electronics has made that difference even more difficult to deal with.
- There is a collision of rights concerning information. Freedom of expression, freedom of information, privacy, and protection of intellectual property often conflict. Sorting out priorities is difficult, especially in electronic environments.
- There is a conflict between our natural urge to communicate and our urge to protect property.

The following are guidelines for actions that should be taken:

1. Make the scope realistic.
2. Make it specific to your organization.
3. Role play.

4. What are you really saying?

5. Ethical codes should guide, not trap.

6. Involve local management directly and extensively.

7. Peer pressure.

8. Commitment by example.

Society vs. Technology

Do our present norms of social behavior carry over to our use of computers? (See Figure 11.7.) As our society becomes more and more computer-dependent, we must address this question. To be effective, an ethic must:

- Be adaptable
- Outline user's rights
- Allow equal access
- Be based on existing laws
- Place time limits on use
- Provide guidelines for use
- Outline manufacturer's responsibilities
- Exert penalties[6]

ACADEMIC INSTITUTIONS INTEGRATING ETHICS INTO CLASSES

With the recent wave of Wall Street scandals, the stories of computer crimes and viruses, and the often-occurring Capital Hill illegalities, one can't help but wonder about the current state of ethics. Countless journal and newspaper articles, books, television programs, and movies are concerned with ethics, or more precisely, the lack of ethics in business, politics, school, and everyday life.

Accreditation bodies such as the American Assembly of Collegiate Schools of Business (AACSB) and the Association for Computing Machinery (ACM), in conjunction with business people, are now placing more rigorous emphasis on the incorporation of ethical issues into the curriculum. Many educational institutions, from grade schools to universities, are drafting formal Ethical Use of Computing Codes for students, as well as orienting students to ethical standards in many of their courses.

"It is unrealistic to put newly graduated IS graduates into an ethical conflict and expect them to make the correct decision if they've never been 'trained' in the issues of professional ethics."[49]

Student Survey

In April 1989, two surveys were conducted at James Madison University, Harrisonburg, Virginia, under the auspices of the Dominion Fellowship Grant by Dr. Karen A. Forcht and Anne Myong to ascertain the level of ethical awareness and practice by college students and practitioners.

FIGURE 11.7

Social Behavior vs. Computer Behavior

Unacceptable Social Behavior	Unacceptable "Computer" Behavior
1. To knowingly infect another person with a communicable disease	To knowingly infect another person's computer with a virus or worm
2. To enter another person's home or drive another person's car without their permission	To enter another person's system without permission
3. To rummage through another person's belongings	To rummage through another person's database(s)
4. To shoplift or steal something that belongs to another	To make copies of copyrighted software
5. To keep the extra money if a store clerk gives us back too much change or to pay a lower price because merchandise is priced wrong	To access another's system or data because the system allows us to do so
6. To lock another person out of his or her own household or car	To deny someone the use of his or her computer

This survey targeted students mainly from James Madison University's College of Business and spanned sophomore students through MBAs. The information was solicited from the participants by utilizing a questionnaire that included key factors such as major field of study, demographics, and other personal information such as career paths, how respondents viewed themselves and their peers morally and ethically, and their personal experience with computer misuse.

The participants in the study ranged in age from 19 to 45, with a heavy concentration in the areas of Accounting, Finance, Computer Information Systems, and MBA candidates. Most of the students were from cities ranging in population from 50,000 to 750,000+ residents. Family income was high, with the heavily weighted median income being $75,000 a year or more.

Most of the students surveyed had previously had computer experience in the workplace, ranging from data entry and word processing to operations and specialized internships in the computer area. When asked if they had engaged in any form of illegal computer use, whether it be software piracy or some form of hacking, almost half of the participants admitted to using the computer for unethical means. Male hackers definitely outnumbered the females, and the majority of these offenders seemed to be in the senior level of college and in a computer-related area of study. It is ironic and perhaps hypocritical that this same age group is adamant about their own morals and ethics, which they judge to be very high.

Students who were majoring in Accounting and Computer Information Systems were more aware of formal ethical statements and honor codes of the university than any other major. This could be attributed to the importance of accurate information produced by these two areas and the means to ensure that the information is indeed correct (i.e., IRS auditors, security officers).

Alarmingly, although CIS majors and MBA candidates are aware of the ethical concerns, they are the foremost group of student hackers of all surveyed. This finding should cause great concern because these future consultants, bankers, and government officials will be working with extremely sensitive information, and yet their ethical standards are lacking at this very early stage in their careers.

A comment from one of the respondents seems to sum up the dilemma quite adequately:

> I think today, more than ever, students are learning that it is more practical and safe to use the business ethics that they are taught while still in school. However, many times when the students get in a real-world situation, they may feel they have to do certain things just to stay competitive.[21]

Practitioner Survey

A questionnaire was mailed by Karen A. Forcht and Anne Myong of James Madison University, Harrisonburg, Virginia, to the Chief Executive Officers (CEOs) of the Datamation 100 companies to ascertain their assessments concerning the ethical standards that have been formally adopted by their organizations and to seek their opinions about the ethical environment that may be present in their organization. The data analysis indicated that, for the most part, the CEOs responding adhere to a very high standard of personal ethical conduct and computer use. Furthermore, and most importantly, they expect (and require) that their employees follow ethical standards. This ethical attitude is reinforced by ethics codes, ethics awareness programs, and sanctions/reprimands of offending employees.

Some of the major survey results follow:

1. When asked whether it was possible to teach ethical behavior in a classroom, rather than being learned "on the job," over 75% felt that ethics could be acquired in a classroom setting.

2. When asked whether companies should require all employees to sign an ethics oath before beginning work, over 50% agreed.

3. When asked whether companies/organizations should develop and administer an ethics awareness program for all employees, over 75% agreed.

4. When asked whether colleges and universities should incorporate an ethical use of computers course in their present curriculum, almost half (46.77%) agreed and 20% strongly agreed.

5. Over 80% of the respondents reported that their organizations have a formal ethics policy. Almost three-quarters (73.3%) were American companies, while only 23.3% of the foreign companies have a formal ethics policy.

6. Most of the respondents, when asked how public figures can best promote good ethics, said "by setting a good example."[17]

Conclusion

These two surveys shed a great deal of light on the ethics awareness dilemma facing education and industry. Even though both groups, students and practitioners, seem to follow a very high personal standard of ethics, morals, and laws, many feel that too often compromise is evident (and necessary) in the workplace in order to stay competitive.

Perhaps if educational institutions and the computer industry work together to foster an attitude of ethical use of computers, the outcome will be a favorable, and acceptable, one. The unique and varied challenges we face in this age of information are truly unprecedented. How we achieve a balance between intellectual/professional growth and ethical compromise—and yet remain "in the ballpark"—is indeed the paramount challenge.[10]

Additional Surveys

David B. Paradise of Texas A & M University offers the results of a similar study of students:

> A survey examining the ethical attitudes of over 125 upper-division undergraduate business students was administered. As this sample contained students from throughout the southwestern United States that will be accepting jobs in a matter of months, this group is representative of entry-level employees in this United States area. Frequently, entry-level employees are naive regarding "accepted" corporate behavior. "Acceptable behavior" may be a particularly ambiguous concept in the information systems field, since the field is still relatively young and is evolving at a tremendously rapid pace. The survey focused on situations containing computer-based systems and investigated motives of obligation, opportunity, and intent. Responses by MIS subjects were compared to non-MIS subjects. Some statistically significant differences between the groups appeared: non-MIS subjects felt programmers should be responsible for the inherent correctness of calculations embodied in programs, MIS subjects indicated stronger notions of professional responsibility, and MIS subjects were more tolerant on the issue of unauthorized software copying. These results indicate entry-level personnel should be given specific guidelines regarding acceptable corporate behavior.[42]

Karen-Ann Kievit of Loyola Marymount University, Los Angeles, California, reports from her study of students that:

> Both majors and non-majors tended to choose the more "ethical" response to most of the choices in each of the scenarios. There was a significant difference between majors and non-majors in four of seven scenarios (job offer from competing firm and taking software; job offer from non-competing firm and taking software; receiving software on a trial basis and making a copy for review purposes due to lack of time to review; finding a password lying around).
>
> This study shows the responses of undergraduate majors and non-majors at two universities in a large, urban metropolitan area. These responses may be different than those of students in rural areas, at different size institutions, geographical location, demographic composition, or class level (undergraduate/graduate). In addition, ethnic/cultural background may play some factor in the response due to the values taught in each culture.[33]

As educational institutions struggle to develop an ethical student who will be, in turn, hired by industry, ethical behavior will, no doubt, continue to be of paramount concern to both academics and practitioners. These studies shed some light on how far we've come and how far we have to go.

SCENARIOS TO PONDER

In each of the following scenarios, consider the legal, ethical, and moral ramifications:

1. A computer system user has given inadequate specifications for an application being developed internally. The analyst and programmers are held responsible for the program's performance according to the specifications agreed on by both parties. Who should be ultimately responsible?

2. A program written in-house contains an obvious weakness, but is generally accepted by the users and programming staff. Who is liable if this weakness causes a problem later?

3. A consultant uses information obtained while working with one company to give a competitive advantage to a second client. Is the consultant or the client at risk?

4. A software product purchased/leased from a vendor contains some flaws (discovered only after possession has taken place). Is the vendor responsible?

5. A log of computer use shows that an employee has attempted on several occasions to obtain access to a confidential file. Should the company institute sanctions against this employee?

6. A computer center employee is "borrowing" computer time to run a private programming service for outside clients. How should the company handle this?

7. A programmer develops a payroll program while being employed by a company and later writes a similar program for a competitor. Is there a conflict?

8. A programmer complains to management that he will not sign a non-disclosure/ethics oath as a condition of his employment. Can the company refuse to continue his employment?

9. An employee is overheard in the lunchroom talking about confidential/private information that he has viewed during the course of his job. What can the company do about this?

10. An employee copies software that the company has purchased so she can take the program home and work on it. Is this acceptable?

11. An employee falsifies his employee records so that he will get a larger raise this year. How should the company react?

12. An employee uses questionable means to obtain data/information about a competitor so that her company can have "a jump on the gun." Who is responsible—the employee or the company?

13. An employee sells time on the company's computer to outside clients—unknown to the company. What should be done about this?

14. A college student falsifies his transcript to obtain entrance to graduate school or to get a job. Should he be dismissed for deception?

15. A hacker enters an unauthorized file and just snoops. Is this illegal since nothing was changed?

SUMMARY

Within the computer industry, production is up, as well as quality. There is heavy debate over the state of ethics involving individual privacy and rights in the industry.

Does the IS department have a special challenge and ethical responsibility because of its close ties with organizational data and information? There are justifiable arguments for both sides of this subject. The IS manager has a critical job in communicating the importance of security and integrity to IS employees.

Individual ethics and organizational ethics must be combined cohesively for a business to operate effectively. There is also much debate over how much organizational ethics can promote ethical behavior. One view is that what really determines the degree of ethical behavior is what type of person each individual is.

In discussing individual ethics, keep in mind that there are responsibilities not only to the company but also to society, the profession, and others, as well as to oneself.

Essentially, the practice of information gathering in business is a standard one. It can involve ethical questions if the methods used are illegal or even in "grey areas" of professional judgment.

The issue of established codes of ethics raises questions of feasibility due to the relative newness of the IS profession. It also brings out questions involving realistic effectiveness of codes that consist of broad, general guidelines instead of specific rules. Along with those questions are concerns of actual enforcement. Today, even professional associations that have chartered codes of ethics have weak mechanisms to deal with reported breaches of their codes.

Ethics issues in the computer industry still lay unresolved and, with its fast-paced growth, the situation is not likely to simplify.

This chapter discusses **normative ethics systems** and applies these principles to the use of computers. **Ethics** is described as the science of social survival. The responsibility for ethical use of computers falls on all users, academia, and practitioners. The Information Age presents ethical issues of privacy, accuracy, property, and accessibility. Many professional organizations, both academic institutions and industry, have adopted formal Codes of Ethics to serve as guidance to users and as sanctions against violators.

REFERENCES

1. Axline, Larry, "The High-Ethics IS Manager," *Information Executive*, Vol. 2, No. 4, Fall 1989, pp. 21-24.

2. Brady, F. Neil, *Ethical Managing . . . Rules and Results* (New York: Macmillan, 1990).

3. *Code of Ethics*, Information Systems Security Association, Newport Beach, CA.

4. *Code of Responsible Computing*, Computer Learning Foundation, Palo Alto, CA.

5. Cohen, E., and L. Cornwell, "College Students Believe Piracy Is Acceptable," *CIS Educator Forum*, Vol. 1, No. 3, March 1989, pp. 2-5.

6. "Computer Ethics," *The Futurist*, August 1984, pp. 68-69.

7. Cougar, J. Daniel, "Preparing IS Students to Deal with Ethical Issues," *MIS Quarterly*, Vol. 3, No. 2, June 1989, pp. 211-218.

8. Ethics Resource Center, Inc., Washington, D.C.

9. "Ethics Violators Often in Midlife," *Wall Street Journal*, October 4, 1988.

10. Forcht, Karen A., "Assessing the Ethical Standards and Policies in Computer-Based Environments," *Ethical Issues in Information Systems*, eds. R. Dejoie, G. Fowler, and D. Paradice (Boston, MA: boyd & fraser Publishing Company, 1991), pp. 56-69.

11. Forcht, Karen A., "Developing An Attitude of Ethical Use of Computers," *IBSCUG Newsletter,* May 1989, p. 21.

12. Forcht, Karen A., "Developing Awareness of Computer Ethics," *Computer Security Institute*, 14th Annual Computer Security Conference, Anaheim, California, November 1987.

13. Forcht, Karen A., "Developing Ethical Attitudes of College Computer-Using Students," *Proceedings of National Computer Security Conference*, Baltimore, Maryland, October 1989.

14. Forcht, Karen A., "Establishing a Code of Ethics in Computer-Based Environments," *Proceedings of Disaster Recovery Symposium and Exhibition*, Atlanta, Georgia, September 1989.

15. Forcht, Karen A., "Establishing an Ethical Environment in Computer-Related Classrooms . . . Results of Recent Research," *Proceedings of WCCE/90, Fifth World Conference on Computers in Education*, Sydney, Australia, July 1990.

16. Forcht, Karen A., "Ethical Use of Computers," *Proceedings of International Conference on Information Systems*, Boston, Massachusetts, December 1989.

17. Forcht, Karen A., "Ethics in Information Systems," *Proceedings of Minnesota Joint Computer Conference*, Minneapolis, Minnesota, February 1990.

18. Forcht, Karen A., Charles Bilbrey, and Richard Christoph, "DP Courses Don't Include Ethics Study," *Government Computer News*, July 4, 1986, p. 11.

19. Forcht, Karen A., Richard Christoph, and Charles Bilbrey, "The Development of Information Systems Ethics: An Analysis," *Journal of Computer Information Systems*, Winter 1987/88, p. 31.

20. Forcht, Karen A., and J. K. Pierson, "Developing Computer Security Awareness," *IRM Bulletin*, November 1987, p. 2.

21. Forcht, Karen A., and Anne Myong, "A Student Assessment of Computer Ethics . . . Findings of a Recent Survey," *IBSCUG Newsletter*, Fall 1989, p. 19.

22. Forcht, Karen A., Ben Bauman, and Joan Pierson, "Developing Awareness of Computer Ethics," *Proceedings of ACMSIGCPR International Conference on the Management of Information Systems Personnel*, College Park, Maryland, April 1988.

23. Forcht, Karen A., Ben Bauman, and J. K. Pierson, "Developing Awareness of Computer Ethics," *Proceedings of Information Systems Security Association 2nd Annual East Coast Regional Meeting*, Timonium, Maryland, August 1988.

24. Forcht, Karen A., J. K. Pierson, and Ben Bauman, "Developing Awareness of Computer Ethics," *Proceedings of the Ninth International Conference on Information Systems*, Minneapolis, Minnesota, November 30, 1988.

25. Forcht, Karen A., J. K. Pierson, and Ben Bauman, "Developing Awareness of Computer Ethics," *PanPacific Business Association Conference*, Sydney, Australia, May 1989.

26. Forcht, Karen A., Joan Pierson, and Ben Bauman, "Halting Computer Crime and Abuse: Management's Newest Challenge," *Journal of Applied Business Research*, Fall 1989, p. 52.

27. Forester, Tom, and Perry Morrison, *Computer Ethics: Cautionary Tales and Ethical Dilemmas in Computing* (Cambridge, MA: MIT Press, 1990).

28. Hilton, T.S.E., "A Framework for Teaching Computer Ethics," *Instructional Strategies*, Vol. 5, No. 3, Summer 1989.

29. Hilton, Thomas, "Information Ethics . . . Content and Pedagogy," *Proceedings of the Annual Convention of the International Association for Computer Information Systems*, October 1991.

30. Hilton, Thomas, "Information Ethics Among the Fortune 500," *Proceedings of the 32nd Mountain Plains Management Conference*, October 1990.

31. Johnson, Deborah G., *Computer Ethics* (Englewood Cliffs, NJ: Prentice-Hall, 1985).

32. Johnson, Deborah G., and John W. Snapper, *Ethical Issues in the Use of Computers* (Belmont, CA: Wadsworth Publishing Company, 1985), pp. 11-12.

33. Kievit, Karen-Ann, "Information Systems Majors/Non-Majors and Computer Ethics," *The Journal of Computer Information Systems*, Fall 1991, pp. 43-49.

34. Kohlberg, L., and Turiel, E., *Moralization: The Cognitive Development Approach* (New York: Holt, Rinehart, and Winston, 1973).

35. Levine, Marilyn M., "Ethical Space," *Information Express*, July 1986.

36. Manning, George and Kent Curtis, *Ethics at Work* (Cincinnati, OH: South-Western Publishing Company, 1988).

37. Mason, Richard O., "Four Ethical Issues of the Information Age," *MIS Quarterly*, March 1986, pp. 5-12.

38. McEwen, J. Thomas, "Computer Ethics," *National Institute of Justice Reports*, January/February 1991, pp. 8-11.

39. Nash, Jim, "Technology Raises Many New Ethics Questions," *Computerworld*, October 14, 1991, p. 87.

40. Neumann, Peter G., "Computers, Ethics, and Values," *Communications of the ACM*, Vol. 34, No. 7, July 1991, p. 106.

41. *New Ethics for the Computer Age?*, The Brookings Institute, Washington, D.C., 1988.

42. Paradice, David B., "Ethical Attitudes of Entry-Level MIS Personnel," *Information and Management*, Vol. 18, No. 3, 1990, pp. 143-151.

43. Parker, Donn B., *Ethical Conflicts in Computer Science and Technology* (Arlington, VA: AFIPS Press, 1986), pp. 1-201.

44. Parker, Donn B., "Ethics for Information Systems Personnel," *Journal of Information Systems Management*, Vol. 5, No. 3, Summer 1988, pp. 44-48.

45. Rifkin, Glenn, "Are Corporate Codes Enough? Maybe Not," *Computerworld*, October 14, 1991, p. 86.

46. Rifkin, Glenn, "The Ethics Gap," *Computerworld*, October 14, 1991, pp. 83-85.

47. Sieber, Ulrich, *The International Handbook on Computer Crime* (New York: John Wiley and Sons, 1986), pp. 1-276.

48. Slater, Derek, "New Crop of IS Pros on Shaky Ground," *Computerworld*, October 14, 1991, p. 90.

49. Smith, Mark W., "Professional Ethics in the Information Systems Classroom: Getting Started!" *Journal of Information Systems Education*, Vol. 4, No. 1, Spring 1992, pp. 6-9.

50. Spiro, Bruce E., "Ethics in the Information Age," *Information Executive*, Vol. 2, No. 4, Fall 1989, pp. 38-41.

51. Van Duyn, Julia, *The Human Factor in Computer Crime* (Princeton, NJ: Petrocelli Books, 1985), pp. 1-162.

52. Vitell, Scott J., and Donald L. Davis, "Ethical Beliefs of MIS Professionals: The Frequency and Opportunity for Unethical Behavior," *Journal of Business Ethics*, May 1990, pp. 63-70.

53. Ward, Gerald M., and Jonathan D. Harris, *Managing Computer Risk: A Guide for the Policymaker* (New York: John Wiley and Sons, 1986).

54. Wood, Wallace A., "A View of Computer Ethics by Managers and Students," *The Journal of Computer Information Systems*, Winter 1991-92, pp. 7-10.

REVIEW QUESTIONS

1. Why is the issue of ethical use of computers becoming important?

2. Briefly describe what constitutes a normative ethical system.

3. Compare laws, ethics, and morals.

4. Why do variations in ethics and morals exist?

5. Describe the stages of moral development.

6. What is a crime of technology and intellect?

7. What is the Information Age?

8. What are the four basic ethical issues of the Information Age?

9. What are some of the issues in a Code of Ethics?

10. What are the objectives of Codes of Ethics?

DISCUSSION QUESTIONS

1. "Ethics are completely intertwined with corporate culture." Can a company influence employees, or is the reality that employees influence the company?

2. "Reestablishing and reaffirming a national commitment to ethics won't happen in a year or perhaps a decade. It has taken many years for us to reach our present state of decline, and reversing this trend will be a long and arduous task." Discuss this statement and its implication on computer users.

3. "Publishers do not want their software copied because they and the creator of the software want to make a fair profit on their investment. When a program sells for $40.00, buyers often think that the price is unfair since a blank diskette costs only $2.50. They say a profit of 625 percent is unfair! It costs much more than $2.50, however, to produce a software program." What are the ethical issues in this statement?

4. "Social psychologists point out that human behavior is influenced by those we perceive to be authorities." How can industry and academia "show by example" the acceptable ethical conduct of computing?

5. Jay BloomBecker, Director of the National Center for Computer Crime Data in Santa Cruz, California, frequently lectures on ethics to computer classes around the country. Recently a student commented during his lecture that "being ethical only allows other people to take advantage of you . . . but I'm not a computer criminal." Comment on this student's statement.

EXERCISES

1. Interview a key individual in your school's computing center and find out if he or she is aware of Codes of Ethics and if he or she has adopted or plans to adopt one for faculty, staff, and students.

2. Locate a periodical or a newspaper article that reports on Codes of Ethics.

3. Interview a person in an organization to determine if his or her company has a Code of Ethics. Does their code include computer use?

4. Develop a short listing of scenarios and ask a group of students to comment on whether the actions are legal, ethical, or moral. Tally your results.

5. Develop a poster or banner that could be posted in your school's computer center to foster computer ethics.

PROBLEM-SOLVING EXERCISES

1. You have purchased a software package to use. You paid for it personally. The license agreement stipulates "you may use the program on a single machine." You want to make a copy to use on your home computer. You will make sure that you are the only person using the package. This approach appears to adhere to the "spirit" though not the "letter" of the license agreement. Is this consistent with a code of ethics?

2. You develop a system for inventory control that is very complex and very costly in terms of the amount of time and effort needed for development. A company, possible a rival, secures a copy from unknown sources. What do you do?

3. A university course requires students to prepare laboratory assignments. Students are given the assignments during class and are required to submit the assignment by the next week's class. The professor discovers several cases where different students submit similar assignment results. When confronted, the students admit to working together, but plead ignorance of knowing that submitting similar copies is wrong. What action should the professor take?

4. State your opinion concerning the following individuals or situations in terms of legal, moral, or ethical issues. (In some cases, all three may apply.) Your responses should represent *your* personal opinion, not necessarily a broader viewpoint (corporation, society as a whole, religious body, etc.).

   a. A recognizable public figure is exposed as having lied about a personal situation

   b. The actions of Bonnie and Clyde

   c. Col. Oliver North's actions

   d. Falsifying one's income for IRS tax records

    e. Discussing with friends information of a confidential nature

    f. Covering up mistakes on the job in order to gain a promotion or raise in pay

    g. Actions of Jim and Tammy Bakker

5. Do you agree or disagree with the following statements? Answer "yes" or "no."

    a. It is acceptable for employees to copy proprietary software to evaluate for future purchases.

    b. In information systems processing, the human factor is considered to be less critical than the technological factors.

    c. It is not really possible to teach ethical behavior in a classroom; rather, it must be learned on the job.

    d. Employees should be allowed to recreate a product/program/design for another company if they change jobs and are no longer employed by the company who paid them to create it.

    e. Computer center employees should be more closely scrutinized than other employees as they have more information access.

CASES

Rockcreek High School

Mr. Scott faces a difficult decision. He is chairman of Rockcreek High School's computer purchasing committee. This year he hoped to purchase four new computer systems at a cost of $1,500 each and to buy $4,000 worth of software. However, his budget was reduced from $10,000 to $7,000.

Two members of a five-member committee have recommended that only two computers be purchased, largely because the school desperately needs software for all teachers. The two members also believe that the school should wait another year before introducing two computers into the English classes.

The three other members, however, have recommended that the software budget be reduced to $1,000 and that four computers be purchased, as requested. One of the members suggested that the school purchase $1,000 worth of software and then make enough copies for all the teachers. This way the school would have both the needed computers and software.

Mr. Scott knows that it is against the law to copy the software, but he also knows that other schools are copying software. By copying the software, the committee would be able to satisfy everyone.

1. What alternatives have been recommended by the committee members?

2. What is Mr. Scott's ethical dilemma?

3. Should Mr. Scott copy the software and keep the same number of computers? Or should he sacrifice one or more computers to buy the software?

4. If you were a student of Mr. Scott, would you report him for violating the law if he copied the software? To whom would you report him?

Basketball Player's Dilemma

Bill, a good friend of yours, is in three of your classes, and computers are used in one of the classes. Students have been told that assigned work on the computer will not be accepted late. Bill is a star basketball player and misses classes when the team plays out-of-town games. He has a computer at home, and he could do the assignments there, but he does not have the software needed to do the work. Bill is behind in his assignments, and if his work isn't turned in on time, he faces getting a D in the class. He will be suspended from the team, and his team is likely to lose out on their bid for the tournament. As a good friend, Bill has asked you to make a copy of the software for him so that he can do his assignments at home and make a passing grade in the course.

1. Would you make a copy of the software for Bill?
2. Is it unethical for Bill to ask you to copy the software?
3. What alternatives could you take instead of copying the software?

MANAGERIAL ISSUES

MANAGERIAL ISSUES

LEARNING OBJECTIVES

After studying this chapter, you will be able to:

1. Understand the basic managerial functions of planning, staffing, organizing, directing, and controlling.
2. Explain why managers must set and meet organizational security objectives.
3. Explain the process of long-range goal setting in relationship to security objectives.
4. Describe the basic tenets of a security policy.
5. Explain information classification, ownership, and valuation.
6. List the desired characteristics and normal functions of the Data Security Officer.
7. Explain the various placements of security in the organization.
8. Describe the various computer center job classifications and the security considerations of the individual positions.
9. Explain the make-up and use of tiger teams.
10. Consider the effect of budget constraints on the security function.
11. Explain how computer security awareness and training can be achieved in an organization.
12. Describe the function of security audit procedures.
13. Consider the impact of computer monitoring.
14. Define the following terms.

TERMS

- Human Resources
- Capital Resources
- Information Classification
- Information Ownership
- Information Valuation
- Data Security Officer (DSO)

- Tiger Team
- Chargeback System
- Orientation
- Compliance Training
- Authorization
- Preprocessing Review

TERMS (Continued)

- Batching
- Control Totals
- Conversion Verification
- Key Verification
- Check Digit
- Logic Tests
- Control Total Balancing
- Input Security Controls
- Correct File Controls
- Programmed Controls
- Validity Test

- Completeness Test
- Matching Test
- Reasonableness or Limit Test
- Error Correction Procedures
- Recovery and Backup Procedures
- Postprocessing Review
- Master File Controls
- Periodic Internal Audit
- Whistleblowing
- Electronic Monitoring

STUDY FINDS MANAGERS FEEL INSECURE

Most managers in most organizations have little confidence that data security requirements are being met. That is the finding of a recent study by Ernst and Whinney, which found that:

- 62% of managers believe security risks are rising
- 75% say they are taking substantial steps to implement security policies
- Only 6% say security measures taken by their organizations are adequate

On the positive side, "Awareness of the issues is on the upswing," according to Toni Fisher, co-director of Ernst and Whinney's Information Security Services Group. "But as is the case in many areas, awareness is one thing; implementation is another."

The focus has shifted from more general management issues to security and contingency planning, however. More management people are involved in IS concerns

Continued

STUDY FINDS MANAGERS FEEL INSECURE

From page 371
because they have become a major part of running a business. Organizations that have addressed security issues such as disaster recovery are usually more apt to have established data security procedures. Unfortunately, Fisher points out, most companies have adopted the "Scarlet O'Hara plan: 'I'll think about it tomorrow.'"

The respondents of the Ernst & Whinney survey point out the following issues as priorities for IS and other managers:

- Protecting data from the competition—whether in government or industry
- Employee security
- Data classification and network security
- End-user authentication

As reported in Alper, Alan, "Management Insecure About Security Measures," *Computerworld*, March 2, 1987, pp. 57+.

Now that we have looked at the legal and ethical issues of computer security, several other issues need to be discussed so that the behavioral side of security balances with the technical side. As stated earlier in this book, computer security is a "people" problem, as well as a technical problem. This chapter will address the managerial issues of computer security—"controlling the monster that computers have created." Every asset of an organization must be successfully managed to ensure the efficient operation and acceptable outcome of the product or service. Information (and the computers and people who create it) should be managed as carefully as any other asset.

DETERMINATION OF GOALS AND PRIORITIES

The emergence of computer security as a major problem has been caused by the relative success of the computer and its proliferation. . . Quite significantly, data processing usually involves a concentration of vital facilities and assets. The rapid growth and acceptance of computer technology has not been accompanied by a parallel growth in the management of this technology. The management log, coupled with the fact of concentration, gives rise to an exposure that can be detrimental to the very existence of even a large organization.[22]

Management Responsibility

Information—its creation, protection, and dissemination—is critical to the ongoing operation of the organization. All managers must use the resources at their disposal to meet corporate objectives by performing the following functions:

- Planning
- Staffing
- Organizing
- Directing
- Controlling

Traditionally, managers have approached productivity improvement through work simplification and employee motivation. Recently, management has been looking to the information resource for competitive advantage. "Management must concern itself with safeguarding the resources under its jurisdiction. The manager's function is essentially the management of resources: human resources and capital resources."[22] In the computing environment, **human resources** are represented by skills needed to operate and control both hardware and software facilities. **Capital resources** are represented by the investment in equipment and operating programs.[22]

Meeting Organizations' Security Objectives

"In all organizations, the implementation of a new or revised system security policy should start with a formal announcement of management objectives and plans."[31] In order to make an effective strategic appraisal of their organization's security policies and procedures, managers must:

1. Identify the business consequences of poor security
2. Evaluate the risk of each of these consequences materializing (by considering each of the threats that could result in the particular consequence)
3. If this level of risk cannot be tolerated by the organization, then identify which of the appropriate control techniques reduce this risk at a reasonable cost

Page Assured Systems, Inc., of Fort Lee, New Jersey, describes management's commitment in the following way.

Modern organizations are increasingly dependent upon the critical resource of information and its supporting data processing activities. Special vulnerabilities arise because of this dependence. Loss or impairment of information processing functions can threaten survival. Inadequate systems or incorrect results can undermine organizational viability.

Managers are individually and collectively responsible for the integrity and security of information-related assets. In today's environment a manager may be held personally liable for security oversights or inadequate control. Government regulations, stockholder lawsuits, and other court cases have underscored the need for improved management attention to security and control—and the painful alternatives. Management today must get it right, get it secure, get it in control, *and* most importantly, **get it going**. The challenge is to ensure the adequate control, effectiveness, and security of data processing functions. **No** manager is exempt from this collective accountability. Proper control of data processing must include:

- Integrity and quality of information
- Security of data processing resources (information, software, hardware, personnel)
- Cost-effectiveness
- Balancing control and productivity

Management Commitment to Security

The security exposures resulting from lack of management commitment can be insurmountable. Even if the highest ranking official within the corporation believes that security is important, a secured computing environment is not ensured. Top management needs to propagate through the organization a definite commitment to security in the form of corporate resources. Some potential exposures resulting from top management's failure to make this commitment include the following:

1. Inadequate control over security software
2. Confusion about data ownership
3. Lack of security awareness

Long-Range Goals

To make security a part of an organization's culture, managers must view security as a cooperative effort that is a part of the "corporate culture." Proper security planning requires a focus on long-range goals rather than quick fixes for short-term security problems. With the many advances in technology, security is an issue that must be kept in the forefront of management concerns.

In his article titled "Everybody's Business," Charles Connolly identifies eight fundamental steps for successfully making security a part of an organization's culture.

1. **Obtain commitment from top management.** Management must take the initiative and ensure that the employees understand the security plan. In doing so, management must "recognize that positive attitudes and enthusiasm are contagious, lead by example, participate in the improvement process, know in detail how to improve security, and continually monitor the program."

2. **Create a security steering committee consisting of senior management and labor representatives** who would be responsible for tailoring the improvement process specifically for the corporation. This committee should be responsible for developing awareness of security issues through conducting educational programs.

3. **Ensure management participation.** Managers must realize that the security program applies equally to all departments and that active employee participation is fundamental to the success of the program. Work is a "cooperative

venture" and management must institute trust in its employees by sharing problem-solving responsibilities with them.

4. **Encourage teamwork.** Encourage active employee participation in the program by forming process improvement teams, task forces, or quality circles.

5. **Get individuals involved** by giving them an opportunity to contribute to the program and to be recognized for these contributions.

6. **Work to improve the system.** Improvements for the program must continue with advances in technology.

7. **Establish a quality assurance mechanism for security.** Evaluate the effectiveness of the security program.

8. **Plan for security.** Establish long-range security strategies supported by short-range plans.[7]

Numerous studies show that the failure of security programs may be attributed largely to management's misuse of the improvement process, lack of participation, unwillingness to make a long-range commitment, and failure to make the program a part of the business. Responsible managers who increase their own knowledge and share this knowledge with their employees are fundamental to the implementation of a successful security program.[7]

Security Policy Goals

All organizations should develop a security policy statement and verse **all** employees on its contents. A policy statement should include main checkpoints that are directed specifically at an individual organization's operations:

1. **Avoidance.** Avoidance involves removing potential threats or eliminating/moving assets away from potential threats.

2. **Deterrence.** To "deter" means to discourage from action. The concept of deterrence is to combat intent before it reaches the degree necessary to carry out a violation of trust. Be alert for possible drug, alcohol, or gambling problems that may force employees to steal vital data in the hopes of converting it into cash. Also, be on the lookout for employees who have a personal or family crisis that may have them in a financial bind.

3. **Prevention.** This traditional security function is most often used and is conditioned on cost-effectiveness. Its three types are:

   a. **Intensive.** Prevent unauthorized acts by even the most sophisticated attackers.

   b. **Confrontation.** Employ enough safeguards to make any perpetrator aware that he/she is performing or attempting to perform an unauthorized or possibly illegal act.

   c. **Avoidance.** Involves removing assets from direct temptation.

4. **Detection.** To have a good preventive measure in a security system means to have an effective detection available. Detection by itself would not be enough. Timely detection is important in determining that an unauthorized

act is impending, is in progress, or has occurred recently enough to prevent or limit losses. Technical or automatic detection is generally preferred over manual detection because computers are ideal detection devices and humans are poor detectors.

5. **Recovery.** Recovery is needed for a wide range of events and activities. Computer production jobs must be designed to recover from hardware failure, software bugs, operator error, and data errors. Recovery from other, less frequent and sometimes more catastrophic events occupies more time of the computer security specialist. Recovery commonly takes place in an environment where people are confused, upset, and not functioning effectively. Security planning should include the testing of threat scenarios to anticipate the failure of safeguards and possible alternative results. The cost of recovery can be significant and should be taken into account in determining the cost of a potential loss.

6. **Correction.** Corrections to security flaws should be implemented immediately after recovery. Overreaction may occur because of the recent, vivid memory of the incident and painful recovery process. The security specialist must approach recovery objectively at such times to maintain a sense of balance and perspective.

In addition, a security policy should include the following points:

- Data is a valuable corporate asset that must be protected with the care and concern shown other valuable assets.

- Business continuity depends heavily on the integrity and continued availability of critically needed data and of the means by which that data is gathered, stored, processed, communicated, and reported.

- Adequate control of data depends on the ability to restrict access to only those individuals who need it and on the ability to hold each person fully accountable for what he or she enters, changes, or reads.

- The functional managers of the business must have primary responsibility for the security of *their* data. The vast majority of the people who create or contribute to security problems work in non-DP areas. Furthermore, the managers of those non-DP areas should be responsible for:

  - Monitoring the integrity of their data to identify problems as early as practicable.

  - Assessing the impact on the organization of security problems involving the data for which they are responsible, and advising the DP organization as to the nature and degree of security protection needed.[8]

DPMA Policy Statement

The Data Processing Management Association's (DPMA) Executive Council has adopted a Model Corporate Computer Security Policy Statement that is intended to serve as an example statement for use by organizations in drafting their own statements. Organizations are cautioned to use the DPMA Statement as a guide as they attempt to tailor their policy to fit the nature of their organization and to retain consistency with other organization policies.

The policy statement is a brief description of the major objectives, value, and intended use of computer resources and information. It is not intended to provide specific procedures for the implementation of the policy.

Procedures should be placed in a separate, and more detailed, document containing such things as:

- The levels of information in the computer system and the types of information requiring protection
- The required user clearance level and the individuals able to access various pieces of information
- Control measures
- Assignments of various security responsibilities

In summary, the basic goals of any policy should include the following:

1. Protecting against the loss or misuse of information.
2. Defining employee responsibility and accountability to maintain protection of information.
3. Preserving the integrity and security of information through appropriate controls.
4. Establishing a basis for measuring security effectiveness.
5. Preserving and supporting audit and legal compliance.[22]

INFORMATION CLASSIFICATION, OWNERSHIP, AND VALUATION

In order to determine security objectives and goals, information must be treated as a corporate asset—something of value. Three methods for classifying data and information are classification, ownership, and valuation.

Classification

Classification of information is its systematic labeling to indicate a specific set of protective controls. Raw material is data that, when processed, becomes information. A clear appreciation of the difference is very important because of security implications. "If information is unavailable, has been modified, is incomplete, or loses its confidentiality, the impact on the organization is far greater than if the subject of such mistreatment were data."[9] Data is collected to establish records. These records are collected to establish files. Files are processed to establish information, which is used to make decisions.

Before the subject of a guide to data and information classification is considered, the following reasons for classification should be considered:

- What is the intrinsic value to the organization?
- Are data/information at higher levels to be treated differently?

■ What analysis should be given to access information at higher levels?

■ Will importance of the information influence recovery and backup planning?

Information can be initially classified using the matrix approach where there is a breakdown according to level, contribution, horizon, form, and evaluation. (See Figure 12.1.)

FIGURE 12.1
Initial Classification of Information

Decisions	Level	Contribution	Horizon	Form	Evaluation
Strategic	Top Management	External Sources	3-5 years	Often Informal	Difficult
Tactical	Middle Management		1 year		
Operational	Lower Management	Internal Sources	Hour - Day	Usually Structured	Easy

To further break down and evaluate data and information, criteria should be established to fit the following categories.

1. **Competitive-edge data/information.** Such data is recognized because of its role in distinguishing your product or service from that of the competition. Examples of such data/information are designs, recipes, methodologies, client lists, and pricing/costing details.

2. **Operational/confidential data/information.** Tactical and strategic future plans fall into this category. If such data/information were to get into the wrong hands, it would seriously affect profitability. Examples of such data/information are lists of prospective clients, takeover strategies, new products/services, advertising campaigns, and new personnel targets.

3. **Privacy-based/confidential data/information.** Clients, customers, suppliers, and employers all expect confidentiality standards to protect their private information. Therefore, such files as payroll and supplier and customer details require security attention.

4. **Operational/critical data/information.** In every factory or office situation, there is some identifiable data that is essential for the running of that particular production unit. It may be a reference file, such as a price list, or it may be an instruction set, such as a word processing computer program.

5. **High-cost data/information.** Some material is the result of extensive investment, particularly research-type information. Its cost to re-create is extremely high and thus justifies careful attention.

6. **Fraud potential data/information.** Employees in particular have access to material with potential financial benefits.

7. **Security data/information.** Access reference files and keys, encryption algorithms, and other security items, such as passwords, have particular relevance.[9]

A formal rating scale can be established and kept on file to record the evaluation process.

Ownership

The owner is that *individual* manager or representative of management who has the responsibility for making and communicating judgments and decisions on behalf of the organization with regard to the use, identification, classification, and protection of a specific information asset.[19]

Information ownership, then, is a *role*, assigned by management to the owner for control of the asset. The specific name used by the organization for "owner" is not the important issue; it is the concept that is important. Due to the legal connotations associated with the term "owner," some organizations have used other names, such as "guardian" or "sponsor." From a legal perspective, the actual owners of the organization's information are ultimately the shareholders, proprietor, and taxpayers.

The point should be made clear to all employees, as part of the education and awareness program, that the organization is the legal owner, not the employee. Ownership flows through the hierarchy of the organization. Thus, all the organization's information is ultimately the responsibility of the CEO. The objective is to place the ownership at the lowest reasonable level possible in the organization—the level at which the most knowledge about the information, along with orientation to management responsibility, is found. A helpful question is, "Who has the 'property rights' responsibility for the information?" It must be an individual, not a function, certainly not a department. It should be a person who is close enough to the day-to-day concerns about the information to make the best informed decisions regarding its use.[19]

Ownership of most of the organization's business information should *not* reside in the Information Systems (IS) function (or DP department). For most of the business information it handles, IS is ordinarily the *custodian*. For example, ownership of data such as system control programs and utilities of payroll belongs to an individual in the payroll department, not in IS.

Recommended ownership responsibilities are as follows:

- Identify information and acknowledge ownership
- Classify the information
- Specify business controls
- Authorize access
- Assign custody

■ Approve application controls

■ Perform or participate in risk assessment and acceptance

■ Develop contingency plans

■ Monitor compliance and review periodically[19]

Valuation

Valuation of information is a method of establishing the worth of data/information to the organization.

Charles Cresson Wood offers the following rationales to the issue of establishing value of information.[62]

Rationale #1—Information has a cost/value.
There are three bases for the worth of information:

1. The *cost* to acquire, develop, or maintain the information

2. The *value* of the information to its owners, authorized users, and unauthorized users

3. The *cost/value* of information already present in the real world

Conclusion: Information will have cost, value, or both, and cost may or may not be an accurate reflection of value. The value of information is almost always composed of a series of attributes that, in the judgment of the relevant parties, are greater than its identifiable costs.

Rationale #2—The cost/value of information can be established or assigned.

■ There are three fundamental components of cost: acquisition (purchase), development or replacement, and maintenance. These costs can readily be determined.

■ The value of information is always at least equal to its costs.

■ The value of information is meaningful only to its owners and authorized users (or those desiring to own or use the subject information).

■ The objective value of an asset is only a mutually agreed-to subjective value that may have little to do with cost or other measurable attributes.

■ Agreement as to value among relevant parties is the key issue in establishing and assigning objective value.

Note that information itself exists in a world of imperfect and incomplete information about virtually every object, and agreement as to value occurs every day despite uncertainty.

Rationale #3—There are worthwhile reasons to establish and assign cost/value.
Reasons for valuing information that are directly related to information security are to:

■ Provide an accepted basis for information security budget cost/benefit analysis

■ Establish better integration of the information security function with widely recognized organizational functions

Other reasons for valuing information that have to do with information security include the following:

- Avoiding regulatory penalty
- Preserving confidentiality and integrity of proprietary information
- Assuring continuity of business/mission functions or human services relying on information

Reasons for valuing information that are not directly related to information security include the following:

- Knowing marketplace value to facilitate buying/selling of information at a reasonable price
- Owner/user perception of information value

Rationale #4—Circumstances dictate whether to establish and assign cost/value.
These circumstances include the following:

- Materiality of the decision
- Time value
- Avoidance of disruption or loss
- Opportunity costs
- Need to support security measures

Rationale #5—There is an ethical obligation to make the best possible decision in any given circumstance.
Managers are obliged by fiduciary responsibility and ethical standards to use all justifiable tools and techniques to secure information. Failure to use best security tools may be irresponsible and may incur culpability under the Foreign Corrupt Practices Act of 1970 or other civil liability.

LOCATING AND TRAINING COMPUTER SECURITY PERSONNEL

Growing concern about computer security has created new jobs—particularly those for Data Security Officers and Directors of Computer Security. It takes considerable time to train DSOs and Directors to become knowledgeable in a company's operation and to assess where security best fits in. The problem is compounded for DSOs as they must locate and train personnel to aid in the ongoing computer security operation.

The DSO's Role

The **Data Security Officer (DSO)** should be recognized in the organization as a data security specialist who, acting similarly to a consultant, makes suggestions, offers alternatives, reviews proposed solutions, and educates team members to raise their level of security awareness.[42]

The DSO's (or Computer Security Officer—CSO) responsibilities should include the following:

- Securing the physical environment
- Maintaining logical access controls
- Establishing the security policy manual
- Training employees in security awareness
- Establishing good security practices

The ideal DSO should have the following characteristics:

- Demonstrated security awareness in both logical and physical domains.
- Experience with computer use, with knowledge of hands-on operations and programming
- Demonstrated capability as an administrator
- Successful problem-analyzing skills
- Ability to communicate well with all levels of staff and management in both a training role and in presentations to others.[57]

Identifying, structuring, and valuing the information resource is the responsibility of the DSO. Protection of business resources is properly assigned to the DSO or Manager/Director of Security, who must respond to the information values assigned. The DSO must develop and apply such security measures as are appropriate to a resource of equivalent value.[48]

Placing Security in the Organization Chart

Very often the business information security effort is hopelessly squandered because security's organizational responsibilities are poorly defined or never made clear. Examples include these corporate scenarios:

1. Where the systems manager has a computer security function but there is no information valuation program
2. Where the director of security has responsibility for protecting the business but the information systems department has its own "special" security program
3. Where computer output reports are marked and protected but the same information in typewritten form is handled in a casual manner[48]

In placing the computer security function in the organization at the most strategic level possible, conflicts often arise. Alan Krull, of the firm Business and Professional Education in Redwood City, California, sums up the situation as having three causes:

1. Scope of security needs to be broader—to include all operations, not just the mainframe
2. Security professionals still haven't attacked the problem of getting the educated opinion and judgment of their most important customers—their senior management

3. Too many security people are demoralized. They still have a great interest in their profession, but they feel unable to do the right job. They feel they only get lip service from their management.[26]

Krull has stated, "Security is still perceived as the responsibility of the data security department . . . you don't get profits by appointing a profit administrator; so too you don't get security by appointing a security administrator."[26]

Placing the computer security function in the organization can take many forms. Figure 12.2 shows the organization of the computer security function, and Figure 12.3 depicts the placement of the computer security function within the corporate organization chart.

FIGURE 12.2
Computer Security
Function

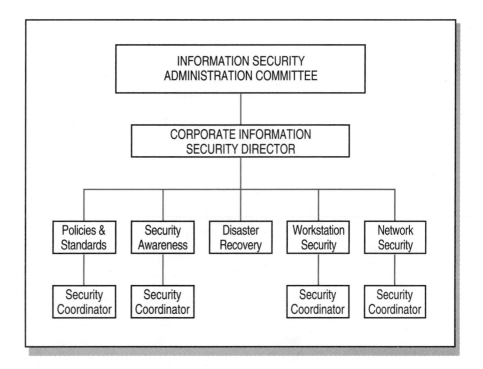

Reporting to Whom?

In a recent issue of *ISP News,* managers of information security were asked the question, "Should information security report to an organization's MIS department or to corporate security?"[40] The responses were somewhat varied, but they consistently emphasized that upper management must support and encourage computer security to the fullest extent possible.

In a small installation, if a company cannot afford to have separate entities, then combining information security and corporate security is perfectly logical. However, if the installation is medium to large size, where the company has a full-time information protection professional, then it makes sense for the information

FIGURE 12.3
Security in the
Organization

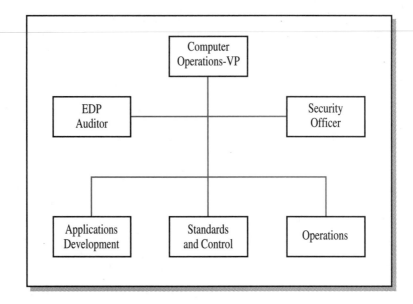

security manager to report to the vice-president for information systems—not to the applications managers, not the systems managers, but directly to the head of data processing.

Stan J. Kiyota, *Manager of IS Security*
The Clorox Company

My observations are from an industry perspective. I see two primary information security functions. First is policy development and enforcement, which I see typically performed by corporate security. Second is the issuance of specific guidelines and secure implementation procedures, such as what software to use on PCs, or what RACF settings should be. This function is the province of the information systems department. The difference between the two is that corporate security determines what is to be accomplished, while IS determines how to do it. Finally, there is a third function, awareness, the two areas need to cooperate on. Corporate needs to set the stage to show upper-level management support, and that information security is a priority for the entire organization. That's what we do at AT&T, and it works well for us.

Catherine W. Englishman, *Technical Staff*
AT&T Bell Laboratories

I tend to go more for information systems, but there is a potential conflict of interest there. They could be persuaded not to put in security measures because they are more costly or less efficient, if they have the same management. On the other hand, I have a real problem with the mentality of guards, guns, and dogs. I think they have less understanding of the needs of data security; they're not as technically oriented. I guess my real choice would be 'Other'—and I'm not ready or willing to define who other is. Overall, I still think I would prefer the information systems area. I've known a lot of people who reported to guns, guards, and dogs, and they've always had problems, both budget-wise and psychologically.

Sally Meglathery, *Past President*
Information Systems Security Association

This is an age-old question, and the answer really depends on the culture of the organization. There is a real conflict of interest around protections; MIS has not succeeded in meeting the overall corporate objectives as well as they should. Also, have MIS and infosecurity been able to expand beyond the concerns of computer systems to address enterprise-wide security needs? If not, then a different approach is clearly appropriate. There is a trend to move it to corporate security. In previous years, the gun-toters and retired FBI agents were uncomfortable with computers and high technology. But these days, more and more physical security is electronically based, and corporate security is much more conversant with technology. If you can marry an understanding of computers and the organization's information with the traditional security attitude—the tendency to question everything, and an understanding of the need for protection—then putting info security under corporate security is a good move. The real key is that, to meet the organization's best interests, information security has to address much more than what's on computers.

Robert P. Campbell, *President*
Advanced Information Management, Inc.

Staffing the Computer Security Function

All personnel who are working with computers as a primary part of their job should be carefully selected and trained to ensure that security is "woven into the fiber" of the entire computer function.

The Federal Bureau of Justice Statistics, with the aid of SRI International, has classified 17 occupations in computer technology and the types of risk/vulnerability that they present to the data center.[2] Because the majority of these job functions, by whatever title, are present in almost all data center environments, they are presented here in brief to provide insight into those areas that could present security problems.

User Transaction and Data Entry Operator. With knowledge of source document content, terminal protocol, and other procedural controls, this position has opportunities to modify, destroy, or disclose information entering the system. While data modification is the greatest danger, destruction of source data is a major consideration, particularly if no backup for that source exists.

Computer Operator. Knowledge of operating system functions, utility programs, work flow, system accounting procedures, console protocol, privileged access procedures, and physical access procedures gives this position opportunities for abuse of equipment, data, applications programs, or internal systems programs resident in main memory or on tape or disk files.

Peripheral Equipment Operator. Because this position has knowledge of input/output operations, vulnerability lies primarily in the area of disclosure or destruction of vital company information.

Job Setup Clerk. Knowledge of work flow, system accounting procedures, media libraries, and physical access procedures can cause vulnerability to destruction or disclosure of data or programs external to the system but in the general operations area. An additional vulnerability is the possibility of media theft.

Data Entry and Update Clerk. Like the user transaction and data entry operator, this position has online access to data, making possible modification, destruction, or disclosure of data within the system.

Media Librarian. With knowledge of file names and labels, library and job accounting procedures, work flow, physical access procedures, and media archiving, the vulnerabilities of this position are destruction, disclosure, modification, and theft of media. Fraud by an individual in this position is unusual because of lack of knowledge of file contents.

Systems Programmer. With knowledge of the operating system internals, programming languages, terminal and console protocols, identification/verification procedures, work flow, system architecture, and almost everything that is vulnerable within the system, this function can disclose, modify, or destroy data that is internal or external to the system. A major area of abuse is unauthorized use of the system and the ability to deny authorized use.

Applications Programmer. Knowledge of programming languages and procedures and terminal protocols makes this function capable of physical, operational, and programming violations. The principal vulnerabilities involve modification, destruction, or disclosure of programs within the system. This destructive ability is usually limited to those systems with which the programmer works or has worked.

Terminal Engineer. This position's electronic, mechanical, and communications engineering knowledge makes possible physical, operational, and engineering violations. The main threat is modification, destruction, or theft of terminal hardware.

Computer Systems Engineer. In-depth knowledge of programming languages, electronic, mechanical, and communications engineering, digital logic design, terminal protocol, and physical access procedures gives this function the ability to pose the greatest threat to security. The principal areas of vulnerability are modification, destruction, or theft of hardware, and the unauthorized use of the system (or denial of authorized use). A secondary vulnerability is modification, destruction, or disclosure of systems programs.

Communications Engineer Operator. Knowledge of electronic and communications engineering, data communications, terminal protocols, identification/verification procedures, and physical access procedures make physical and operational abuse possible. The principal area of vulnerability is destruction or disclosure of data during transmission. In addition, there is the threat of modification, destruction, or theft of hardware.

Facilities Engineer. Knowledge of electromechanical engineering and physical access procedures allows physical violations. The two primary areas involved are denial of access/service and destruction or theft of system equipment.

Data Center Operations Manager. Knowledge of work flow, hardware configuration, operating procedures, media storage, job accounting, physical access, system integration, system maintenance, and software causes physical and operations vulnerabilities. Several serious threats are posed by this function, primarily in the areas of destruction or disclosure of data, applications programs, or systems software, destruction or theft of hardware, and unauthorized use or denial of authorized use of the system. In addition, this function is in a position to destroy or disclose data files or programs stored in the media libraries, as well as to modify parametric data internal or external to the system.

Database Administrator. Knowledge of database concepts and languages, file contents and layout, work flow, security access controls, terminal protocol, and identification/verification procedures gives this function the ability to violate physical and operational security. The two most serious vulnerabilities are internal and external access to all data maintained in a database and illegal file modification. Thus, there is opportunity for destruction, modification, and disclosure of data.

Programming Manager. With the same skills as an applications programmer, the programming manager's vulnerabilities are much the same. Access is usually limited to those systems under the manager's direct control.

Security Officer. Knowledge of security, including identification, DP hardware and software technology, industrial security products, procedural, operational, and personnel practices and policies, together with privileged access to all areas and all systems functions, creates vulnerabilities of all types for an individual holding this position.

EDP Auditor. Knowledge of audit techniques, controls, safeguards, system design, software organization, computer applications, and facility security, together with privileged access, allows this function opportunity to commit violations in all areas. Detection of those violations is virtually impossible; thus, great care must be taken to ascertain the competence and trustworthiness of individuals performing this function.

Because all the functions listed above are part of an average data center, great care must be taken to ensure separation of duties and responsibilities. Separation of responsibilities begins with the principle of least possible privilege—a person should have only enough information to perform the job at hand.

Security Considerations for Organizational Structure

As a starting point, the following should be considered when developing an organizational structure:

- Separate the duties and responsibilities of data entry personnel, computer operators, applications programmers, systems analysts, and technical staff.
- Permit only those operations personnel whose jobs involve direct contact with the hardware into the computer room.
- Consider physically separating the mainframes from their peripherals.
- Allow the operators in the printer room only to stop and start the printer and change the forms. All other printer control should be performed in a separate room.
- Institute logging and reporting procedures for entry to the data center. Management should review these logs.
- Establish an operations control group to monitor overall operational procedures and controls.
- Separate the functions and reporting structure of machine operators, peripheral operators, media librarians, and job setup clerks. While operators and librarians may report to the same person, scheduling and job setup clerks should not.

- Separate the development function from the maintenance function within applications programming and establish a separate systems testing and installation group.
- Within technical programming, create separate functions for operating system maintenance, data communications maintenance, and database administration.

DP Personnel Practices

To ensure security in DP personnel practices, the best place to begin is the hiring process. Most computer abuse, fraud, theft, and sabotage would probably not happen if organizations followed better hiring practices.

The best time to determine if a person is going to be an asset to the organization is at the point of hiring. At no other time is information more readily available about a person, and it is much easier—and much less costly to the organization—to turn down a job applicant than to deal with a problem employee. Personnel departments must understand that the DP operation is necessary for corporate health and that security is a major consideration. Written guidelines serving this purpose should be available to those persons involved in initial recruitment interviews.[37]

Hiring Practices

It is important that the members of the personnel department responsible for hiring be made aware of the overall security practices of the organization and of the specific emphasis that should be taken with highly sensitive positions in the computer area. When screening applicants, the following should be thoroughly reviewed:

- Reference checks
- Educational records
- Military records
- Prior employment
- Criminal records
- Law enforcement records

Probationary Period

After an employee has been hired, he or she should be placed in a probation period for the purposes of training and indoctrination to acquaint the employee with the organization's procedures. After this initial period, the employee should sign a nondisclosure statement, a conflict-of-interest statement, and a statement that continued employment is based on the employee's adhering to these agreements.

Employment Termination

Termination is one area in which personnel practices often fail to meet security criteria. When an employee leaves for any reason, an exit interview should be conducted, during which all physical company property should be retrieved, especially keys, badges, magnetic cards, and other physical access mechanisms. The former employee's passwords and access codes should be immediately canceled so that access to the computer system can be revoked. While many organizations accept the standard two weeks' notice, it may be better to adopt a policy of providing severance pay in lieu of notice, especially if an employee is suspected of harboring negative feelings toward the company.[37]

Most companies use a hiring checklist, which prescribes the forms to be filled out, orientation to be supplied, and other details. A similar checklist should be established for termination and should include:

- Surrender of keys
- Surrender of identification cards
- Exit interview
- Return of all company materials
- Changes in passwords
- Changes in locks
- Changes in access or authority codes[22]

After the hiring is completed, employees should be versed on the need to establish and maintain security practices. The following measures should be considered when establishing personnel practices in the data center environment:

- Rotate department managers. Not only does this practice give the organization greater flexibility, it can point out areas of abuse and need.
- Rotate computer operators among shifts.
- Insist on at least a full week's vacation in one consecutive period. In the case of employees who receive three or more weeks' vacation, insist that they take at least two weeks consecutively.
- Review all job descriptions to ensure that employees are performing the job described.
- Establish quarterly rather than yearly preformance reviews. These should not be confused with salary reviews. Rather they are operational reviews to discuss specific problem areas that require management attention.
- Establish both formal and informal channels of communication within and among departments to discuss security risks. Employees and managers should be able to seek resolution to security problems without fear of repercussion. Regularly scheduled meetings and memoranda should also address security risks.
- Use job rotation and cross-training. Consider promotions or lateral transfers if regular rotation is infeasible. This practice ensures that more than one person can perform each function.
- Establish a policy stating that members of the same family may not work in the same or interdependent functional areas. (Many types of computer abuse and fraud require collusion.)[37]

Size of Security Staff

Charles Cresson Wood offers this insight regarding the size of the computer security staffing issue:

> Information security staffing ratios vary by size of organization and by industry type. The results of this informal survey, however, fall within narrow windows reflecting a standard-of-due-care. For example, the number of information security staff should be from 1% to 5% of the data processing department staffing level. Similarly, information security staff is expected to be from 0.05% to 0.28% of total employment.[62]

Staffing level ratio calculations provide a starting point for a conversation about what constitutes a reasonable level of staffing for information security.[62]

Investigative Team-Building Approach to Security

Computer-related crimes have become one of the most challenging problems encountered by organizations today. Unfortunately, a majority of these crimes are committed by insiders who earn the trust of management and use it to access sensitive information that they are not authorized to see. In fact, computers have become so user-friendly in recent years that they are also vulnerable to criminal activity. The best way for management to combat these problems is to develop an investigative team within the organization whose goals are to effectively prevent, detect, and investigate computer crimes.

This team should be composed of the following members:

- *A coordinator* should be the director or manager of security who is responsible for ensuring that the team work together in the investigation of computer crimes, by scheduling team meetings so that the colleagues will maintain communication with each other.

- *An investigator* is responsible for conducting interviews, interrogating suspects, and conducting background investigations of suspicious individuals.

- *An information system specialist* is knowledgeable about all computer systems involved in the organization. He or she must be reliable and objective, able to maintain confidentiality, able to operate through procedures, and be suitable as an expert witness. This individual will be responsible for explaining the technology if the organization were ever to appear in court.

- *An attorney* is responsible for establishing an educational relationship with the other team members by informing them of the legalities involved in the collecting of investigative evidence and the use of this evidence to establish a criminal case.[30]

Robert F. Littlejohn, Director of Corporate Security for Avon Products, Inc., in New York, provides these keys to an investigative team's success:

- Closely supervise technicians. Don't permit technicians to work by themselves. If you assign a technical person a task, a member of the investigative team who understands evidence should closely supervise the work.

- Have the team's attorney advise the team on how to collect acceptable evidence, safeguard it, and maintain a chain of custody. The smallest error can destroy a case.

- Understand what the information systems specialist is doing. If the specialist can't explain exactly what he or she is doing so that you can understand, he or she will not be able to explain it to an investigator. You will not be able to sell the case to a prosecutor and surely not be able to present it to a jury effectively.

- Insist on documentation; it's essential. By putting it in writing you maintain a chain of custody. You must do everything with computer evidence that you would do with any other evidence.

- Identify other specialists before an incident occurs. Spend some time with the internal electronic data processing auditors and the data processing personnel. Know who handles internal control.[30]

Companies such as IBM, Honeywell, Control Data, Apple, and others understand that computer crime does happen, and they are prepared to help under appropriate circumstances.

Tiger Teams

To test and evaluate whether all personnel adhere to the security policies, a team may be organized to test the system. A **tiger team** is a team of experts who try to "crack" the system to identify and exploit flaws. Members of the tiger team are trained to think like bandits by putting themselves into the hacker's or criminal's shoes to try to beat the security system in every way possible. "While most of them have been trained to know how the computer system should operate, tiger team members must concentrate on how the system should not operate."[10]

> The effectiveness of the tiger-team approach was clearly demonstrated in 1973, when a U.S. Army tiger team penetrated the Pentagon's supersecret Defense Intelligence Agency (DIA) and obtained access to the agency's top-secret computer system. It was the first documented use of the tiger-team concept against a computer system within the Department of Defense.

> This particular unit served as a model for the eventual formation of the U.S. Army Automatic Data Processing Systems Security Detachment (ADPSSD), which is a subordinate element of the 902nd Military Intelligence Group, U.S. Army Intelligence and Security Command (INSCOM), which has base headquarters in Fort Mead, Maryland. The ADPSSD tests the security of hundreds of Department of Defense (DOD) classified computers each year.[10]

The following are just a few of the more important hints for conducting a tiger-team attack. The tiger team should use imagination and ingenuity to discover other weaknesses.

- Determine if the system has dial-up ports. If so, obtain the telephone number of the port.

- Learn about the on and off times of the system. Determine the best time to attack the system in order to minimize risk of discovery.

- Learn the layout of the network's wire lines and determine if the lines can be tapped with little risk of discovery.

- The systems personnel, those people who operate the computers, generally are one of the weakest links in the security system. They tend to be careless in safeguarding their passwords, and hackers take advantage of this. The tiger-team member should therefore concentrate on this area first. Since many people have trouble memorizing numbers or passwords, they frequently use social security numbers, birth dates, or house numbers. These are the first numbers that the tiger-team member should try. The codes often are jotted down on calendar pads, under desk blotters, and other "secure" places around a person's desk.

- When trying to access the system through a dial-up port, first try the common password that most computer companies deliver with their machine. Frequently, users fail to change this password. This weakness was taken advantage of by the "414 gang."

- Contact a former employee for his password. Managers frequently forget to change passwords when someone leaves.

- Check the trash for passwords that have been written on a scrap of paper and then thrown into the waste baskets.

- Watch the screen as someone operates the personal computer. Unless the security program suppresses the printing of the password, he or she may be able to obtain it in this way; a pair of binoculars can be helpful.[10]

BUDGET CONSTRAINTS

> Convincing senior management to shell out the money for a security system can be difficult for MIS. The trouble stems from the fact that although computer crime is a serious problem, it has, by and large, been swept under the carpet.[53]

One of the reasons MIS has had a tough time convincing senior managers to take computer crime seriously is that most publicized examples of security breaches are of young hackers or petty criminals. Management is often reluctant to spend thousands of dollars protecting against minor breaches. Another argument senior management may use is that purchasing a system is unnecessary when the company is already insured against losses.[53]

Resource Allocation

"Not only does the computer security problem have grave potential impact upon organizational resources: so does the cure."[61] Security is not just a data processing problem—it is an organizational problem. The resources to resolve the problems must come from general organizational resources—not squeezed out of an already strained DP budget.

> There is no such thing as perfect security. Most organizations can achieve a level of protection appropriate to their needs. The objective of a data security program is to cut the risk and probability of loss to the lowest affordable level and also to be capable of implementing a full recovery program if a loss occurs.[45]

Donn Parker states, in *Computer Security Management,* that "Poor budgeting and responsibility assignment in organizations can also lead to serious security deficiencies and suboptimizations, particularly when security responsibilities are shared by departments with separate budgets."[39]

Billing for Security Functions

A **chargeback system** is a billing system that multiplies the number of units by the predetermined price and accumulates these charges throughout a suitable charge period (e.g., a month).

There are a variety of billing systems ranging from no- or low-chargeback (overhead charging) to full-chargeback (cost center) operations. Charging internally for the use of a central computer facility involves allocating the costs of the operations to its users.

A chargeback system would be based on the following criteria:

- CPU operation time
- Storage
- Input/output channel time
- Peripheral devices use
- Priorities utilized by each program
- Proprietary programs provided
- Auxiliary services supplied
- Terminal and communication charges
- Personnel involved
- Security measures necessary[13]

Selling Management on Budget Allocation

Management must be convinced that a financial outlay for security is a necessity, not a luxury. Some basic points that should be realistically presented to management are these:

1. What degree of risk does the company face from poor security?
2. How does security relate to other business risks?
3. How is the "information has ownership" concept stressed?
4. What are the results of a formal, complete risk analysis?[1]

Security in Tough Economic Times

In tough economic times, such as we are now facing, many organizations are forced to lay off employees. Along with the pink slips come hard feelings. Some disgruntled employees might be moved to vent their anger and frustration on the company's information or information systems technology. What should an organization do to protect itself?[40]

It is vital that information security policies be in place as soon as possible—before any layoffs are announced or budgets are slashed.

Some of the positive steps that should be taken to ease tension and alleviate employee's fears are the following:

- Support should be reaffirmed from the top.
- Published policies and procedures should be in place.
- All media and programs should remain in the building.
- Security awareness training should be emphasized.
- Security and audit staffs should remain highly visible.
- Counseling and employment assistance should be provided to departing employees.

"Call it downsizing or rightsizing, workforce reduction creates a host of new security problems and intensifies virtually all the old ones."[40]

SECURITY TRAINING AND AWARENESS

Security awareness depends on cooperation and feedback of all computer users. One method of maintaining or improving morale in the organization is to provide professional development and continued education. If employees fully understand the need for good security practices, they will be more likely to cooperate to reach desired objectives. Another method is to demonstrate the cost of poor security—the risks, damages, and costs associated with breaches. "In effect this type of training can appeal to the employees' solidarity and to their sense of belonging to the organization. It can be shown that if the organization is damaged significantly, so will they be."[22]

Basic Training

All new hires must undergo orientation sessions covering required job skills and practices. Orientation sessions are often followed by both formal and on-the-job training. "The orientation and training should follow well-defined procedures or steps and stress good housekeeping and personnel security practices to prevent the loss or disclosure of information."[13]

In most cases, employees will cooperate with security requirements only if they understand why security is appropriate in each specific situation. "Many users, however, are unaware of the need for security, especially in situations in which a group has recently undertaken a computing task that was previously performed by a central computing department."[41]

There are two kinds of training that are significant from the security standpoint: orientation and skills. **Orientation** is training whose intent is to set a climate, convey initial information about a situation, and so on. Employee orientation at the beginning of employment is of interest here. The second kind of training is specific skills training: courses of a few days or weeks' length which help employees upgrade their knowledge and capability or add new knowledge. From the security perspective, orientation relates to employee awareness and organizational culture; and skills training relates to reinforcement of the culture, and the critical need to have competent, knowledgeable people.[4]

Orientation should include the following:

- Written corporate policies regarding confidentiality and conflict of interest
- Formal reporting relationships
- Disclosure policies
- Security and control procedures

It is highly recommended that employees review policies regarding confidentiality and security and sign new acknowledgments each time there is a material change in employee duties and responsibilities. In any case, employees should review these policies periodically (perhaps each six months or yearly).[4]

Awareness Sessions

The two concepts that should be covered in awareness sessions are *necessity* and *persuasion*. Falling under the category of necessity are:

- Security policies and procedures
- Meaningful approval(s) by meaningful individual(s)
- Communicating data security expectations to *all* members of the organization

   User friendly "persuasions" include items such as:

- Restricted workstation access
- Protection of media
- Back-up and storage
- Restricted system(s) access
- Solutions to the "first you see it, now you don't" problem(s) in unsecured environments

It is extremely important, says Morrison, to obtain "feedback" from end users in the form of a data security acknowledgment. This document should be a written statement signed and dated by the individual users when they are first introduced to the system.

The acknowledgment might read as follows:

> *I understand that my (organization name) password is confidential information. I understand that disclosing my own password or attempting to discover another user's password is a violation of (organization name) data security policy. I understand that I am responsible for all system activity conducted under my user ID. I understand that all work performed on (organization name) information processing systems is for (organization name) business purposes only. I understand that violation of these provisions may be cause for disciplinary or legal action or termination of my employment with (organization name).*
>
> *I have read and understand this data security acknowledgment.[60]*

Security awareness should foster these attitudes on the part of all personnel:

1. Awareness of what security problems exist and what opportunities for improvement are available

2. Awareness of the employee's own security responsibilities as well as those of his coworkers

3. Awareness of what to do when security violations are detected[45]

Compliance Training Program

As an aid to assist federal agencies to comply with Public Law 100-235 (commonly referred to as the Computer Security Act of 1987) all federal agencies are now conducting security training. The Computer Security Act stipulates that **all** government employees and contractors shall receive annual computer security awareness training that is commensurate with their level of computer usage. Since many employees rely on computers for a majority of their work, it makes sense for upper management to make these courses available to all employees.

Compliance training defines security responsibilities for federal employees who use, manage, or administer computer systems. This training should be designed to reflect the needs of users in multiple environments, as well as various levels of expertise. The courses should be designed to be informative, explain concepts and logic of assurance principles, as well as serve as a seed for causing users to take protective steps for their information.

EVALUATING AND UPDATING SECURITY PROGRAMS

One of the most effective tools used today to evaluate and update security systems is the use of an audit—by both internal and external auditors.

Management should budget for and promote advanced EDP audit training activities. Current state-of-the-art data security courses should be required for those individuals auditing complex systems. With technology changing at breakneck speed, internal auditors should also attend technical computer courses to ensure their familiarity with changing technology trends.

Hiring internal auditors having a computer-oriented background—whether by academic training or by prior experience in a data-processing environment—and mandating continuing education are also essential to effective computer security.

Both external (contracted) auditors and internal (employee) auditors are particularly helpful in performing a thorough review of an organization's security procedures. The audits can take several forms:

1. Periodic (on a scheduled) basis

2. Unannounced (on the spot) basis

3. Discrete (ongoing) basis

According to the National Institute of Justice, Washington, D.C., Certified Internal Auditors (CIAs) have been certified by the Institute of Internal Auditors (IIA). Certification includes subscribing to a code of ethics, holding a baccalaureate degree or equivalent work experience, and passing an examination based on a Common Body of Knowledge for Internal Auditors. The CIA rating was established to promote and increase the professional standing of internal auditors but is not a requirement for being an internal auditor.

The issues of detecting and investigating fraud and other irregularities have varied over the years and from one organization to another. Some organizations do not charter their internal audit function with responsibility for detecting fraud, justifying this decision on a cost/benefit basis. Other organizations view the internal audit function as both detecting fraud and acting as a deterrent to fraud. The consideration for fraud detection is directly addressed in the IIA's "Standards for the Professional Practice on Internal Auditing." This is not limited to any one area such as EDP, but is a general standard dealing with due professional care.

In exercising due professional care, internal auditors should be alert to the possibility of intentional wrongdoing, errors and omissions, inefficiency, waste, ineffectiveness, and conflicts of interest. They should also be alert to those conditions and activities where irregularities are most likely to occur. In addition, they should identify inadequate controls and recommend improvements to promote compliance with acceptable procedures and practices.

Due care implies reasonable care and competence, not infallibility or extraordinary performance. Due care requires that auditors conduct examinations and verifications to a reasonable extent, but it does not require detailed audits of all transactions. Accordingly, the internal auditor cannot give absolute assurance that noncompliance should be considered whenever the internal auditor undertakes an internal auditing assignment.

When an internal auditor suspects wrongdoing, he or she should inform the appropriate authorities within the organization. The internal auditor should recommend whatever investigation is considered necessary in the circumstances. Thereafter, the auditor should follow up to see that the internal auditing department's responsibilities have been met.

The Bank Administration Institute (BAI), also concerned with standards of internal auditing, has issued a statement on the internal auditor's responsibility for detecting fraud. The statement appears in the BAI's "Statement of Principle and Standards for Internal Auditing in the Banking Industry."

Audit proficiency includes the ability to evaluate fraud exposures. Sufficient information is available in the literature on auditing concerning how frauds may be committed in banking. The auditor should be familiar with that literature.

The systems of control and not the internal audit function provide the primary assurance against fraud. Internal auditors, however, must evaluate the capability of the systems to achieve that end. When in doubt, the auditor should consider applying additional procedures to determine if fraud has actually occurred.

In fixing the internal auditor's responsibility for detecting fraud, it should be recognized that the internal auditor cannot be responsible for detecting irregular transactions for which there is no record, e.g., an unrecorded receipt of cash from a source for which there is no evidence of accountability; an isolated transaction that does not recur, e.g., a single fraudulent loan; or irregularities that are well concealed by collusion. However, in the usual course of the audit cycle, the internal auditor should detect irregularities that significantly affect the financial statements, repeatedly follow a suspicious pattern of occurrence, or can be detected by a reasonable audit sampling. Internal auditors must also accept responsibility for those irregularities that result from their failure to report known weaknesses in the systems of control.

In judging the preventive capacity of the control systems and the internal auditor's responsibility, the principle of relative risk should not be ignored, namely, costs must be balanced against intended benefit.

The EDP auditor can be an excellent source of information. Because the function is based on (or may actually be a part of) an internal audit, important professional standards and principles dictate how work is performed. Specific information on controls, weakness in security, recommendations for strengthening controls, and general information on elements of the EDP environment should be readily available from them. In addition, the EDP auditor often has computer tools specifically designed to assist investigators in reviewing, testing, and evaluating computerized records and computer systems. Some EDP auditors may even be experienced investigators of computer-related fraud or abuse.

EDP auditors should be familiar with these topics:

- **Basic topics:** introduction to data processing, computer hardware overview, computer programming overview, computer documentation overview, introduction to data processing application controls, and introduction to general data processing controls
- **Advanced topics:** on-line systems controls, data communication controls, continuous operation controls, storage media/device controls, audit trace considerations, and special audit software

EDP auditors use numerous tools and techniques to audit the computer environment. The tools and techniques can be classified by the function that they perform:

- **Auditing systems development and change control:** code comparison, system acceptance, and control group
- **Computer application control testing:** test data method, basic system evaluation, integrated test facility, and parallel simulation
- **Selecting and monitoring transactions for compliance, testing, and data verification:** transaction selection, embedded audit data collection, and extended records
- **Data verification:** generalized audit computer program
- **Analysis of computer programs:** snapshot, tracing, mapping, and control flowcharting
- **Auditing computer service center:** job accounting data analysis

Auditing Software

The most widely used tool is the Generalized Audit Computer Program package. The other frequently used tools and techniques are test data method, transaction selection, and control flowcharting.

The following major strong points of EDP auditors make them valuable in detecting abuse:

- **Level of confidence.** Because of the nature of their profession, auditors are highly respected as analysts and evaluators; the standards, principles, and codes of ethics that dictate how auditors conduct their work are well established; to a degree, auditors have a responsibility to their profession as well as to their employer.

- **Technical expertise.** With proper training and experience, EDP auditors provide a high level of EDP technical knowledge, both for the data processing profession in general and the specific computer environment within their organizations.

- **Tools and techniques.** Because EDP auditors must regularly use EDP audit tools and techniques, they are often available for testing and investigation; the EDP auditor should have some of these tools ready for immediate use (especially a generalized audit computer program package that can be used for retrieving and analyzing computerized records). However, the admissibility of evidential data obtained using these tools as an ordinary business practice is doubtful.

- **Independence.** Because auditors have no direct responsibility for nor authority over any of the activities that they review, they have a broad mandate, and they report to top management. Their independence is well established, a critical factor in any investigation.

The following major weak points of EDP audit must also be considered:

- **Relationship in organization.** Because audits are evaluations of the organization, they often cause the EDP audit and audited groups to disagree; this conflict can result in an adversarial relationship that may compromise cooperation.

- **Immaturity of profession.** Because of the relative newness of the EDP audit specialization in contrast to the general field of audit, generally accepted EDP audit principles, standards, guidelines, and tools and techniques are still developing.

- **Training.** Even with formal education and certification programs, the level of EDP auditor expertise varies widely; some auditors have excellent EDP and audit backgrounds, others are much stronger in one area than the other, and some have entered the profession with little EDP audit knowledge.

The adequacy of security controls may be enhanced by involving EDP auditors in evaluating applications program controls and consulting them to determine needed tests and checks in handling sensitive data. Audit trails built into computer programs can both deter and detect computer fraud and abuse.

The basic functions of security software should cover these principles:

- Security audit trails should be available to track the identity of users who update sensitive information files.
- If the sensitivity of information stored on microcomputers requires audit trails, then both physical and access controls are essential.
- In a computer network, the host computer, not the terminal, is where audit trails should be located.
- Audit trails should not be switched off to improve processing speed.
- Audit trail printouts should be reviewed regularly and frequently.

Internal Audits

The types of controls in an EDP system can be conveniently classified into two categories:

- **General controls**—relate to all parts of the EDP system; must be evaluated early in the audit
- **Application controls**—apply to the specific use of the system; must be evaluated specifically for every audit area in which the user accesses the computer so that the auditor can reduce assessed control risk

General controls break down to form four categories:

- Plan of organization and operation of EDP activity
- Procedures for documenting, reviewing, and approving systems and programs
- Hardware controls
- Controls over access to equipment, processes, and data files

Auditors usually evaluate the effectiveness of general controls before evaluating application controls. If general controls are ineffective, there may be potential for material error in each computerized accounting application. The auditor should be concerned with the potential for fictitious transactions or unauthorized data and omission accounts such as sales, purchases, and salaries.

Application controls that are related to the audit function can include the following:

- **Authorization.** Specific authorization of input documents by appropriate personnel is required before they can be processed.
- **Preprocessing review.** Someone from either the user department or EDP control group reviews the documents for such things as completeness and correctness.
- **Batching.** Input documents are collected, controlled, and processed by discrete groups.
- **Control totals.** These totals are used to determine whether all data put into the systems were processed.

- **Conversion verification.** This process checks to see that all data put onto machine-readable media (disks, tapes) are free of entry errors. The methods used to verify are:

    1. **Key verification.** Different operators repeat the keying process for all or part of the entry.

    2. **Check digit.** Part of an identification number, the check digit is used to determine whether a recorded ID number is correct.

    3. **Logic tests.** These tests involve various types of computerized comparisons of input data with programmed criteria that determine the acceptability of the input data.

    4. **Control Total Balancing.** Equipment can be programmed to produce totals from data entered for use in balancing.

- **Input security controls.** These controls prevent improper access to devices and files.

- **Correct file controls.** These controls verify that the correct file is processed.

- **Programmed Controls.** Written into computer programs to detect erroneous input, processing, or output, programmed controls can take four forms:

    1. **Validity test.** This test compares identification number with an existing list to determine if the number exists. (Is employee number included in a list of all employees?)

    2. **Completeness test.** This test examines every field in a record to determine whether all are complete. (Is employee number, name, number of regular hours worked, number of overtime hours worked, and department number included and complete?)

    3. **Matching test.** This test compares two fields to determine if they match. (Is employee in the same department as the one to which he or she is assigned?)

    4. **Reasonableness or limit test.** This test compares contents to see if amounts are within defined limits. (Is employee's gross payroll less than or equal to 60 hours or $999 for the week?)

- **Error correction procedures.** The procedures include specific steps to follow for investigation, correction, and resubmission if data is rejected.

- **Recovery and backup procedures.** The procedures involve maintaining backup copies of programs and files so that destroyed files can be recovered.

- **Postprocessing review.** User or EDP control group reviews the output for errors.

- **Master file controls.** These controls ensure the integrity of the master file at different times.

- **Periodic internal audit.** Internal auditors review input, output, and processing controls on an ongoing basis.

External Audits

Independent certified public accounting firms audit corporations and certify the accuracy of corporate financial information (for example, the statement in a company's annual report). These audits are performed under the provisions of the federal securities laws. When acting as the independent auditor of a publicly owned corporation, the external auditor has public responsibilities and must satisfy requirements of the federal and state governments regarding performance of those responsibilities. The objective of the ordinary examination of financial statements by independent auditors is the expression of an opinion on the fairness with which they present the financial position, results of operations, and change in financial position in conformity with generally accepted accounting principles.

CPAs (certified public accountants), certified by state examining boards as having met stringent qualifying requirements to practice accounting, may serve as independent auditors for publicly owned corporations. Noncertified accountants may engage in some of the audit work, but a CPA is required to direct the effort and to sign the opinion.

The American Institute of Certified Public Accountants (AICPA) is the national association that guides and directs the auditing profession. Various AICPA committees are chartered to issue pronouncements and rules on auditing matters; for example, Statement on Auditing Standards Number 3, The Effects of EDP on the Auditors' Study and Evaluation of Internal Control. In addition, a code of professional ethics supports the standards and provides a basis for their enforcement. Starting in January 1989, CPAs were required to audit for fraudulent activity.

The major purpose of external auditing is to attest to the accuracy of company's financial statements, not to audit internal controls per se (e.g., controls involved with data processing). However, a number of CPA firms have developed audit tools to assist in EDP auditing. The major tool, the Generalized Audit Computer Program Package, is used to retrieve and analyze data stored in computer files.

From an EDP perspective, external auditors typically do not perform a detailed review of the full computer environment—the financial attestation does not require that type of effort. Nonetheless, they usually have staff with EDP expertise and use them as needed, typically for helping to extract computerized financial records or for management consulting on special projects (other than the financial audit function).

CPAs normally produce two reports, the opinion letter and a management letter. The opinion letter is a short statement of the scope and date of the audit, an opinion of the accuracy and fairness of the financial statements, any expectations, and whether the financial statements are presented in accordance with generally accepted accounting principles that have been consistently observed over the preceding periods. The management letter includes findings regarding weak or missing controls and recommendations for corrective action. In addition to producing these formal reports, external auditors have well-defined standards of field work that include the compilation of sufficient evidential matter (in work papers) to support the rendered opinion.

DISCLOSURE OF SECURITY VIOLATIONS

Even with the increase of computer-related crimes being made known to the public, organizations are not always willing to report these violations. Organizations are often reluctant to make full disclosure of violations because they do not want the vulnerabilities of their operation to be known for fear of losing the public's or customer's confidence.

Another reason organizations fail to report crimes is that hard evidence of computer crime is difficult to gather. "It is estimated that businesses report only about 6% of criminal acts aimed against their computer systems."[32] The consequences of reporting computer crime have traditionally been perceived as being most costly to the victim.

At the same time that companies fail to report computer crimes, there is a rapid rise in the number (and dollar volume) of these crimes. "Estimates of loss due to high-tech thieves range as high as $3 billion to $5 billion annually in the U.S."[32]

In a 1989 study reported in *Sloan Management Review,* Jeffrey Hoffer and Detmar Straub report that "information about large-scale losses is often suppressed because of management embarrassment. The effect is to promote a false sense that systems are more secure than they really are."[20]

The study reinforced earlier findings that *computer abuse is a serious and underreported problem.* Of the 1,211 organizations that returned the survey, 211 firms reported 259 separate incidents of abuse. Five of the 211 organizations reported dollar losses of more than $100,000 at the local level; in the pilot study, one organization reported a $2 million loss. These results, which are similar to results from earlier studies, indicate that one out of five organizations experiences one or more security breaches in a three-year period. This figure is undoubtedly an underestimate, since no single manager is aware of all the abuse uncovered in an organization. Based on validation studies, a single manager may be aware of less than 50 percent of the abuse actually discovered in a firm. Uncovered abuse, moreover, is probably only a portion of actual abuse. Several studies have found that half of all known computer-related crimes are discovered by accident.[20]

Whistle-blowers

Another factor that organizations must consider, along with the willingness to publicly disclose violations, is employees' crimes or misuse. **Whistle-blowing** involves employees reporting illegal, immoral, or illegitimate practices under the control of their employers to parties who can take corrective action. The whistle-blower is usually seen as an adversary who "rocks the boat" and is often isolated, shunned, fired, or suffers in some other manner for his or her actions. The whistle-blower's role can be seen as a dichotomy. On one hand, whistle-blowers have the ability to help organizations correct unsafe products or working conditions or wasteful practices and, thereby, avoid substantial adverse consequences. On the other hand, whistle-blowers may threaten the organization's authority structure, cohesiveness, and public image.

John P. Keenan of the School of Business, Management Institute, University of Wisconsin, Madison, states that

Research on upward communications in organizations indicates that openness, trust, and receptivity tend to have beneficial effects in terms of organizational performance.

If there is mistrust, lack of openness, and suspicions [about] reciprocals, the willingness to [report] illegal or wasteful activities might be expected to be reduced. . .

Communication climate plays a major role on whether employees choose to speak up . . . [The] role of defensive and supportive communication climate indicates that a direct correlation exists between these items and whistleblowing.[27]

There are currently over a dozen federal laws to protect whistle-blowers, along with numerous state laws. Even with the legal protections in place, many employees are reluctant to report violations for several reasons:

- The risks of whistle-blowing and the motivations behind it are not easily influenced by statutory enactments.
- The statutes as enacted are not perceived as being effective or are not adequately understood.
- The statutes have a beneficial but unanticipated effect—causing employers to change their policies by encouraging internal whistle-blowing and refraining from retaliation against whistle-blowers.

Management must not discourage whistle-blowing entirely, but instead set up an effective reporting system as an open avenue to employees that includes:

- Personal communication
- Confidential advice and counseling
- Investigation, conciliation, and mediation
- Adjudication
- Upward feedback

CRITICAL MANAGEMENT ISSUES IN COMPUTER SECURITY

To have a fully functioning, efficient security system, managers must apply the same solid management practices that they use for any other asset. The "people" skills of these managers are the essence of success or failure.

Balance

After an intrusion by pranksters at the *Calgary Herald* in Calgary, Alberta, Canada, management was forced to look at the security of its computer system. One lesson the *Herald* learned was the importance of good management. While many companies might have looked to computer-based prevention, the *Herald* faced a basic reality.

It was primarily a management problem . . . We should have been able to identify this before it affected our computer operations . . . The *Herald* didn't ignore the need for additional security requirements, however, we added new levels of security, such as passwords and restrictions on functions various users can perform. We had to balance security and access . . . If we restricted our people too much, we'd slow down the writing, editing, and production. We'd hamstring our operations.[2]

Controls should be built into each application system to ensure that all data entered into the system is completely and accurately processed into output format. This is a vital part of a well-designed control framework. Present technology makes it possible to automate virtually all aspects of system balancing, so that the only time manual intervention is needed is after the diagnosis of an out-of-balance condition by the system. A system should be manually balanced during initial parallel, and after maintenance modifications or system enhancements. Any time there is a program change, balancing must be retested. This will ensure that known changes to the system have not created an out-of-balance condition where data is not completely processed or inaccurately processed. To ensure that unknown or unrecognized changes have not created an out-of-balance condition, the system should be retested on a periodic basis.

Balancing controls must ensure that all data is successfully transferred between system interface points. Controls for interface balancing should be established wherever data are processed by or passed between multiple applications and application systems. When data are passed between application systems that are under the responsibility of two different managers, frequently a responsibility void (and communication void) will develop with respect to balancing controls. Management should establish a clear balancing control policy and periodically assess their personnel for adherence to it.

Electronic Monitoring of Employees

Electronic monitoring is the process whereby employers collect, store, and analyze information on employee performance using computer and/or telecommunications technology. Through currently available forms of electronic monitoring, employers can analyze employee performance and productivity. Computers, for example, can record when workers turn their video display terminal (VDT) off and on, count the number of keystrokes per second, track the number of operator errors per day, and monitor employees' telephone conversations. More than 60,000 companies use computer and phone technology to check work pace, rank performance level, record break times, and measure tone of phone conversations.

As electronic monitoring of workers spreads through the workplace, however, employee advocacy groups and unions have raised concerns about the stress imposed by this kind of monitoring and the potential for invasion of privacy. They have compared the monitoring of the workplace to George Orwell's "Big Brother" where workers are under constant surveillance and are constantly scrutinized for their performance. On the other side of the spectrum, there are groups defending management's right to rely on computerized work-measurement systems, noting their relevance and importance to effective quality control.

It has been speculated that the dispute over workplace privacy and electronic surveillance will be the hottest employment-law topic of the 1990s.

According to the Office of Technology Assessment (OTA), a congressional research agency, electronic work measurement is most commonly focused on office workers who occupy positions in which a limited number of standardized tasks are performed. This type of monitoring can take many forms: closed-circuit television, phone monitoring, the monitoring of computer workstations, and metal detectors at plant entrances. Phone monitoring is common in the airline reservation and phone customer assistance businesses, while computer monitoring is more common in the word processing and customer service areas. Similar techniques, however, have also been applied to higher level, skilled positions in technical, professional, and managerial fields as well. Things commonly looked at with managers include the number of appointments they have and the number of hours spent on various tasks.

Employers use enhanced monitoring for a variety of reasons: to reduce theft, to study and improve efficiency, to ensure that customer relations are being handled correctly, to pace work, and to evaluate employees. If used correctly, electronic monitoring can be conducted without interfering in any way with performance of the work and can collect vast amounts of incredibly detailed work measurements. Airline reservation computers, for example, closely measure how long individual clerks take to handle each customer and the amount of time the employee spends between calls. The computer takes note of any idle moment and measures lunch hours, coffee breaks, and even trips to the bathroom. All this is done without the need for a supervisor to be physically present (on-site) with the employee. Most observers agree that certain data generated by such a system can provide valuable and detailed information to an employer to evaluate and compare one worker objectively with all the others. Another reason for increased use of electronic monitoring is that the installation of such a system is relatively easy, requiring only specially written software. However, the amount of information readily available to employers presents a host of other issues and concerns—for example, the response time to customer calls and the number of inquiries abandoned because of delays. It is still questionable whether most electronic monitoring is any more invasive of privacy than the sorts of personal monitoring that have always been allowed. The problem of electronic monitoring is that it is impersonal and often unobservable by the employee. Both the employee's inability to know when the monitoring is occurring and the use of monitoring to constantly pace and gauge performance increases workplace stress. Critics contend that these advanced systems could become counterproductive, leading to work-related stress and high turnover rates. Finally, with employers so concerned about the quantity of worker output, some companies may begin to overlook a factor more difficult to measure: quality.

Congress has considered legislation to protect employees from other kinds of mechanical and electrical devices used by employers to investigate their workers. Traditionally, employers have monitored employee performance visually. Today, however, employers rely increasingly on machines of various sorts to enhance or supplement personal visual monitoring. Advances in electronics on computers have given employers increasingly efficient methods of gathering vast amounts of employee information silently and without the employee's knowledge. The most common types of monitoring devices in the workplace today relate to video, phone, and computer usage.

Video Monitoring. Visual monitoring through the use of cameras and television has been a topic of growing concern. However, it has received little public attention. A few court decisions have held that use of such equipment can be justified as a legitimate monitoring device. Consequently, monitoring places such as bathrooms, changing areas, and private offices (where there is great expectation for privacy) could lead to successful suits that employees could file against the company unless the employer can show a great need for surveillance.

Some examples of video monitoring include these:

- General Electric Co. uses tiny, fish-eye lenses installed behind pinholes in walls and ceilings to watch employees suspected of crimes.

- DuPont Co. uses hidden, long-distance cameras to monitor its loading docks 24 hours a day.

Phone Monitoring. Fifteen million employees work in industries where monitoring occurs, and an estimated 1.5 million of those employees had their conversations monitored in 1987. Phone monitoring has received special attention because of the reasonable expectation of privacy that many employees and third parties have regarding their phone conversations. Although phone monitoring is subject to control under the federal wiretapping law, many exceptions arise in relation to employee phone monitoring. One exception allows the employer, "in the ordinary course of business," to use another telephone to monitor employee calls. An employee would be protected only if private phone calls are allowed by the employer, and if the employer had not given notice of phone monitoring. To date, the courts have ruled that monitoring that occurs through the use of extension phones is permissible if a firm suspects an employee is discussing business matters with a competitor or verbally abusing customers. An employer can also monitor for personal calls only long enough to determine if a call is personal. In April 1987, Representative Don Edwards (D-CA) introduced H.R. 1950, a bill that would require identification of phones that might be monitored and a regular audible warning tone (i.e., a "beep") when an employer is listening. Thus, the stress of not knowing whether someone is listening to your conversations would be absent. California and West Virginia both passed similar "beep" bills, but these laws are no longer in effect.

Here is an example in which phone monitoring has paid off for a company:

Soon after installing software that keeps track of employee telephone calls, managers at a Los Angeles insurance company noticed a puzzling pattern. Each day at noon, someone at the firm dialed a number where the conversation lasted for 58 minutes. A file clerk stationed at that extension was asked to explain the matter. Her answer: Every day she called her mother, who put the phone down beside the television set in her home so that her daughter could listen to her favorite soap opera.

Although this is an extreme case, it illustrates that telephone monitoring can pay off by quickly locating inefficiencies and waste in a corporation.

Computer Monitoring. Computer monitoring is the process that connects the workstations in offices to computers that keep a close watch on the individual worker's production. Those whose jobs involve desktop terminals are the most vulnerable: central mainframe computers keep track of input speed, time spent working, and even the total mistakes made. What sets computer monitoring apart from all other similar devices is its ability to monitor work as it takes place. At Giant Food store chains, optical scanners at checkout counters eliminate pricing guesses by employees, improve inventory control, aid in work scheduling, and track the workers' speed—all of which produce savings in excess of $15 million. Recently Congress has focused on employees' rights and privacy issues relating to monitoring in the private sector. The following are some examples of legislation that would safeguard employee rights:

- *Privacy for Workers & Consumers Act (Bill H.R. 2168)* states that the employer would have to provide the following to their employees if monitoring were to occur:
    - Prior written notice that it is engaging in electronic monitoring
    - Explanation of what information it is trying to obtain
    - Statement of how it will use the information
    - List of what performance standards will be monitored
    - Results of the monitoring
    - Assurance that the only information collected is that which is relevant to the employee's work performance

    This legislation applies to any electronic means to observe employees, including telephone, computer, wire, radio, or optical system.

- Illinois Senator Paul Simon introduced a bill that would require employers to notify workers when their telephone or computer use is being electronically monitored. This bill would prohibit employers from using monitoring data as the sole basis for production quotas and would provide civil fines of up to $10,000 per violation. Surprisingly, several companies have found monitoring counterproductive and have eliminated its practice.

    Federal Express. In 1984 it decided to discontinue a telephone monitoring system tracking its 43,000 employees in the customer service division. The management felt monitoring productivity was sacrificing quality.

    AT&T, Tempe, Arizona. AT&T operated a hotel billing office without any monitoring or supervisors. Service was guaranteed only by following up on specific customer complaints. Despite the absence of monitoring, the 135 employees showed increased levels of service and productivity.

The following are some guidelines on how to use electronic monitoring effectively:

- Employees should be informed when work is monitored and be shown how the data are used.
- Subliminal messages should not be flashed across the screen.
- A written policy should be given to employees and supervisors.
- Employees should understand how and when their work is being measured.

- Employees should have access to their records and receive regular performance feedback.

- Individual differences in need for breaks or in work speed should be anticipated.

- Individuals should be rewarded appropriately for performance.

With new technology present in the workplace, productivity has improved substantially. Yet, along with this growth in productivity is the employers' ability to monitor the work of their employees through electronic means. Management must balance increased productivity against the possibility of alienating employees and causing undue stress. Monitoring of employees in itself is neither good nor bad. Rather, it's how managers use these technologies that determines their acceptance in the workplace. Employers can utilize current technology to advance the goals of the firm while at the same time improving the quality and skills of their workers.

The 1990s will deal with many technological and legal issues concerning electronic employee monitoring. Lack of protection from undue monitoring in the workplace is slowly being challenged and will continue to be challenged in the future. The harshest critics of electronic monitoring argue that it should be banned from the workplace, while others have lobbied for legislation limiting its use. At a minimum, the critics desire policies that will prevent the outright abuse of this technology. Rules and policies alone will not ensure effectiveness of monitoring. By managers focusing their attention on sound principles of performance appraisal and feedback, the negative impacts of electronic monitoring for employees can be minimized.

Promoting Security

When assessing security's role, it is important to keep in mind an obvious point: Resources are limited. Each security recommendation must compete for funding and support with all the other recommendations bubbling through the organization. Some security recommendations—including many of those described here—are more important than others. If they are not carefully presented and properly justified, their *significance* may not be recognized by top management and they will not be funded. Here are four basic business realities you must clearly understand before you embark on the preparation of a proposal.[38]

- Phrase recommendations to get attention—increasing revenue or reducing operating costs.

- Write recommendations that will make upper management look good and avoid adverse publicity.

- Be people-oriented.

- Include present cost and benefits, both tangible and intangible.[38]

"Budgetary presentations are, in essence, sales presentations. Each department manager 'sells' the financial decision-makers on why that department should receive a bigger slice of the corporate budgetary pie."[57]

SUMMARY

This chapter presents the security issues facing managers in the computer environment. While senior management teams in most organizations are becoming more aware of data security requirements, few managers are confident that the safeguards they have already taken are adequate. Information—its creation, protection, and dissemination—is critical to the ongoing operation of the organization. The management functions of planning, staffing, organizing, directing, and controlling apply to the management of the security function. **Human resources** and **capital resources** must be considered in order to meet corporate security objectives. Long-range goals must include a commitment from top management, creating a steering committee, ensuring management participation, encouraging teamwork, getting individuals involved, working to improve the system, establishing a quality assurance mechanism, and planning for security.

Security policy goals should address avoidance, deterrence, prevention, detection, recovery, and correction. Information, in order to meet security objectives and goals, must be evaluated according to **classification**, **ownership**, and **valuation**.

Locating and training computer security personnel should involve selection of a **Data Security Officer (DSO)**, placing security at the appropriate and most effective level in the organization structure, establishing reporting policies, staffing the security function in all computer-related positions, establishing hiring practices (including probationary periods and termination policies), and building investigative teams.

Budget constraints that will impact the security function should address resource allocation, billing for security functions, selling management on security, and economic difficulties.

Security awareness and training sessions should be established that include all employees who create or use information.

Evaluating and updating the security program can be accomplished by conducting audits—both internal and external.

Organizations do not disclose security violations to the public for various reasons. **Whistle-blowers** are employees who report illegal, immoral, or illegitimate acts.

The critical management issues that should be considered when developing a fully functioning, efficient security system are balance, electronic monitoring (video, phone, and computer), and promotion of security throughout the organization.

REFERENCES

1. Baker, Richard H., *The Computer Security Handbook* (Blue Ridge Summit, PA: Tab Books, 1985), pp. 1-281.

2. Ball, Michael, "To Catch a Thief: Lessons in Systems Security," *Computerworld*, December 14, 1987, pp. 73-74.

3. Berg, H.K., W.E. Boebert, W.R. Franta, and T.G. Moher, *Formal Methods of Program Verification and Specification* (Englewood Cliffs, NJ: Prentice-Hall, 1982).

4. Brebner, Kratz, and Fites, *Control and Security of Computer Information Systems*, Computer Science Press, 1989.

5. Caldwell, Kenneth, "Answer Fundamental Questions About Security Planning," *Government Computer News*, December 5, 1986.

6. Cashin, Jerry, "Forewarned Is Forearmed," *Software Magazine*, April 1988, pp. 69-70.

7. Connolly, Charles P., "Everybody's Business," *Security Management*, Vol. 34, August 1990, pp. 45-47.

8. Courtney, Robert H., Jr., "Contemporary Data Security: A Leadership Vacuum," *Computer Security Journal*, Vol. IV, No. 2, December 21, 1987, pp. 7-15.

9. "Data Classification—Or Should It Be Information Classification?" *Computer Control Quarterly*, Spring 1986, pp. 63-69.

10. Delmage, Sherman, "Tiger Teams," *PC Week*, February 11, 1986, pp. 55, 60, 61.

11. Dooley, Ann, "Crime Time," *Computerworld*, June 5, 1989, pp. 31-32.

12. Dworkin, Terry Morehead, "Protecting Private Employees from Enhanced Monitoring: Legislative Approaches," *American Business Law Journal*, Vol. 28, Spring 1990, pp. 59-85.

13. Enger, Norman L., and Paul W. Howerton, *Computer Security . . . A Management Audit Approach* (New York: AMACOM, 1980).

14. Farber, Ken, "Integrity Security Throughout an Enterprise," *ISP News*, September/October 1991, pp. 40-41.

15. Farhoomand, Alif, and Michael Murphy, "Managing Computer Security," *Datamation*, January 1, 1989, pp. 67-68.

16. Freeman, Layton, and Peter P. Mykytyn, "Computer Crime and Data Copyright: Issues for IS Management," *Journal of Systems Management*, November/December 1991, p. 18.

17. Frenzel, Carroll W., *Management of Information Technology* (Boston, MA: boyd & fraser publishing company, 1992), pp. 1-283.

18. Gelbspan, Ross, "Keeping a Close Eye on Electronic Work Monitoring," *The Washington Post*, December 13, 1987, p. 4.

19. *Good Security Practices for Information Ownership and Classification*, IBM Corporation, G360-2705-01, November 1986.

20. Hoffer, Jeffrey, and Detmar W. Straub, Jr., "The 9 to 5 Underground: Are You Policing Computer Crimes?" *Sloan Management Review*, Vol. 3, No. 4, Summer 1989, pp. 35-43.

21. Hollocker, Charles P., *Software Reviews and Audits Handbook* (New York: John Wiley and Sons, 1990).

22. Hutt, Arthur E., Seymour Bosworth, and Douglas B. Hoyt, *Computer Security Handbook* (New York: Macmillan, 1988), pp. 1-399.

23. *Information Technology Resource Management Standard, Information Security*, Commonwealth of Virginia, COV ITRM Standard 91-1, Richmond, VA, December 9, 1991.

24. Johnston, R.E., "From Security Staff to Protective Parents," *Infosystems*, February 1987, p. 38.

25. Johnston, Robert E., "Don't Blame Management for Lack of Security Commitment," *ISSA Access*, 1st Quarter, 1989, pp. 16-40.

26. Kay, Russell, "Alan Krull on Getting Things Done: The Art of Productive Politics," *ISP News*, July/August 1992, p. 50.

27. Keenan, John P., "Sex, Home Office, and Whistle-blowing: Implications Regarding Managers' Response to Fraud, Waste, and Mismanagement," Proceedings of SWFAD, New Orleans, LA, March 1989, pp. 371-375.

28. Kesim, Susan N., "Securing Computers: A Checklist," *Security Systems*, p. 42.

29. Laurance, Robert, "Managing Security," *Journal of Data and Computer Communications*, Summer 1988, pp. 77-79.

30. Littlejohn, Robert F., "Teaming Up to Fight Computer Crime," *Security Management*, July 1990, pp. 37-38.

31. Lobel, Jerome, *Failing the System Breakers* (New York: McGraw-Hill, 1986).

32. Loch, Karen D., Houston H. Carr, and Merril Warkentin, "Why Won't Organizations Tell You About Computer Crime?" *Information Management Bulletin*, February 1991, pp. 5-6.

33. McKibbin, Wendy, "Who Gets the Blame for Computer Crime?" *Infosystems*, July 1983, pp. 34-36.

34. Makely, William K., "Computer Security's Worst Enemy: Management Apathy," *The Office*, March 1987, pp. 115, 116, 120.

35. Martin, James, *Security Accuracy and Privacy in Computer Systems* (Englewood Cliffs, NJ: Prentice-Hall, 1973).

36. Menkus, Belden, "Crime Prevention in System Design," *Journal of Systems Management*, May 1991, p. 19.

37. Moore, Warren S., "Secure Data Center Operations," *Data Security Management*, Auerbach Publishers, Inc., New York, NY, 1984, pp. 1-12.

38. Moulton, Rolf, "A By-Product of Effective Security-Improved Organizational Productivity," *Computer Security Journal*, Vol. V, No. 1, pp. 31-37.

39. Parker, Donn B., *Computer Security Management* (Reston, VA: Reston Publishing Company, 1981), pp. 1-308.

40. Peltier, Thomas, "Hard Times: Infosecurity and Downsizing," *ISP News*, July/August 1992, p. 25.

41. Pfleeger, Charles P., *Security in Computing* (Englewood Cliffs, NJ: Prentice-Hall, 1989), pp. 1-538.

42. Prause, Peter N., "The DSO's Role in Designing and Implementing Application Systems," *Data Security Management*, Auerbach Publishers, Inc., 1984, pp. 1-8.

43. Ross, Steven J., and Michael Pinna, "Defeating the Would-Be Abuser," *New Accountant*, January 1992, pp. 31-34.

44. Rothfeder, Jerrey, "Is Your Boss Spying on You?" *Business Week*, January 15, 1990, pp. 74-75.

45. Rullo, Thomas A., *Advances in Computer Security Management* (Philadelphia, PA: Heyden, 1980).

46. Russell, Deborah, and G.T. Gangemi, Jr., *Computer Security Basics* (Sebastopol, CA: O'Reilly and Associates, Inc., 1991), pp. 1-441.

47. Schweitzer, James, "Tangible Losses," *Computerworld*, January 18, 1988, pp. 75-81.

48. Schweitzer, James A., *Computer Crime and Business Information . . . A Practical Guide for Managers* (New York: Elsevier, 1986), pp. 1-195.

49. "Security: Striking a Balance," *Infosystems*, January 1986, p. 22.

50. "Selecting an Effective Manager," *Security Systems*, Vol. 16, No. 12, December 1987, p. 5.

51. Smith, Cherie, "Marketing Security Acquisitions to Management," *ISSA Access*, 1st Quarter, 1989, pp. 10, 42.

52. Steinberg, Richard M., and Raymond N. Johnson, *Journal of Accountancy*, August 1991, pp. 60-68.

53. Stevens, Larry, "Security Systems: Getting Management to Sell Out," *Computerworld*, December 14, 1987, pp. 75-77.

54. Straub, Detmar, Jr., and William D. Nance, "Discovering and Disciplining Computer Abuse in Organizations: A Field Study," *MIS Quarterly*, March 1990.

55. Sweet, Frank, "How to Build a Security Chain," *Datamation*, February 1, 1987, pp. 69-70.

56. Teodoro, Reynaldo S., "DP Technology Security Impact Statement," *Data Security Management*, Auerbach Publishers, Inc., New York, NY, 82-02-10, pp. 1-16.

57. "The Computer Security Officer," *Computer Control Quarterly*, Spring 1986, pp. 52-54.

58. "The Security Importance of Personnel Recruitment," *Computer Control Quarterly*, Spring 1986, pp. 1-3.

59. *Using Microcomputers in GAO Audits: Improving Quality and Productivity*, United States General Accounting Office, Information Management and Technology Division, March 1986.

60. Weber, Austin, "Effective Security Programs Start with Awareness," *Data Management*, November 1985, p. 34.

61. Wofsey, Marvin M., Editor, *Advances in Computer Security Management* (New York: John Wiley and Sons, 1983), pp. 1-268.

62. Wood, Charles Cresson, "How Many Information Security Staff People Should You Have?" *Computers and Security*, September 1990, pp. 395-402.

63. Yerespej, Michael, "How Much Can You 'Buy' Employees?" *Industry Week*, Vol. 238, August 7, 1989, pp. 65-66.

REVIEW QUESTIONS

1. Contrast the human resources and capital resources that are part of an effective computer security program.

2. List some of the necessary long-range goals that corporations must include in a corporate security plan.

3. State a brief description of information classification, ownership, and valuation.

4. What is the role of the Data Security Officer (DSO)?

5. Briefly explain the need for proper placement of the security function in the corporate reporting structure.

6. Why are established hiring practices (including probationary periods and termination) critical to the effective functioning of computer security?

7. What is the role of a tiger team?

8. What is a chargeback system of billing for computer security operations?

9. Contrast computer security awareness and training. Which employees would be involved in each?

10. Compare an internal vs. an external audit.

DISCUSSION QUESTIONS

1. "Security managers must constantly resell value to top management due to rapid turnover." How could this reselling be best accomplished and yet maintain consistency?

2. "Wrongful and abusive discharge is the major labor issue today." How can computer security managers avoid this situation or better protect themselves against charges?

3. "The only thing worse than an installation totally devoid of security measures is one that believes its security measures are adequate." How can organizations avoid this complacency?

4. "Many managers think that security gadgets and gimmicks will prevent computer crime." What is the best way to convince people that computer security is a "people problem" that a "technical fix" does not cure?

5. Explain the statement, "The cost of security should not exceed its value."

EXERCISES

1. Prepare a list of the pros and cons of internal vs. external audits of computer security functions.

2. Locate an article in a non-computer journal that addresses the issue of management's responsibility to computer security.

3. Interview an individual in an organization to gain insight to where computer security is placed in that corporation's structure.

4. Find out where computer security functions are placed in the administrative structure of your college or university.

5. People often say, "Why mark something as valuable? It just shows the bad guys what to steal." With hacking increasing on college campuses, could labeling or classifying information "send up a red flag"?

PROBLEM-SOLVING EXERCISES

1. Develop a one-page policy statement (using the DPMA Statement as a guide) that could be adopted by a corporation as their official policy statement relating to computer security.

2. Develop a 10-15 minute presentation that you would present to top management to "sell" the need for computer security resource allocation.

3. If you were on the committee that was screening applicants for the position of Data Security Officer (DSO), what criteria for this position would you check?

4. "Once major activities are completed, security is often in a maintenance mode. This is when the parental attitude of security managers begins to surface." Design a policy that would avoid the parent-child syndrome and would foster the adult-adult way of approaching security.

5. The use of the power of the computer to uncover suspicious patterns is very prevalent today. Develop a list of ways the computer could perform security procedures more effectively than humans.

CASES
King Aviation

Joe King, President of King Aviation, an independent for-hire flight service head-quartered in Chicago, Illinois, recently has become more concerned about computer security due to several breaches in King's system. Even though the breaches were somewhat minor, King is concerned about the vulnerability of the data and information as his clients expect complete discretion. Also of concern is the competition—there are numerous other flight companies in the Chicago area.

A security consultant, on her initial visit to King, mentioned that personnel policies must be addressed first, then technical issues. The consultant observed three distinct personnel areas that need to be rectified:

1. Employees consistently arriving early and remaining late—without visible signs of extra work being produced.

2. No debriefing of and no policies related to terminated or resigning employees.

3. Employees with access to sensitive data who take computers to and from their office in order to do extra work.

Mr. King feels these policies should not be so stringently enforced that employees feel threatened. How could he clean up some of its loose policies so the vulnerability of the system is lessened, yet continue to foster a productive, highly motivated organization?

Bishop Financial Services

BFS is an independently owned financial advisory servicing clients in the Dayton, Ohio area. John Bishop, president of BFS, feels that to ensure the safety of clients' information, internal audits should be performed periodically. He also feels that external audits should be conducted at least yearly to verify the results of the internal audits.

1. What types of operations and transactions should the audit strive to check?

2. Who should perform the internal and external audits?

3. Should both of these audits be visible or discrete? Why?

DISASTER RECOVERY AND CONTINGENCY PLANNING

LEARNING OBJECTIVES

After studying this chapter, you will be able to:

1. Explain the need for disaster planning and recovery.
2. Identify the elements of an effective disaster recovery plan.
3. List the basic steps in a risk analysis.
4. Explain several analytical risk analysis procedures.
5. Discuss how vulnerabilities can be identified and corrected.
6. Explain how a contingency plan is developed and utilized.
7. Discuss some types of application software used for risk analysis and contingency planning/disaster recovery.
8. Discuss basic back-up procedures and how they are organized and used.
9. Describe an uninterruptible power supply (UPS) and how it is used.
10. Explain the insurance involved in contingency planning.
11. Describe how training of employees is conducted.
12. Explain testing procedures for disaster recovery.
13. Define the following terms.

TERMS

- Risk Analysis
- Vulnerability
- Annualized Loss Exposure (ALE)
- Courtney Risk Assessment
- Delphi Approach
- Security Plan
- Safeguards
- Contingency
- Contingency Plan
- Disaster
- MTBU
- On-Site Backup
- Off-Site Backup

- Hot Sites
- Warm Sites
- Cold Sites
- Relocatable Shells
- On-Site Local Backup
- Off-Site Local Backup
- Off-Site Remote Backup
- Archival Storage
- Uninterruptible Power Supply (UPS)
- Spikes
- Oscillatory Voltage Surge
- Checklist Test

WORLD TRADE CENTER BOMBING CHANGES DISASTER RECOVERY FOREVER

Infosecurity experts believe that the February 1993 terrorist bombing of New York's World Trade Center has changed the way we think about disaster recovery planning.

On February 26, a van packed with explosives blew up in an underground parking garage. People throughout the 110-story building felt the blast. It left a 60- by 100-foot by 3-story-deep hole and knocked out electrical power and telephone communications throughout the building. Over 1,000 people were injured, six people killed, and thousands evacuated down dark and smoky stairwells.

From a disaster recovery planning standpoint, the most crucial result of this disaster was not so much the physical damage to office space, equipment, or data; but rather, the fact that access to offices and, therefore, data was denied to so many businesses. After the building was evacuated and the extent of the damage realized, the owners of the complex (the Port Authority of New York and New Jersey) gave limited access to its tenants. Security guards escorted groups of workers to their offices, many having to wait for up to five hours. Once there, workers had only half an hour to retrieve any files, tapes, disks, and equipment.

To give an historical perspective, this bombing was the fourth largest disaster since the disaster recovery industry was born, based on the number of data centers that became inoperable, according to John Nevola, manager of IBM's Business Recovery Services facility in Franklin Lakes, New Jersey. The three disasters with greater impact included Hurricane Andrew and the Chicago flood in 1992, and the New York power outage in 1990. The after-effects of the World Trade Center bombing continued because of the extended recovery period.

All in all, the disaster affected approximately 50,000 employees of 350 companies, many with large data centers right in the building. The fact that so many organizations were affected is also a factor that forced us to change the way we think about disaster recovery.

As reported in "Terror in the Towers: Bombing Shapes the Future of Disaster Recovery," by Russell Kay, *Infosecurity News*, May/June 1993, pp. 45-47.

The February 21, 1992, issue of *Computerworld* offers this brief scenario and summary of the need for disaster planning and recovery operations:

"Imagine this. You get a phone call at 3 a.m., telling you that your main computer room has burnt down. What do you do?

"Do you immediately contact your backup facility, getting the operation going again—or do you reach for your resume and start looking for a new job?"

Will Davis, senior consultant with the Amdahl Corporation, cites one recent world survey which showed that half the responding companies saw themselves as operating for no more than two to three days in the event of a system breakdown.[28]

This survey was part of a worldwide report commissioned by Amdahl in 1990. Some of the disasters that occurred—that could have been overcome or alleviated by a backup system—were:

- In 1989, hurricanes blew the roof off an English telephone sales company. The computer still functioned, but the staff could not work.

- A European computer services company office was flooded when a nearby stream burst its banks. The computer hall and its computers were flooded in 15 minutes.

- A fire outside a European bank disrupted the telecommunications and electricity supply to the computer center for four weeks.

- Data incorrectly input, in miles instead of kilometers, led to an unmanned satellite having to be aborted.

- The 1989 California earthquake caused a major computer manufacturer to lose both primary and secondary power as well as its cooling systems, affecting its ability to exchange and process information.

During the Newcastle, Australia, and San Francisco earthquakes, one international airline with offices in both cities received a double blow within the space of four months. Fortunately, a backup recovery system was in place in Australia, and the airline managed to restore communications there within half an hour of the earthquake. The same facility was not operating with the same efficiency in San Francisco, where 12 hours elapsed before calls were rerouted through the Australian office.

Davis stressed that off-site backup is a topic that must be covered as early in the planning effort as possible.

"While backup of the critical systems is important, correct backup of all systems is necessary since they will all have to be restored at some stage after a major disaster," he said.

"The backup scheme should be one that responds effectively to the most likely problem situations while still providing a tolerable solution in major recovery situations."[20]

CRISIS MANAGEMENT

Managing and operating any computer facility not only involves the day-to-day operations but should also include planning for disasters, contingencies, and recovery. Ongoing operation is the prime objective; the steps needed to reach that goal should be carefully planned, organized, implemented, and tested.

A thorough analysis of any computer center's disaster operation plan should include a risk analysis, a disaster plan, recovery and resumption procedures, a systematic plan for training key employees, and periodic testing of the entire plan. (See Figure 13.1.)

FIGURE 13.1
Disaster Recovery Plan

☑ Risk analysis
☑ Disaster plan
☑ Recovery and resumption
☑ Training employees
☑ Testing of plan

The key question is, "How long can our computer center be down before profits and service are adversely affected?"

"Computer misfortunes arise from causes as diverse as human error, deliberate sabotage, acts of God, and impeccably executed computer crime."[1] The threats of natural disasters, viruses, hackers, and human error have already been discussed at length. This chapter will focus on recovery from these threats to computer systems. "Don't wait for a crisis to access your security. Learning now to outsmart hackers, survive floods and fires, and fend off viruses can help you save your company thousands of dollars in lost data."[1]

RISK ANALYSIS

Data security risk analysis is an effective means for convincing management of the importance of data security. One of the most critical and thorny roadblocks information processing managers face when they implement a data security program is convincing upper-level management of this importance. Some risk analysis controls include:

1. Organizational and administrative controls designed to provide effective segregation of duties and restrictions on accessing data, supplemented by the tests of the effectiveness of security protection procedures

2. Authentication of system users supplemented by the additional verification procedures designed to validate users

3. Physical security measures designed to provide for continuity of data processing services in the event of natural and man-made disasters, and to control access to computer equipment

4. Communication line protocol provisions specifically incorporated to provide for data security

These controls are labeled as general controls because they apply to all application systems processed within a computer installation.

The data security risk analysis should be organized to evaluate the effectiveness of the general controls within the organization, and to analyze the related risks associated with four general types of security penetration or breach.

Other experts suggest the list of controls should include the following:

1. Internal controls
2. Administrative and physical controls
3. Qualifications and training of staff
4. Data integrity
5. Software integrity
6. Communication controls
7. Cost processing controls
8. Interactive controls[17]

Risk analysis is the process of identifying the risks to an organization, assessing the critical functions necessary for an organization to continue business operations, defining controls that are in place to reduce organizational exposure, and evaluating the cost of such controls. The risk analysis often involves an evaluation of the probabilities of a particular event. Risk analysis is often referred to as risk assessment, impact assessment, corporate loss analysis, risk identification, exposure analysis, or exposure assessment.

John O'Mara of the Computer Security Institute states that ". . . there are a lot of companies out there that won't know what computer security is until disaster strikes. Then it will be too late."[1] Essential to the success of risk assessment and analysis is a written charter from top management stating its commitment to the purpose of the study and its resolve to implement as many of the recommended courses of action as is fiscally and physically possible.[12]

Some of the primary goals of the task force performing the risk analysis should be to:

■ Perform a preliminary risk analysis to identify major problem areas and select urgent "quick fix" security measures as needed to correct major problem areas

■ Estimate potential losses to the computer facility and its users from

   ■ Physical destruction or theft of physical assets
   ■ Loss or destruction of data and program files
   ■ Theft of information
   ■ Delay or prevention of computer operation

■ Estimate the probability of occurrence for potential threats and their effect on the computer facility in terms of the above four classes of loss potential

■ Combine the estimates of loss potential and threat probability to develop an annual loss expectancy

■ Select the array of remedial measures that effects the greatest reduction in the annual loss expectancy at the least total cost. Remedial measures will include changes in the environment to reduce exposure, measures to reduce the effect of a threat, improved control procedures, early detection, and contingency plans.[12]

Benefits of Risk Analysis

Some of the benefits of a thorough risk analysis are the following:

- Improved security awareness among employees
- Identification of all assets, vulnerabilities, and controls
- Improved decisions on controls needed
- Justification of expenditures for security measures

Basic Steps in Analysis

Risk analysis is an orderly process adapted from practices in management. The six basic steps should include the following:

1. Identify assets.
2. Determine vulnerabilities.
3. Estimate likelihood of exploitation.
4. Compute expected losses.
5. Survey applicable controls and their costs.
6. Project annual savings of control.[27]

Uses of Risk Assessment

There are a number of reasons why an organization would want to identify or measure risk in a computer or communications environment:

- To justify implementation of controls
- To provide assurance against unacceptable risk
- To assess compliance with regulations, laws, and corporate policy
- To balance controls to risk

Quantitative Risk Techniques

Initial risk assessment often uses quantitative techniques that require the estimate of probability or occurrence rate of given threats and the dollar amount of loss per occurrence. A fundamental equation might be:

$$R = F \times L$$

where risk (R) is the product of a given threat frequency (F) times the single-time loss that occurs as a result (L). The output is usually expressed as an **annualized loss exposure** (**ALE**) for purposes of comparison.

Courtney Risk Assessment

A quantitative method of numerical risk assessment recommended by Robert Courtney of IBM in 1977 is one approach to safeguarding. It is based on estimates of the expected frequency and amount of loss from each particular actualized threat. The expected frequency of threats occurring per year (P) is calculated (in most cases approximated) by the formula $P = 10 (P - 4)$ where P is assigned a value from this scale:

P = 0	Practically never
P = 1	If once in 1,000 years
P = 2	If once in 100 years
P = 3	If once in 10 years
P = 4	If once a year
P = 5	If once a month (10 times a year)
P = 6	If twice a week (100 times a year)
P = 7	If three times a day (1,000 times a year)

Delphi Approach

The Delphi approach is a technique in which several raters individually estimate the probable likelihood of an event. The estimates are collected, reproduced, and distributed to all raters. All raters are then asked if they wish to modify their ratings based on the initial ratings. After revisions, all values are collected. If the values are reasonably consistent, the final value is assumed. If the values are inconsistent, the raters need to discuss the inconsistencies and attempt to select a final value.

Identifying Vulnerabilities

The three basic goals of computer security are ensuring secrecy, integrity, and availability. A **vulnerability** can be identified by considering situations that could cause loss of the object. One vulnerability can affect more than one asset or cause more than one type of loss. The following should be considered when assessing risks and vulnerabilities:

- What are the effects of natural and physical disasters (fires, storms, floods, power outages, component failures)?
- What are the effects of outsiders (network access, dial-in access, hackers, people sifting through the trash)?
- What are the effects of malicious insiders (disgruntled employees, bribery, curious browsers)?
- What are the effects of unintentional errors (incorrect commands, wrong data, using incorrect master files, disposing of output insecurely)?

Risk Analysis Software

A number of software products on the market today can be used to calculate the risk assessment process. The advantages of using a software product rather than the manual method are as follows:

- A software product will automate the tedious calculations, thus saving time and money.
- Consistency and accuracy are more likely as human error is lessened.

- Nontechnical managers and users are more easily involved.
- A "what-if" modeling can be facilitated for decision making.

There are basically three types of software products that can be used (note: some products will incorporate more than one method):

- **Spreadsheet models** usually run on micros and employ standard software, such as Lotus 1-2-3.
- **Qualitative assessment packages** that usually present an easy-to-use questionnaire to the user, who is led through the session focusing on his or her own environment or system. The answers are weighed and scored, and the results shown on a scale. Analysis or corrective action may be indicated.
- **Quantitative packages** run the gamut from a simple analysis to the use of actuarial statistics and complex statistical routines that take confidence factors into consideration.

Once the mechanisms for cost-effective control have been identified by assessing risks, the next step is to develop a disaster or contingency plan. In the event of an emergency, what actions are to take place? All key areas should be addressed and included so that there are no "holes" in the procedure.

SECURITY PLAN

A **security plan** is a document that describes how a company will address its security needs. The plan is subject to periodic review and revision as the security needs of the organization change. The aspects of the security plan should include:

- What the plan should contain
- Who writes the plan
- How to acquire support for the plan

The issues to be addressed in the plan should include:

- Policy—achieving goals
- Current statistics
- Recommendations
- Accountability
- Time table
- Periodic review schedule
- Allowances for revisions

The following outline for a security plan details common components:

- A brief descriptive inventory of the critical applications and types of sensitive information. The description should include the purpose, use, criticality, and sensitivity of each application or type of information; the physical, operational, and technical environment supporting these applications and information; and the general type of risk to which each application or type of information may be exposed. In describing risks the plan should classify each source of risk by the level of current concern—for example, whether the risk presents a major, minor, or minimal source of concern.

- A general description of the information security safeguards currently in place to address the defined risks. Safeguards may include, but are not limited to:
    - Management controls, including assignment of an information security officer or other responsible individual, procedures for conducting risk assessments and keeping the results current, and personnel administration procedures
    - Systems development controls, including adherence to standards for project management, documentation, systems maintenance, and use of development methodologies
    - Information security controls, including software controls installed on hardware or used in shared processing environments
    - Operational controls, including production, I/O, and programmed controls
    - Security awareness and training, including brief descriptions of training programs
    - Physical security controls, including access, surveillance, and environmental controls
    - Contingency management, including provisions for backup and recovery of critical application systems
- A general description of information security safeguards not currently in use but planned and a schedule for their implementation
- A brief description of any deficiencies in information security safeguards as may have been noted in recently completed audits performed by an internal auditor. If management has adopted a set of corrective actions in response to audit findings that are not otherwise addressed above, these would be described.

Safeguards are physical controls, software controls, mechanisms, policies, or procedures that protect resources from risk.

"After the plan is written, it must be accepted and its recommendations carried out. Acceptance is a function of sensibility, understanding, and management commitment."[27] After the security plan is in place, a contingency plan should be written to fully outline the actions to be taken if and when the security plan fails.

Contingency Planning

A **contingency** is a condition in which information resource(s) are unavailable, as a result of a natural or manmade occurrence, that is of sufficient duration to cause significant disruption in the accomplishment of objectives of the organization. A **contingency plan** (or disaster recovery plan) is the advance planning and preparations that are necessary to minimize loss and ensure continuity of the critical business functions of an organization in the event of a disaster. A **disaster** is a sudden, unplanned calamitous event that causes great damage or loss. Any event that creates an inability on the company's part to provide critical business functions for some predetermined period of time is a disaster. (Contingency planning is also called outage planning, catastrophe planning, business resumption planning, corporate contingency planning, business interruption planning, or disaster preparedness planning.)

The intent of contingency plans is to ensure that users can continue to perform essential functions in the event that information support is interrupted. End users of information, as well as computer installations that process applications, should be required to have contingency plans.

Policies

Contingency plans must be written, tested, and regularly communicated to staff. Contingency plans must take into account backup operations (i.e., how information will be processed when the usual computers cannot be used) and the recovery of any information that is lost or destroyed.

With small computers and word processors especially, the contingency plan should address selected equipment breakdowns, such as a single printer servicing many stations. Procedures and equipment should be adequate for handling emergency situations (fires, floods, etc.).

Backup materials, including the contingency plan, should be stored in a secure and safe location away from the computer site. Contingency procedures must be adequate for the security level and criticality of the information. Employees should know what to do in case of an emergency and be familiar with the contingency plan.

The contingency plan may be needed at a time of great stress, when key personnel are not available. Training of staff is vital.

Goals

"The initial phase of contingency planning must be to define and establish the goals that are expected for the activity."[18] The goal statement should contain:

- A statement of importance of business resumption following a disruption, disaster, or outage
- A priority rating of the functions or applications to be performed
- A listing of the persons responsible for implementing the plan
- A statement about urgency and timing

Written Procedures

Geoffrey H. Wold, National Director of Information Systems and Technology Consulting for the CPA consulting firm of McGaldrey and Pullen, states that "the contents of the plan should follow a logical sequence and be written in a standard and understandable format . . . A standard format for the procedures should be developed to facilitate the consistency and conformity throughout the plan."[41] Wold recommends following this standard format:

- Purpose of the procedure
- Scope of the procedure (e.g., location, equipment, personnel, and time associated with what the procedure encompasses)

- Reference materials (i.e., other manuals, information, or materials that should be consulted)
- Documentation describing the applicable forms that must be used when performing the procedures
- Authorizations listing the specific approvals required
- Particular policies applicable to the procedures

Instructions should be developed on a preprinted form. A suggested format for instructional information is to separate headings common to each page from details of procedures. Headings should include:

- Subject category number and description
- Subject subcategory number and description
- Page number
- Revision number
- Superseded date

Writing Methods. Procedures should be clearly written. The following writing tips should be used in the procedures:

- Be specific. Write the plan with the assumption that it will be implemented by personnel completely unfamiliar with the function and operation.
- Use short, direct sentences, and keep them simple. Long sentences can overwhelm or confuse the reader.
- Use topic sentences to start each paragraph.
- Use short paragraphs. Long paragraphs can be detrimental to reader comprehension.
- Present one idea at a time. Two thoughts normally require two sentences.
- Use active verbs in the present tense. Passive sentences can be lengthy and may be misinterpreted.
- Avoid jargon.
- Use position titles (rather than personal names of individuals) to reduce maintenance and revision requirements.
- Develop uniformity in procedures to simplify the training process and minimize exceptions to conditions and actions.
- Identify events that occur in parallel and events that must occur sequentially.
- Use descriptive verbs. Nondescriptive verbs such as "make" and "take" can cause procedures to be excessively wordy. Examples of descriptive verbs are:

Acquire	Back Up	Count
Activate	Balance	Create
Advise	Compare	Declare
Answer	Compile	Deliver
Assist	Contact	Enter

Explain	Log	Replace
File	Move	Report
Inform	Pay	Review
List	Print	Store
Locate	Record	Type

Planning Assumptions. Every disaster recovery plan has a foundation of assumptions on which it is based. The assumptions limit the circumstances that the plan addresses; the limits define the magnitude of the disaster the organization is preparing to address. The assumptions can often be identified by asking the following questions:

- What equipment/facilities have been destroyed?
- What is the timing of the disruption?
- What records, files, and materials were protected from destruction?
- What resources are available following the disaster:
    - Staffing?
    - Equipment?
    - Communications?
    - Transportation?
    - Hot site/alternate site?

The following is a list of typical planning assumptions to be considered in writing the disaster recovery plan:

- The main facility of the organization has been destroyed.
- Staff is available to perform critical functions defined within the plan.
- Staff can be notified and can report to the backup site(s) to perform critical processing, recovery, and reconstruction activities.
- Off-site storage facilities and materials survive.
- The disaster recovery plan is current.
- Subsets of the overall plan can be used to recover from minor interruptions.
- An alternate facility is available.
- An adequate supply of critical forms and supplies are stored off-site, either at an alternate facility or in off-site storage.
- A backup site is available for processing the organization's work.
- The necessary long distance and local communications lines are available to the organization.
- Surface transportation in the local area is possible.
- Vendors will perform according to their general commitments to support the organization in a disaster.

Team Approach. Assigned teams have specific responsibilities to allow for a smooth recovery.

Potential teams include these:

- Management team
- Business recovery team
- Departmental recovery team
- Computer recovery team
- Damage assessment team
- Security team
- Facilities support team
- Administrative support team
- Logistics support team
- User support team
- Computer backup team
- Off-site storage team
- Software team
- Communications team
- Applications team
- Computer restoration team
- Human relations team
- Marketing/customer relations team
- Other teams.[40]

Figures 13.2 and 13.3 show results of several studies dealing with the impact of disaster recovery planning.

Donn Parker states, "Many organizations are becoming so dependent on continuous availability of their computer services that loss of a few days can sometimes be fatal. The amount of time that an organization can go without computer services is referred to as **MTBU** (Maximum Time to Belly Up)."[26]

Emergency Response Procedures

When an emergency occurs, decisions regarding the choice of action must be made swiftly and effectively. Some of the guidelines for emergency response are as follows:

1. In the event of any life-threatening emergency (fire, earthquake, harmful chemicals, or explosives) the supervisor in charge is responsible *first* to start evacuation of personnel in an orderly fashion. Established evacuation routes and procedures will be followed with emergency assistants guiding the evacuation.

2. After evacuation, all individuals known to be on the premises must be accounted for (including vendors, suppliers, repair persons, visitors).

FIGURE 13.2
Industry Penetration

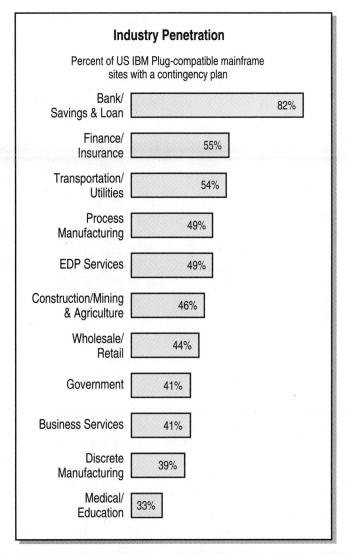

Source: Peter D. Anderson, "Disaster Planning . . . The Need for an Integrated Approach," *Computer Control Quarterly*, Vol. 9, No. 3, 1991, p. 37.

3. The supervisor is responsible for prompt notification of the emergency to appropriate authorities:

- Fire department
- Police department
- Emergency rescue groups
- Medical emergency groups
- Internal security
- Management, vendor support staff, etc.

FIGURE 13.3
Intangible Costs of an Extensive Computer Outage

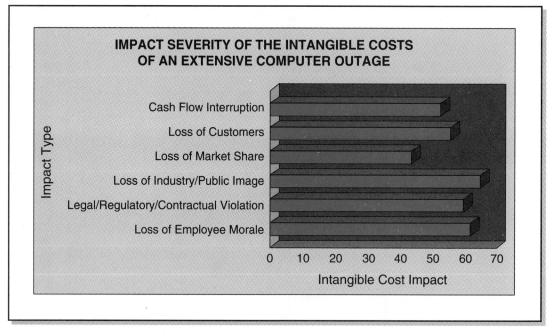

Source: Kevin J. Fitzgerald, "Planning for Disaster," *Computer Control Quarterly*, Vol. 9, No. 3, 1991, p. 53.

4. If possible (without endangering anyone's life), remove critical data files during evacuation and shut down equipment and utilities in a prescribed sequence.

5. If the emergency is *not* life-threatening, arrange for an orderly termination of jobs in progress prior to equipment and utility shutdown.

6. Establish damage control coordination to guide recovery and repair activities and to facilitate notification.

7. Activate emergency control center procedures if the scope of the emergency or the expected duration is sufficiently serious.

8. Recall personnel for special assignments, initiate recovery and repair procedures, and notify users and customers of conditions.

Disaster Recovery Software

As with risk analysis, there are also software programs that aid in the disaster recovery process.

Few security pros today would approach contingency planning without using an automated package. Since the early 1980s, PC-based disaster recovery tools have assisted numerous organizations in developing planned responses to interruptions in their normal business operations. These tools have evolved from simple text and word-processing applications to link data and information from a variety of sources and platforms.[2]

These software tools vary widely in functionality, complexity, and ability to meet a particular organization's resumption needs. Costs range from a few thousand dollars for a simple tool to as much as $25,000 for a highly sophisticated plan.

Some of the factors that should be considered before purchasing a disaster recovery software application relate to software, methodology, and model plan.

Software factors include the following:

- User-friendly interface
- Extensive customization capability without source code
- On-line help system
- Full-featured word-processing interface
- Relational database management system
- Integrated text processing and database functionality
- Standard and custom report generation
- Consistent functionality throughout the system
- Facilitation of plan testing and maintenance
- Automated testing and recovery management module, which provides for simulated disaster scenario analysis and on-line recovery
- Complete documentation, including planning methodology, user's guide, and action plan
- Integral backup and restore functions

Methodology issues include the following:

- Company-wide business resumption planning capability
- Team or department orientation, which provides extensive sample procedures for plan activation, notification, emergency response, reaction, crisis management, command and control center, and others
- Flexible methodology and phased approach to planning
- Flexibility to adapt to unique requirements
- Risk analysis and business impact analysis through an expert system interface

Model plan factors include the following:

- Comprehensive model plan with generic procedures for business resumption
- Disaster avoidance strategies and problem escalation strategies
- Standard and user-defined data collection forms
- Address of all areas of data processing and office equipment
- Address of communications and network recovery strategies and information

BACKUP PROCEDURES

Recovery from a contingency depends on having the required resources available. Generally, backups are necessary for:

- Data and documentation for critical applications
- Hardware (including peripherals and communication links)

- Software (including operating systems, utilities, and application programs, along with necessary documentation)
- Supplies (e.g., customized forms, paper, tapes)
- Personnel
- Facilities to house equipment and operation and support personnel

Hardware Backup

Backup for physical equipment is an important element of the contingency plan. It is also very visible and can serve as a focal point for other protection and backup schemes.

Procedures to back up computer hardware should include both on-site and off-site procedures. **On-site backup** is contained within the computer center (or nearby) so resumption is quickly achieved with the least amount of disruption. **Off-site backup** is the process of storing records and alternate hardware at a location removed from the normal place of use. There are three types of off-site backup facilities:

1. **Hot sites** are alternate facilities that have the equipment and resources to recover the business functions affected by the occurrence of a disaster. Hot sites may vary in types of facilities offered (such as data processing, communications, or any other critical business functions needing duplication). These sites are usually fully configured and are ready to operate within a few hours. The equipment and systems software must be compatible with the primary installation being backed up. Usually, the only additional needs are staff, programs, and data files.

2. **Warm sites** are partially equipped backup sites, usually with selected peripheral equipment, such as disk drives, tape drives, and controllers but without the main computer. Usually, a warm site is equipped with a small CPU—one large enough to run critical programs only.

3. **Cold sites** are alternate facilities that are void of any resources or equipment except air conditioning and raised flooring. Equipment and resources must be installed in such a facility to duplicate the critical business functions of an organization. Cold sites have many variations, depending on their communications facilities, UPS systems, or mobility (sometimes known as **relocatable shells**). These shells are computer-ready cold sites that can be transported to a disaster site so equipment can be obtained and installed near the original location.

The major distinctions between the three types of sites are activation time and cost. The hot site offers a reasonably fast response to a disaster situation but at a fairly high cost. The warm site is less expensive but not as quick to activate (several days to several weeks). The cold site is the least expensive approach but is usually not as responsive; it is generally used for long-term needs.

Software and File Backup

In addition to the hardware and other physical devices necessary to get back into operation, the computer software, programs, and essential files/data must also be backed up. "One or preferably two sets of duplicate records should be stored with a third party specializing in off-site storage."[36] Off-site storage is advantageous for these reasons:

- **Cost.** Outside vendors are less expensive.
- **Risk reduction.** Separate locations lessen the chances of the same disaster striking both places.

Some of the considerations to review before choosing an off-site storage location are these factors:

- Most backup records need to be close enough to be accessible but far enough away to prevent loss from the same disaster (two to four hours or 15 to 100 miles).
- The more secure and protected the backup facility, the better. The backup should have the same stringent security as the primary location.

When deciding which records to store in off-site facilities, a classification of records should be performed, putting records in priority order. In determining how to rank information, three loss situations should be considered:

1. Destruction (denial of timely use)
2. Modifications (use of undetected false data)
3. Disclosure (unauthorized use)

Records are then classified into three categories:

- **Class I**—data that require off-site storage of backup copies under secure vault conditions with timely restoration in case of operational information loss from (1) denial of timely use, (2) use of undetected false data, or (3) unauthorized use.
- **Class II**—data that require off-site storage of backup copies under conditions that resist accidental damage with timely restoration in case of operational information loss from (1) denial of timely use or (2) use of undetected false data.
- **Class III**—data that require no off-site storage of backup copies because the data could be regenerated if needed at a cost that is less than the cost of off-site storage, or because loss of the data would result in another cost less than the cost of off-site storage.[36]

Special File Access Problems

Special file access problems complicate file retention and backup procedures. For example, if customer, classified, trade secret, and sensitive data are processed in an on-line environment, an automated security software package must limit file access. The following issues address the access considerations for such files.

- **Customer files.** If an organization performs processing for other entities, it must access their files and programs. Customer agreements should be reviewed to determine the organization's contractual liability. Standards and procedures should have safeguards that minimize the possibility of lost or altered files. The customer should agree in writing to the proposed backup and retention schedules.

- **Classified data.** Files with a government security classification must be stored and transported according to government regulations. The Data Security Officer (DSO) and auditor should review these limitations to verify their enforcement.

- **Trade secrets.** Trade secrets are programs, formulas, and processes that give an organization a competitive edge. Methods of handling and protecting trade secrets are specified by law and therefore vary. Several requirements, however, are basic to maintaining trade secrets:

  - A trade secret and its associated material must be physically secured and designated confidential.

  - The information cannot be published or made available to the public.

  - If the material must be disclosed to someone outside the organization, this individual must be advised of the trade secret and must agree to maintain its confidentiality.

  - A permanent record should be maintained of the material's location when it is not in its usual storage location.

Senior management must designate which files or programs represent trade secrets, and the DSO and auditor should ensure that an organization's retention and backup standards and practices meet state and internal requirements. The standards and practices should be reviewed with legal counsel to ensure adequate control.[38]

Types of Software and File Backups

Storage facilities can consist of several types of operations. **On-site local backup** is usually a fire-resistant safe or vault located in or near the computer center. It is used to house the most recently created backup files until they are replaced by the newer generation.

Off-site local backup is a fire-resistant vault or safe located in another building within a close radius of the computer center. It is used to store backup files for up to one week; files are usually rotated daily.

Off-site remote backup is a fire-resistant vault or safe located 5 to 10 miles from the computer center. It is used to retain remaining backup files in active use for more than one week; it is accessed weekly.

Archival storage is an underground, fire resistant, and earthquake resistant storage facility located at least 40 to 50 miles away from the computer center. It is used to house permanent records to be retained for several years, including microfilm, printed reports, and selected backup files such as year-end data.

Without software the computer hardware is of little value. Contingency planning must provide for protection of software and information in an adequate and thorough manner in order to achieve meaningful security.

Uninterruptible Power Supply (UPS)

An uninterruptible power supply (UPS) is a backup power supply with enough power to allow safe and orderly shutdown of the central processing unit should there be a disruption or shutdown of electricity. "The cost of a UPS is often less than the potential cost of an electrical disturbance damage and/or a sudden blackout situation."[10]

Computers suffer from several types of power inconsistencies or disturbances and from power outages. **Spikes** are caused by circumstances related to the utility power source outside the building. Lightning is the most common cause of spikes as the voltage suddenly increases. **Oscillatory voltage surges** (noise) are brownouts and spikes that are caused by other equipment in the building—elevator motors, air conditioning motors, office equipment.

You can get absolute insurance against these potential problems with an uninterruptible power supply (UPS) system that smooths out variations in utility power, protects against more localized power disturbances, and most important, supplies battery backup in case of a complete power failure.

Batteries supplied with a UPS system are usually capable of operating the protected equipment for several minutes to an hour or more, depending on the user's requirements. This is long enough to bridge the customarily short power outage and allows sufficient time for a backup engine generator to get on line. If there is no generator or if it fails to start, the time is sufficient for the computer to shut down in an orderly fashion without loss of data.

A UPS system receives alternating current (AC) power from a utility's supply lines. The AC power is fed into the UPS system's rectifier, which changes it into direct current (DC). From the rectifier, some of the DC power goes to the batteries to keep them charged, but most of it goes to the inverter, which changes the DC back to AC. The output of the inverter is clean computer power, free of spikes, surges, noise, or other potentially damaging disturbances. In the United States, UPS systems operate on 60 Hz AC. When needed, the inverter can also act as a frequency converter to supply 400 Hz for those systems that require it.

In case of electrical overload or component failure, and during maintenance, a static bypass switch in the UPS transfers the protected equipment from the inverter to a reserve power source without interruption.

INSURANCE

Even though a risk analysis is performed and contingency plans are in place, a carefully designed computer insurance program can help reduce the financial impact on the organization if computer services are disrupted. "The right computer insurance program can also mitigate the financial consequences of crime, fraud, inadvertent error, malpractice, and other damages."[18]

Because each organization and its data processing facility usually has unique requirements and considerations, the appropriate insurance coverage will have to be negotiated during discussions with the insurance representatives serving the organization. These meetings should take place after management has completed its risk analysis to identify threats, countermeasures, and costs.[12]

Computer center risks are somewhat unique compared to other corporate assets for several reasons:

- Computer equipment is expensive and can take some time to replace.
- High dollar values are concentrated in a small area or room.
- Information is very concentrated and, in some cases, all corporate files are located in one place.
- Computer equipment is very vulnerable due to its sensitivity to heat, smoke, and humidity.
- Organizations are very dependent on computers.
- Malpractice can result from major errors and omissions.
- Millions of dollars in damage can result from crime and fraud.

Considerations

Insurance is a risk transfer method to reduce the effects of an event, called a loss, on the insured. Characteristics of an insurable event include these:

- The loss must be sufficiently large so as to require that it be averaged over either a large group or over an extended period of time.
- The circumstances of loss should not affect a significant portion of the total insured population (small population within a large group theory).
- The event causing loss must have a relatively small chance of occurrence.
- The event of loss should have the likelihood of equally occurring on any member of the group or class making up the population.
- The event causing the loss should be unexpected and unpredictable in time.
- The event of loss should be undesirable as an end result and generally not controllable.
- The event of loss should be clearly definable.
- The event of loss should be statistically identifiable in terms of mathematical relationships between claim experience, premium income, and profit margins.

Comprehensive Coverage

To provide the organization with more comprehensive coverage, if necessary, several considerations should be addressed:

- Coverage should be extended to include electronic funds transfer (EFT) systems, telecommunications systems, clearinghouse systems, computer networks, databases, and any other special processing needs.
- Employees should be bonded, and all computer fraud should be covered by employees working within the system.

- Transfer of funds over the telephone by voice instruction should be covered.
- Independent contractors performing services for your organization on location or remotely, using a modem connection, should be covered.
- Service bureaus under contract should be addressed in your policy's wording.
- Acts of former employees who have critical knowledge about your system should be covered.

TRAINING OF EMPLOYEES

"There can be no question but that employee morale is a significant factor in the data processing installation's security."[18] Even though the computer center staff and security officers will bear the ultimate responsibility for security, risk analysis, and contingency planning, empowering **all** employees with security awareness will ultimately result in a safer, more workable plan.

Many employees are required to have access to sensitive information and critical applications in the course of their daily activities. In addition to the obvious job classifications for information systems professionals, a variety of positions must be considered as introducing additional risk to information security. These include, but are not limited to, data-entry positions, fiscal and personnel technicians, and customer/client service employees who use terminals to view and update individual records. Risk assessments must consider that the simple act of viewing a record introduces the risk that violations of confidentiality and proper use of information could occur.

One of the most effective general means of reducing this risk is a program of training and awareness for all employees who may have access to sensitive information or critical applications. Training should be used to reinforce the value of security consciousness in all employees. Information security topics can be included that use general orientation programs for new employees; for others, less formal orientation may be appropriate. Training and awareness programs should stress the individual employee's responsibility for information security, proper use of sensitive information, and procedures for reporting potential violations. Training and awareness can be enhanced by periodic updates that inform employees of changes to laws, regulations, policies, or procedures, and that reinforce the employee's understanding of the importance of good information security practices.

A Sample Training Session

This session is designed to be interactive. The trainer gives the logical structure for the discussion and the basic definitions of terms that the participants will need. The participants will be asked to provide the practical examples from their work environments.

Using the risk assessment process as a model, the trainer should ask the participants to list assets; identify the impact of partial or total loss; assess the vulnerability of the assets; identify threats (someone or something) that could exploit the vulnerabilities; and, finally, determine the measures or countermeasures needed to eliminate or limit the effect of these damaging occurrences.

At the end of the session the trainer should have made a chart that looks like this:

Assets	Impact	Vulnerability	Threat	Countermeasures

A good risk assessment model depends on a solid foundation of believable assets that are important to operations or whose confidentiality must be maintained. An example of such a scenario is the payment of entitlements (annuities, bills, travel claims, social security, education and housing subsidies, farm supplements, etc.). Here are some assets in that situation:

- The paperwork submitted by the claimant
- The workers who process the paperwork
- The database of claimant information

The personnel resources needed in an emergency are critical. It is certain that **all** employees are going to be affected, and any part they can play in the ongoing effectiveness of security and recovery from emergencies will be to their benefit.

Successful implementation of a contingency management plan requires that all involved personnel understand the plan and are capable of carrying out their assigned responsibilities. A program of training for contingency management accomplishes three objectives: (1) it develops the knowledge and skills required to execute the contingency management plan; (2) it increases the likelihood of a successful recovery; and (3) it demonstrates commitment to contingency management planning and protection of resources.

Training, testing, and maintenance of the contingency management plan are all related. Training may reveal aspects of the plan that are unclear or inadequate. Testing provides feedback on the plan itself and the effectiveness of training. A failure during training, for example, may be due either to faulty procedures or to insufficient training on a sound procedure. Revisions indicated by the results of training and testing become part of the plan maintenance process.

Specific tasks associated with this component include the following:

- Design of the overall plan for contingency recovery training
- Development of specific training activities
- Development of evaluation techniques and tools

TESTING THE PLAN

Testing is necessary to determine whether (1) recovery procedures are complete and workable; (2) materials and data are available and usable for alternate processing of critical applications; (3) backup software, documentation, and data are adequate and current; and (4) personnel have been suitably trained.

There are three levels of testing for the contingency management plan; they differ only in how closely they simulate actual recovery conditions.

- **Level 1**—audit of off-site data and documentation. At this level the test checks the adequacy of off-site storage, concentrating on the availability of the fields and documentation needed for recovery. Essentially, Level 1 examines whether the off-site storage facility actually has items identified for off-site storage, as previously identified.

- **Level 2**—system restoration using in-house computer and off-site files and documentation. At this level the test checks the workability and adequacy of recovery procedures as well as management's ability to control and direct the recovery process.

- **Level 3**—system restoration using backup computing, off-site files, and documentation. At this level the test checks the ability to restore operations and processing of critical applications at the backup site. It checks adequacy of the backup facility and management's ability to control and direct the recovery process outside the normal setting.

One method often used to test the plan is a **checklist test**. This method is used to test a computer disaster recovery plan and to determine if information such as phone numbers, manuals, and equipment in the plan is accurate and current.

Areas to Be Tested

Some of the components of the disaster recovery plan that should be tested are:

- Notification procedures
- Personnel availability and awareness
- Damage assessment procedures
- Inventory of off-site vaults
- Readiness of the backup site and backup user environment, including appropriate communications to the site
- Restoration/re-creation of system and application software
- Production scheduling of critical applications
- Alternate processing procedures by users
- Management procedures
- Recovery team procedures

Rolf Moulton, manager of information systems security for a large international corporation where he is responsible for security policy development, offers these procedures for reviewing the disaster recovery plan:

- Check that your disaster recovery test and implementation requirements (hardware, software, and communications) are current; make sure that your recovery vendor can still service them. Given the current trend to distributed processing, make sure that your recovery services contract doesn't call for more than your actual requirements.

- Determine whether a local or a distant recovery facility will better meet your needs. A local vendor will help keep staff travel costs down and will make testing easier. However, a larger but more remote recovery center may be a better choice, particularly where the vendor provides substantial technical skills. Also, a distant recovery center is less likely to be overwhelmed by a local disaster, where other organizations in your area may also initiate recovery activities.

■ Take full advantage of new opportunities created by technological changes and aggressive vendors. Evaluate new recovery services, such as mini-recovery centers located nearby, or specialized communication services that can be installed in your offices or data center. Both can substantially reduce the time and money spent on employee travel, and they may actually provide a more realistic recovery testing environment.

■ Don't overlook the possibility that, with the increased use of distributed processing, your disaster recovery processing and communications requirements may have been significantly reduced. It might be possible that they can be serviced using *internal* resources.[25]

SCENARIOS OF ACTUAL DISASTERS

First Interstate Building, Los Angeles, California

On the evening of May 4, 1988, the nightmare began for the tallest building west of the Mississippi River. The 62-story First Interstate Building in Los Angeles, California, had caught fire.

At 10:37 p.m. the Los Angeles Fire Department's Operation Communication Dispatch Section (OCD) received three separate 911 calls from persons reporting a fire on the upper floors of the First Interstate Building. At 10:38 the initial fire companies were dispatched. While en route the Battalion Chief observed and reported a large "loom-up" in the general area of the Interstate Bank Building. On arrival at the scene he requested an additional 15 fire companies and five Chief Officers. The fire department fought the fire successfully and confined the fire to five floors of the 62-story building. Even so, the fire had caused multimillion dollar damage—the water damage was extensive and the smoke contamination was almost total.

One only has to pause for a moment to realize the catastrophic results that a fire such as this would cause to a company like the First Interstate Bank Corporation, one of the largest banking corporations in America. In just a matter of hours their corporate headquarters and one of their main banks were shut down, totally out of service. All employees, computers, and day-to-day operations conducted in this building were terminated for an undetermined period of time.

In this case, the costs were minimized because the First Interstate Bank Corporation had a plan. Within minutes after arriving on the scene, the bank managers initiated their disaster plan, which took into account catastrophic events such as earthquakes, fires, and other types of disasters that could affect the bank's operations. The main objective was to get the bank back into service as soon as possible.

Blackmon-Mooring-Steamatic Catastrophic Incorporated (BMS-CAT), whose headquarters are located in Fort Worth, Texas, was contacted immediately and had supervisors on the scene by 7 a.m. the next day. Within the first five days after the fire, BMS-CAT responded with the balance of 300 supervisors from their offices throughout the United States, 15 major pieces of rolling equipment, and 900 people hired and trained locally.

BMS-CAT had one specific goal—to put the building in a pre-fire condition; this would be a daunting task. It had a 62-story building with five stories destroyed by fire. The 16th floor downward was water soaked, and the entire building had major contamination from the products of combustion. Each floor was approximately 17,500 square feet in area, with over a million square feet of area to be cleaned. BMS-CAT planned to make the building smell, look, and feel cleaner than it was before the fire.

Cleaning up after a disaster is not anything new to BMS-CAT. Blackmon-Mooring, the original company, started in the carpet, furniture, and drapery business over 40 years ago. Over the years it developed unique chemical and cleaning methods—thus, the beginning of the Steamatic Company. The first major job for the Catastrophic Division of BMS was the clean-up of the Las Vegas Hilton after a fire in 1981. Since that time it has cleaned up after floods, fires, earthquakes, and other large disasters in the United States and Canada. Another example of such a clean-up was one done for the U.S. Postal Service headquarters in Washington, D.C., after a fire.

Initially the parking garage of the bank building was used as BMS-CAT's command post, where it literally set up a task force with major logistical responsibilities. The clean-up entailed the use of many chemicals, seven tons of cleaning cloth, 20,000 cases of Q-Tips (used to clean electronic equipment), and over 500 people per shift. They worked two shifts per day, five days a week, with a maximum of 10 people per supervisor. This enabled the supervisors to observe what each person was doing each minute, making sure that maximum productivity occurred and that security was maintained.

One of the many obstacles facing BMS-CAT was the restoration of an estimated 7,000 items of electronic data processing equipment. This equipment ranged from personal computers to mainframe computers and printers. It was this equipment that was worked on first. If the equipment was wet, it was immediately opened up and a special protective oil was sprayed on the interior computer contents to prevent any corrosion from occurring until workers had a chance to clean and dry the equipment properly. Equipment that was not wet but had soot contamination was inventoried, then taken to the 27th floor where it was put on a priority hold and the serial number was entered into BMS-CAT's main computer. Priority-one equipment was identified for immediate restoration. Priority-two was equipment that could be restored on an "as they get to it" basis.

A service center was set up on the 27th floor. Over 200 technicians, students from service technology schools in the Los Angeles area, were hired to clean equipment. Each piece of electronic data processing equipment was taken completely apart, although no soldering was undone. A Q-Tip, dipped into a special solvent, was used to wash each part, then the part was vacuumed and wiped dry. After the equipment was reassembled, it was given to a representative of the company that serviced that brand of equipment for inspection and recertification. The reassembled equipment had a failure rate of less than 5 percent. The equipment was then sealed in plastic and returned to one of two places: another location set up somewhere in the city to get the bank back into operation, or an inventory holding place, where it was prepared to be returned to its original location.

Computer hard disks were among the few things that could not be salvaged. The cost of cleaning them could not be justified so the memory, which usually was not lost, was removed and replaced on new hard disks. In most cases small electronic items such as calculators, clocks, and radios were also disposed of due to the cost of cleaning them; this was not the case, however, for hundreds of thousands of floppy disks. The disks were salvaged by being vacuumed and wiped clean with special materials. In most cases they retained all their memory.

Another obstacle in the cleaning process was the monumental amount of paper materials that had to be sifted through and cleaned. Bank personnel felt sure that the documents in their major vaults would be safe from contamination, but on opening the vault they found out differently. It seemed that even the vaults were not safe from smoke contamination.

Thousands of pieces of paper had to be sorted. Each shelf, file cabinet, vault, and drawer throughout the building had every paper removed. If the materials were not wet, they were cleaned with special sponges that removed soot from both sides. Wet papers could be handled in two ways: if their value did not justify the cost of salvaging, the papers were discarded, or they were freeze-dried immediately. The freeze-dried materials—over ten 18-wheel truckloads—were then transported to BMS-CAT's main recovery facility in Fort Worth, Texas, where the clean-up process was finished.

After all this was taken care of, the real clean-up effort started. The five floors that were destroyed by the fire had to be completely cleaned of all debris. Some 125 truckloads of debris were removed from the building. At the same time that BMS personnel were working on those five floors—cleaning or removing everything down to the bare metal or concrete—contractors started reconstruction. The building as a whole suffered no real structural damage, probably due to excellent construction features such as spray-on fireproofing. On the other floors, every inch of carpet and every piece of upholstered furniture was steam-cleaned two or three times. Every inch of woodwork, every wall covering, every cabinet, every piece of furniture had to be wiped off and deodorized inside and out. Every section of ductwork had to be cleaned, and all the ceiling tiles had to be removed and replaced. And all these activities took place simultaneously so that nothing was spread from floor to floor.

During the clean-up reconstruction, security consisted of a metal detector and off-duty Los Angeles police officers. All employees were screened daily as they went in and out of the building for the entire 12 weeks that BMS-CAT crews were on site. No major security problems developed during this time.

On September 12, 1988, the First Interstate Bank was reopened. Due to appropriate planning on the part of the bank, and the work that BMS-CAT did, the disruption to the bank's operation was kept to a minimum. Why? Because they had a plan.[37]

Bluebonnet Savings Bank, Dallas, Texas

On the evening of Wednesday, January 15, 1992, Bluebonnet Savings Bank (BSB) in Dallas, Texas, got to demonstrate first-hand a key DR maxim: a "disaster" should not be thought of only as an external event that strikes computer operations.

Rather, a disaster is anything that interrupts the continuity of business operations. And when disaster struck, Bluebonnet was ready.

That evening, at this multibillion dollar bank (with 34 branch offices spread around Texas and a mortgage servicing company in Atlanta), MIS operations came to a halt. An attempt to re-IPL the bank's IBM mainframe failed when the 3725 communications controller would not load. In addition, operations were experiencing problems with bad tracks on the disk drive.

Like most financial institutions, BSB held that communication with branches and customers was key to continuous effective business operations. Anything that removes that communications link is disastrous. "We have to be able to allow customers to withdraw money, get information on account balances, and the like. You just can't tell people that they can't withdraw money because you don't know how much they have in their accounts," said Chuck Littleton, Disaster Recovery Planner for the Bank. "So it is standard policy for us to declare a disaster on anything that will knock us out for 24 hours or more."

Therefore, when it became obvious that the problem was not going to be fixed immediately, that is exactly what the bank did. Bluebonnet Savings Bank declared a disaster with its IBM hot site in Tampa, Florida, and activated its business contingency plan, automated with Strohl Systems' LDRPS software, at 4:15 p.m. on January 16.

By 8:00 that evening, key bank employees were on a plane to Tampa; by 12:15 a.m. they had begun recovery operations. At 6:00 a.m., the Tampa alternative site system was up and running successfully with all databases loaded.

Back in Dallas, recovery was in progress. By 3:00 a.m. the same morning, the communications controller had been brought back up. "After testing it and solving some communication problems with a few of the branches, we were able to determine that we could switch operations back to Dallas, and we did so at 9:00 a.m. In fact, we were only running live at the hot site for about three hours," said Littleton. "But if the problem in Dallas hadn't been solved, we were ready that Friday morning to be in full operation in a way that would have been transparent to our branches and customers and in a way that would have preserved the continuity of business operations."

Having the hot-site agreement in place was key to Bluebonnet's ability to react and recover quickly. But just as important, noted BSB's Disaster Recovery Coordinator Patti Smith, was having an automated business continuity plan that the bank had developed last September.

"We realized that in the event of a disaster, there was a lot of information that we would need to access quickly," said Smith. "Things like the names and phone numbers of people we needed to contact, organizational plans, task plans, equipment inventories, and the like. That kind of information is critical to have at your fingertips if you are going to keep doing business and servicing customers."

So last fall, using plan development software, Smith and the unit managers automated the bank's recovery plans. They analyzed the needs and functions of their business units and collected the information necessary to ensure the continuity of each key business function in the event of a disaster. "It was the availability of this data from the database that allowed us to react so quickly and efficiently," said Smith.

"Because we were actually up and running again by 9:00 a.m. Friday in Dallas," said Littleton, "this experience served as a thorough test of our disaster recovery and business continuity plan." And there are several key lessons that both Littleton and Smith point to as a result of the experience.

"First," said Smith, "you absolutely have to have an automated planning tool in order to maintain the data that is needed to effect the recovery process efficiently. There is simply no way, realistically, that anyone could control and update that much information in a simple written plan."

Littleton added, "The second lesson we learned is that it is so critical that the data in the database be current and valid that we will now update our continuity plan on a *daily* rather than a weekly or monthly basis. All staff changes, CPU or other equipment configuration changes, etc., will be input to the database immediately. It *has* to be current."

Finally, both agreed that the position of Disaster Recovery Coordinator, Smith's function, must be made clear and the lines of communication kept open for all who are in any way involved in the recovery. "It is really important in order to minimize confusion," said Littleton. "We had far too many people calling all over the place to ask questions when they should have been dealing directly with Patti. But we've cleared that up now. If anything like this ever happens again, everyone knows that Patti is 'central control' for all information regarding recovery operations." In fact, Bluebonnet Savings Bank now deems the position so important that Smith has been assigned an assistant.

Although Dallas was back up and running Friday morning, the disaster recovery team that had flown to Tampa stayed on over the weekend to troubleshoot the problems with the modems and communications lines. They returned on Sunday night, tired but justifiably proud of a job well done.

This time, the disaster was short-lived. But the experience was an important one. It allowed Bluebonnet Savings Bank to test and refine, under fire, the value and quality of its contingency plan. If there ever is a "next time," they will be prepared.[30]

SUMMARY

Managing and operating any computer facility not only involves the day-to-day operations but should also include planning for disasters, contingencies, and recovery. An ongoing operation is the prime objective that should be carefully planned, organized, implemented, and tested.

Risk analysis is the process of identifying the risks to an organization, assessing the critical functions necessary for an organization to continue business operations, defining controls that are in place to reduce organizational exposure, and evaluating the cost of such controls. Quantitative techniques are often used, rather than manual analysis, such as **annualized loss exposure (ALE)**, **Courtney Risk Assessment**, and **The Delphi Approach**.

The three basic goals of computer security are ensuring secrecy, integrity, and availability. A **vulnerability** can be identified by considering situations that could cause loss of the object.

The types of software products that aid in the process of risk analysis are **spreadsheet models**, **qualitative assessment packages**, and **quantitative packages**.

A **security plan** is a document that describes how a company will address its security needs. A **contingency plan** (or **disaster recovery plan**) is the advance planning and preparations necessary to minimize loss and ensure continuity of the critical business functions of an organization in the event of a disaster.

Recovery from a contingency depends on having the required resources available. Backups are vital for hardware, software, files, and personnel.

A **UPS** is a backup power supply with enough power to allow safe and orderly shutdown of the central processing unit should there be a disruption or shutdown of electricity.

Even though a risk analysis is performed and contingency plans are in place, a carefully designed computer insurance program can help reduce the financial impact on the organization if computer services are disrupted.

Training and awareness should be held for **all** employees who would be affected by a disaster that results in loss of computer operations.

Testing of recovery plans should be conducted to determine whether (1) recovery procedures are complete and workable; (2) materials and data are available and usable for alternate processing of critical applications; (3) backup software, documentation, and data are adequate and current; and (4) personnel have been suitably trained.

REFERENCES

1. Aaland, Mikkel, "Preventing Computer Disaster," *Working Woman*, November 1988, pp. 88-92.

2. Adibi, Sina, "Disaster Recovery Software: Selecting the Right Tool," *ISP News*, March/April 1990, pp. 49-50.

3. Anderson, Peter S., "Disaster Planning . . . The Need for an Integrated Approach," *Computer Control Quarterly*, Vol. 9, No. 3, 1991, pp. 30-37.

4. Arnold, Richard, "Underground Flood Hits Chicago's Loop, Shutting Down Business for Weeks," *Disaster Recovery Journal*, Vol. 5, No. 2, April/May/June 1992 Special Report, pp. 1-4

5. Baker, Richard H., *The Computer Security Handbook* (Blue Ridge Summit, PA: Tab Books, 1985).

6. Baker, Robert A., "Guidelines for Securing an Existing Computer Facility," *Security Management*, September 1985, pp. 72-73.

7. Bush, Mike, "Network Recovery and the Illinois Bell Central Office Outage," *ISSA Access*, 4th Quarter, 1989, pp. 28-29.

8. *Computers: Crimes, Clues, and Controls . . . A Management Guide*, prepared by the Prevention Committee, President's Council on Integrity and Efficiency, Washington, D.C., March 1986.

9. Dawson, Gregg, "Emergency Planning for the Disabled," *Disaster Recovery Journal*, Vol. 5, No. 2, April/May/June 1992, pp. 47-52.

10. DeVenny, James E. J., III, "Power Protection," *Government Data System*, September/October 1986, p. 54.

11. "Disaster Recovery to New York Clearing House Standards," *Computer Control Quarterly*, Vol. 10, No. 2, 1992, pp. 32-33.

12. Enger, Norman, and Paul W. Howerton, *Computer Security* (New York, NY: AMACOM, 1980).

13. Fitzgerald, Kevin J., "The Role of Public Relations During a Computer Disaster," *Computer Control Quarterly*, Vol. 9, No. 3, 1991, pp. 4-7.

14. Fitzgerald, Kevin J., "Planning for Disaster . . . The Need for Planning Your Recovery From Disaster," *Computer Control Quarterly*, Vol. 9, No. 3, 1991, pp. 49-53.

15. Gallegos, Frederick, "Risk and Control of the Software Maintenance Process," *Quality Data Processing*, January 1987, pp. 12-13.

16. Hankins, Joseph, "Choosing the Right Automatic Sprinkler," *Disaster Recovery Journal*, Vol. 5, No. 2, April/May/June 1992, pp. 56-60.

17. Hodge, Bartow, Robert A. Fleck, Jr., and C. Brian Honess, *Management Information Systems* (Reston, VA: Reston Publishing Company, Inc., 1984).

18. Hutt, Arthur E., Seymour Bosworth, and Douglas B. Hoyt, *Computer Security Handbook* (New York: Macmillan, 1988).

19. *Information Technology Resource Management Standard . . . Information Security*, Council on Information Management, Commonwealth of Virginia, COV ITRM Standard 91-1, Richmond, Virginia.

20. Kesim, Suzan N., "Securing Computers: A Checklist," *Security Systems*, December 1987, p. 42.

21. Knapp, Thomas J., "Selling Data Security to Upper Management," *Data Management*, July 1983, pp. 22-25.

22. McDermott, Sue, and Kirk Lowery, "We Knew What We Needed to Do . . . ," *Disaster Recovery Journal*, Vol. 5, No. 2, April/May/June 1992, pp. 18-19.

23. Moore, Pat Williams, "Don't Let Vital Records Go Up in Smoke," *Computer Control Quarterly*, Vol. 10, No. 2, 1992, pp. 51-53.

24. Moran, Robert, "Cutoff!" *Computer Decisions*, August 1988, pp. 65-69.

25. Moulton, Rolf, "A By-Product of Effective Security–Improved Organizational Productivity," *Computer Security Journal*, Vol. V, No. 1, pp. 31-37.

26. Parker, Donn B., *Computer Security Management* (Reston, VA: Reston Publishing Company, Inc., 1981).

27. Pfleeger, Charles P., *Security in Computing* (Englewood Cliffs, NJ: Prentice-Hall, 1989), p. 458, 462.

28. "Prepare—Or Meet Thy Doom," *Computerworld*, February 21, 1992.

29. Rash, Wayne, Jr., "Be Secure, Not Sorry," *Byte*, October 1988, pp. 129-132.

30. Roberts, Mary Lou, "The Test That Wasn't a Test," *Disaster Recovery Journal*, April/May/June 1992, p. 29.

31. Rogers, Michael, "A Data Survival Guide," *Disaster Recovery Journal*, Vol. 5, April/May/June 1992, pp. 14-16.

32. "Taking Steps to Decrease the Risk of Office Fire Losses," *Disaster Recovery Journal*, Vol. 5, No. 2, April/May/June 1992, pp. 8-12.

33. Tartaglia, Benjamin W., "Telecommunications DRP for Natural Disasters," *Disaster Recovery Journal*, Vol. 5, No. 2, April/May/June 1992, pp. 53-54.

34. Teodoro, Reynaldo S., "DP Technology Security Impact Statement," *Data Security Management* (New York, NY: Auerbach Publishers, Inc., 1984), pp. 1-16.

35. "The Big Hurdle to a Recovery Plan," *Pacific Computer Weekly*, February 7, 1992.

36. Walton, Charles, and Ashley Durham, "Information Systems Liability," *Journal of Systems Management*, October 1988, pp. 36-41.

37. Weller, Reginald H., "Off-Site Data Storage: Computer Age Insurance," *Security Management*, September 1985, pp. 70-76.

38. White, David, "After the Fire . . . ," *Computer Control Quarterly*, Vol. 10, No. 2, 1992, pp. 30-31.

39. Wilkinson, Bryan, "File Retention and Backup," *Data Security Management* (New York, NY: Auerbach Publishers, Inc., 1984), pp. 1-12.

40. Wilkinson, Bryan, "Security Standards Guidelines," *Data Security Management* (New York, NY: Auerbach Publishers, Inc., 1985), pp. 1-16.

41. Wold, Geoffrey, "The Disaster Recovery Planning Process," *Disaster Recovery Journal*, Vol. 5, No. 2, April/May/June 1992, pp. 32-34.

REVIEW QUESTIONS

1. What basic elements should be included in a disaster recovery plan?

2. What is the purpose of a risk analysis?

3. What are some of the benefits of a risk analysis?

4. What is the design of the Courtney Risk Assessment?

5. How is the Delphi Approach used in a risk analysis?

6. What is a vulnerability?

7. What items should be included in a security plan?

8. What is a contingency plan?

9. How does a disaster recovery plan function?

10. What are the purposes of backup procedures?

DISCUSSION QUESTIONS

1. It has been said many times by proponents of disaster recovery planning that what we really need is a decent disaster to shake management out of its lethargy. Comment on this statement.

2. Ideally, information security should be managed to encompass:

   a. availability

   b. integrity

   c. confidentiality

   The disaster recovery plan addresses the major part of the "availability" facet. How is this accomplished? What is the relationship of DRP to integrity and confidentiality?

3. Disaster recovery planning is not present in many organizations for two primary reasons:

   a. Management is not convinced that their computing service could suffer a disaster.

   b. Management is convinced that if a disaster did occur that they could survive it without the need for planning.

   Could these assumptions be false? Why?

4. If management does not plan its approach to disaster recovery in a logical manner, it will fall for the trap of implementing solutions before having a real understanding of the problem. Discuss this statement.

5. Computer facilities should not be located under an airline flight path, on a flood plain, on a gas supply line, or on an earthquake fault line. How could these unique vulnerabilities be circumvented economically?

EXERCISES

1. Interview an individual who is a key member of your university's computer facility about its disaster recovery plan.

2. Locate an article in a magazine or newspaper concerning some actions that were taken to recover computer facilities in the aftermath of Hurricane Andrew in August 1992 in Louisiana and Florida.

3. Find out if there are any underground storage facilities, cold sites, or hot sites in your area that are used for computer operations.

4. Prepare a basic disaster recovery plan for a computer lab on your campus.

5. What vital records at your university should have backups? What records could be considered non-essential and not require backups?

PROBLEM-SOLVING EXERCISES

1. Develop a risk analysis matrix for a small business of your choice. This matrix should show the various categories of risks.

2. Develop a brief plan to be presented to upper management that would show a need to perform a risk analysis. (Objectives, process, and outcome should be addressed.)

3. Explain how the quantitative technique of $R = F \times L$ can be used as a risk assessment tool.

4. Develop a hypothetical scenario and analyze the risk analysis using the Courtney Risk Assessment method.

5. Prepare a brief listing of the criteria to be used when choosing a hot site or cold site.

CASES

Patton Automotive, Inc.

George Patton, president of Patton Automotive, Inc., in Nashville, Tennessee, is facing economic hardships due to decreased sales of domestic automobiles in the United States. Patton Automotive is a small manufacturing facility that produces parts for the major automobile companies in the Detroit area. Layoffs may become a necessity in the very near future due to the slowdown. Mr. Patton recognizes that the potential risk to computer centers is heightened due to economic crisis and disgruntled employees. The two key issues in tough economic times become reduced staffing levels and increased productivity demands—both of which could have a devastating effect on the computer operation.

How can Mr. Patton protect his company from a potential disaster?

Bayou Enterprises

In August 1992, Louisiana (along with Florida) suffered unprecedented damage due to Hurricane Andrew. Along with power outages, destroyed property, and lack of food, the public was faced with telephone outages and the closing of essential businesses. Since many of these facilities rely on computers for their operation, their recovery was vital to the overall communication flow. Bayou Enterprises has volunteered its entire computer staff to aid in this relief effort.

What functions would be considered most vital? Least vital? Who would coordinate this effort so that confusion does not reign?

NEW TECHNOLOGIES AND FUTURE TRENDS

LEARNING OBJECTIVES

After studying this chapter, you will be able to:

1. Identify the changing demographic factors and emerging trends that affect the information systems arena.
2. Describe the computer crime patterns that will have a major effect on the future.
3. Explain the major international issues affecting information flow in the global economy.
4. Describe the crime patterns evident in computer terrorism, software infringement, commodity transfers and export licensing, and transborder data flow.
5. Identify the privacy concerns of information collection, transfer, and use.
6. Describe the ergonomic factors that affect the health and safety of information workers.
7. Consider the security implications of new technologies of artificial intelligence, electronic data interchange, electronic imaging, card-based financial transaction systems, and private branch exchanges.
8. Define the following terms.

TERMS

- HERF Transmitter
- Berne Convention
- Universal Copyright Convention
- Buenos Aires Convention
- Transborder Data Flow (TDF)
- Ergonomics
- Artificial Intelligence (AI)
- Expert System (ES)
- Electronic Data Interchange (EDI)
- Image Processing System
- Electronic Funds Transfer (EFT)

- Automated Teller Machine (ATM)
- Point-of-Sale Terminal (POS)
- Financial Transaction Card (FTC)
- Personal Identification Number (PIN)
- Wire Transfer (WT)
- Telephone Bill Paying Systems (TBP)
- Home Banking Systems (HB)
- Private Branch Exchange (PBX)

USING AI FOR AUDIT TRAILS

Artificial intelligence (AI) has long been playing a role in manufacturing—for instance, robot welders in automobile plants. Now there is interest in AI monitors that can wade through audit trail reports to find exceptions that indicate threats to system or procedural integrity. Some of these monitors go far beyond the IS manager's traditional audit trail reports, sorting through deep piles of business details, correlating remote events, and sniffing out creative misuse of the system.

For instance, at Manufacturers Hanover Trust Company, an application called Inspector uses an expert system shell to gather and analyze a record of daily transactions from the bank's 23 offices worldwide. Senior management at Hanover saw a need for a tool like Inspector to spot high-risk or fraudulent currency deals. The expert system uses a 75-rule "knowledge engineering" model, developed with the aid of veteran traders and bank auditors, to spot deviations from standard procedure or violations of bank rules in a nightly four-hour review of the day's international business.

AI has a home at the U.S. Air Force as well. The USAF Cryptologic Support Center in San Antonio uses a custom expert system package to review the audit trails of staff work on the Center's computer system. This tool reduces what would be a daily seven-foot mound of paper to a manageable eight- to nine-page report.

As reported in "Eyeing the Business and All the Businessmen," *Security Dynamics Bulletin*, Vol. 4, No. 2, Fall 1990.

As we conclude this text, having covered the technical issues, physical protection, legal and ethical issues, management and employee concerns, disaster recovery, and contingency planning, looking into the future is appropriate. Key questions to be addressed are: Where is the computer industry headed in the next decade? What privacy issues are becoming most paramount? What effect do computers have on the nature of work performed? As the world "shrinks," how does globalization affect security issues? What new technologies and applications present even greater security challenges?

By viewing computer security as a "moving target" that must be constantly reassessed, only then can we be somewhat comfortable with the systems we create. The computer industry, as a whole, is undergoing constant change and growth, so it is only reasonable that security issues are also in constant flux.

THE FUTURE IS NOW

> Ten or fifteen years from now the security management profession may not seem all that different to managers than it is today. Like today's managers, they will be accustomed to dealing with the day-to-day realities of organizational behavior dictated by the times. The changes predicted for a new century may seem like nothing more to them than other ordinary operational decisions. For now, however, the trends shaping corporations and security organizations are cause for uncertainty.[38]

In the next century, a focus on team management, changes in the work force, and an increase in educational demands will challenge the flexibility of security professionals. In the past, management allowed employees to have only minimal participation in problem-solving responsibilities. Yet, to remain competitive in the year 2000 and beyond, corporations will have to change management techniques by encouraging teamwork among all levels of corporate employees.

To have an effective team, management will have to communicate with its employees more effectively. By allowing its employees more freedom with work patterns, schedules, and techniques, management will receive more creative ideas from its employees and thus harvest the greatest benefits from this valuable asset. A manager will be regarded as a facilitator for the team, rather than a higher level being responsible for the delegation of remedial tasks. As a result, productivity of workers will greatly increase as they will take pride in their enhanced responsibilities. In the future, the job responsibility of managers will be to encourage communication among the team members, provide sufficient training for their employees, and share information about the business with their employees.

Demographic factors will also have a strong impact on the workforce in the year 2000 and beyond. In the years between 1965 and 2000, approximately three-fifths of the entrants into the workforce are expected to be women, a scenario that will pose great changes for security managers. Also, by the year 2000, approximately 29 percent of the workforce will be non-white. These changes will allow corporations to be more selective about its employees as there will be a strong workforce from which to choose.

These demographic changes will also require changes in the educational background of prospective employees. In fact, in the year 2000, almost 90 percent of all jobs will require a postsecondary education. Security personnel will be much more closely examined, and the most qualified will be chosen for the jobs. The primary educational concerns for the future will be literacy, initial job training, and retraining. In addition, the organization must retrain its employees to ensure that they are aware of technological changes and advances.

The combination of these factors will greatly influence the role of management in the security organizations of the future. There will be less emphasis on a hierarchical structure and more emphasis on team management, which is hoped to increase the productivity of the employees, and thus provide the mold for successful operations.[38]

The impact of these changes on the security environment is summed up in "2000 and Beyond" as follows:

> The changing demographics and educational demands of the new work force are significant trends that can strengthen or weaken a corporation. To maintain greater control over their destinies, security organizations will have to increase

the variety, scope, and difficulty of their initial training efforts . . . Security professionals must get involved in a cooperative effort among themselves and within their organizations to develop solutions for the future.[38]

Securing the Future

John deButts, retired chairman, American Telephone and Telegraph, offers three aspects of the current temperament of American society that should be of vital concern to our technical environments:

1. A growing estrangement from technology on the part of a substantial number of our fellow citizens

2. An increasing readiness to acquiesce to mediocrity, to settle for second best

3. A disposition to believe that, in an increasingly complex world, it doesn't make much difference what one person does or fails to do[21]

The public's disenchantment with technology may stem from simply being turned off by the complexity that technology has brought to modern society and the perceived hazards to health and safety that technology introduces (e.g., pollution, nuclear power). The scarcity of resources, setting priorities, and the concern for the quality of life have come to the forefront of our technological advancements. "The roots of . . . estrangement . . . spring from the abruptness with which we have been reminded that we have entered an 'era of limits'."[21] In short, it will take more technology, not less, to assure our civilization.

A new call for technological accomplishment is needed to answer the call for further technological accomplishment. "Ultimately, our technological progress depends upon public acceptance, and that depends on public understanding," added deButts.[21]

deButts postulates that we have retreated from and are now returning to excellence due to the public's acceptance to mediocrity and the belief that no individual's actions mattered to a more interdependent society that values individual performance. "In the final analysis, societies, nations, and institutions exist for no other reason but the fulfillment of the individuals who make them up."[21]

Computer Crime Patterns

The time period 1990-2010 has introduced (and will continue to do so) new waves in sophisticated, computer-aided crimes. As our need for information grows, so, too, does the criminal's methodology in manipulating the data and information held in our computers.

Joseph F. Coates cites a few of the methods currently being used to perpetuate computer crime and abuse:

1. Manipulating data to change image (both positive and negative) of selected people or groups

2. Altering the order of data to give false sense of production and profit

3. Counterfeiting schemes that include altering bar codes and falsifying social security cards, birth certificates, medical records, employment and education histories, immigration reports, etc.

4. Creating false propaganda that can help or hinder political candidates at all levels of government

5. Synthesizing false data to cause panic ("false" hurricanes, tornadoes)

6. Gaining a competitive edge by falsifying data (yours and your competitors') to gain strategic advantage

7. Holding a computer system hostage to gain publicity for a "common cause"

8. Victimizing point-of-sale terminals to transfer funds to illegal accounts (e.g., faking returns of expensive merchandise)

9. Hacking's growing number of active participants. All predictions hold that this practice will continue by taking on more sophisticated, wide-ranging acts.[15]

Analyzing the Risk Factor

A growing number of corporations and individuals are being exposed to a new computer-age phenomenon—information age vulnerability.

> The use of information systems and related technology has provided many benefits to corporations over the years, but the incresing dependence on IS has brought an accompanying need to analyze factors that go beyond traditional computer security, such as organizational, management, and market conditions. The emergence of this third factor in justifying the use of IS has yet to ascend to the forefront of thinking in many companies. The need to do so, however, is becoming more important, as illustrated by the 1987 Wall Street crash, and management must take up the challenge.[67]

> The issues that lie behind corporate vulnerability are not limited to the traditional subject of computer security. They go beyond site, data, and network security issues and cover a broad landscape of corporate concerns: the integrity, volume, and flow of information; the resulting organizational weaknesses; deficiencies in management skills; mutually dependent systems; and the pitfalls of changing business markets around the world.[67]

The key issue to be addressed as we head into the next century is that organizations have become dependent on IS systems for their survival. When you become dependent on any resource, you become vulnerable.

INTERNATIONAL ISSUES

> Modern economics are complex, interconnected systems that can be changed by applying information technologies (IT). Whether information is viewed as a resource or as a commodity, governments, multinational corporations (MNCs), and individuals are increasingly affected by the 'rules' that govern its flow and use.[40]

There are many issues facing IT today as the business arena becomes increasingly globalized and the world "shrinks." Doing business in a globalized economy presents many new opportunities and challenges. Computer security issues are certainly on the forefront as sensitivity to transborder data flow, commodity transfers, software piracy, terrorism, and technology policies become heightened.

Terrorism by Computer

The cozy image of the computer hacker as a bored but benign student has been permanently changed by the discovery that international terrorists are moving out into the networks. Reports of sustained and sophisticated attacks on military sites in Europe and North America have been reported. Hackers have planted software bombs in VAX systems, disguising them to look much like routinely requested classified information using computer commands to search for key phrases such as "Desert Storm" and "Patriot."

> Speculation is growing in the United States that a brokerage service is in operation that will put terrorist groups in touch with suitably skilled hackers. Part of the problem, in this international market scene, is the lack of coordination between different investigation agencies, even between agencies within the U.S. At least 14 different federal agencies were involved in the previously mentioned attempts, resulting in a great deal of redundant work and infighting.[68]

Winn Schwartau, Executive Director, Inter-Pact, Nashville, Tennessee, addresses the issue of computer terrorism by stating, "As we recognize our growing dependence on computers, we begin to see them as potential targets. However, we may not be aware of the kinds of attacks to which our systems are open . . . In today's world, computers are both weapons and targets."[59]

Some of the terrorists' weapons that are now well-known include the following:

- Malicious software
- Intercepting signals from telephones and faxes
- Controlling public telephone systems
- "Listening" to signals transmitted by computers
- Capturing keyboard signals, using spectrum analyzers and digitizing oscilloscopes
- Using a **high energy radio frequency (HERF) transmitter** composed of a signal generator, or amplifier, and an antenna to destroy or cause a glitch to a computer system[59]

Software Infringement

The proliferation of inexpensive, "look-alike" software is well documented by travelers to countries around the globe. Software is virtually sold on the street corner, along with local goods, food, and souvenirs. Much of this software has been found to be contaminated with viruses that will, when used, infect large numbers of computers.

> Legal protection for software depends on local laws and legal precedents and international conventions and bilateral agreements . . . The exclusive rights of a copyright holder that are recognized and protected by most copyright laws are the rights to reproduce, adapt (i.e., prepare derivative works), distribute, and perform the work. The precise nature of these rights, however, often differs between countries.[32]

Certain jurisdictions recognize moral rights (the right to be known as the author of the work) and the right to prevent others from making deforming changes (the right of integrity). These jurisdictions are acknowledged under the Berne Convention (adopted March 1, 1989). The **Berne Convention** members constitute a union open to all countries of the world.

The following protective requirements must be satisfied by the members:

- Granting of certain moral rights to authors with regard to exploitation of their works

- Granting of certain economic rights (e.g., exclusive rights of translation, reproduction, performance, adaptation, arrangement, or alteration) with respect to protected works

- Adoption of certain minimum terms of protection (generally the life of the author plus 50 years) for various works[32]

Prior to the Berne Convention, the United States (and some Latin American countries) adhered to the **Universal Copyright Convention** (**UCC**), which required a copyright notice on published works. Copyright laws become complicated as these laws operate territorially, usually providing protection only for a country's national works or for works first published in that country. In addition to the Berne Convention and UCC, the **Buenos Aires Convention** has been adopted by the United States and most Latin American countries; it provides for national treatment and requires the use of All Rights Reserved or equivalent in the copyright notice.

> Although legal protection for software is increasing rapidly around the world, the scope and effectiveness of that protection varies significantly from country to country . . . these multilateral efforts are intended to promote effective protection of intellectual property rights while ensuring that individual countries' laws enforcing such rights do not themselves become barriers to trade.[32]

Commodity Transfers and Export Licensing

In 1986, the Office of Export Administration (OEA) of the U.S. federal government proposed to establish a new general license under which an exporter could obtain telephone authorization for the export of certain controlled commodities to precertified end users without having to obtain an individual validated license. The proposal to create general license G-CEU was dubbed the "Gold Card Proposal" because of its similarity to credit card authorization procedures.

The Export Administration Regulations would then include the following statement:

> A *Certified End-User* must use and retain the commodities obtained under G-CEU at its own facilities or dispose of them only to other Certified End-Users. Any other use or disposition requires prior individual authorization from the Office of Export Licensing of the Department of Commerce.[25]

Commodities eligible for export to a CEU would *exclude*:

- Nuclear weapons or materials

- Commodities used for wiretapping

- Commodities or technical data controlled by another U.S. government agency

- Commodities listed as under short supply control

- Technical data regarding civil aircraft and electronic navigational equipment
- Commodities or technical data intended for delivery or use by the military or police to disquieted countries

To qualify for certification as an end-user, a business enterprise must be a legal entity with an established place of business in one of the eligible countries. An eligible country's governmental agencies may also be certified. Each candidate for certification will be reviewed by the Commerce Department to determine whether there is a "high expectation that such candidate would be a reliable end-user under this procedure." A CEU candidate must submit to the Office of Export Licensing a Certification Statement and a Comprehensive Narrative Statement prepared in accordance with the regulation's instructions.

Under the proposal, an exporter seeking to ship controlled commodities to a CEU would call a special number to confirm (1) that the proposed recipient is still an authorized end-user; and (2) that the commodities intended for export are within the CEU's authorization. Once the exporter confirms the eligibility of both the end-user and the commodities, it will be given an "approval code" number to place on the necessary shipping documents.[25]

The facilitation of U.S. export licensing procedures is critical to ensure U.S. leadership in high-technology trade. This procedure will aid in lessening this negative effect over time as relative improvement in foreign technology availability continues and as foreign customers' sensitivity to U.S. export controls increases. Control and security measures must be considered in this CEU agreement so that misuse and fraud do not occur and jeopardize both the private businessperson supplying the products but also the release for vital technology to unapproved countries.

Transborder Data Flow

Airlines, hotels, banks, credit card companies, insurance companies, and commodities and stock exchanges all depend on international communications systems to move their financial data around the world. Financial data that crosses international borders is, however, subject to legal and sovereignty constraints, a situation complicated when companies collect data in one country, move it through communications lines in another country, process it in a third, and store it in a fourth. Moving financial data across borders requires respect for host countries' laws and politics, as well as participation in the intergovernmental organizations that regulate international communications.[33]

The term **transborder data flow** generally denotes the flow of data across national boundaries. Much of this data comes from multinational companies that transfer data between headquarters in one country to subsidiaries in other countries. This exchange takes place in many forms, from simple (e.g., oral communication, television, and postal services) to complex (e.g., satellite and microwave transmission). Transmission of computerized data in its original digital format lets a company transmit large volumes of data faster and more efficiently than ever before. The distinction between transmitted voice and data has blurred because both can be transmitted simultaneously through the same transmission line.

Regarding the free flow of and access to information, countries are divided into two camps: those that support it and those that want to restrict it. Advocates of the free flow of information and trade argue that it benefits all parties; opponents argue that it only increases the advantage of countries that already dominate the information market.

Nations opposed to unregulated transborder data flow cite the following justifications:

- **Personal privacy**—protection of the rights of private citizens
- **Corporate privacy**—protection of information from competitors
- **National privacy**—controlling information of national interest (financial plans, economic trends, defense strategy, emergency plans, technical data)
- **National sovereignty**—fear of loss of cultures due to the influx of Western information
- **Data access**—possible unavailability of data stored in other countries during emergencies or wars
- **Data value**—decrease in data flow across national borders if multinational companies find that tariffs and taxation make it financially prohibitive
- **Effect on employment**—prevention of unemployment by processing of data inside borders (before it is transferred)
- **Protection of home industries and markets**—fear that unregulated transborder data flow can inhibit the growth of their countries' own communications and information industries.[33]

Effective management of international information flow depends on how the legal implications are handled. In addition to regulations stipulated by individual countries, many countries have joined intergovernmental organizations that define and limit transborder data flow. The major organizations involved in this debate are:

- The Council of Europe (COE)
- The Organization of Economic Cooperation and Development (OECD)
- The General Agreement on Tariffs and Trade (GATT)
- The Intergovernmental Bureau of Information (IBI)
- The International Telecommunications Union (ITU)
- The United Nations Educational, Scientific and Cultural Organization (UNESCO)
- The European Community (EC)
- The International Telecommunications User Group (INTUG)

Leading experts in transborder data flow (TDF) cite the following problems if restrictions are placed on TDF:

1. Conforming to the myriad TDF laws in every country will result in increased overhead costs. It will also require the creation of additional departments and the assignment of extra personnel, often withdrawing them from revenue-producing activities.

2. Inconsistent tariffs and volume- or content-sensitive pricing will also result in increased costs to multinational corporations (MNCs). Prices of products in various countries will increase since data flows account for an increasing share of the cost of providing goods and services today.

3. TDF legal sanctions, if applied, will hinder operations and may disrupt them completely. In some cases, data protection laws are vague and leave room for discriminatory application. They may, therefore, discourage TDF use because MNCs may fear the penalties for innocent violations of complex regulations.

4. Given that ready transmission of data across national borders may not be possible, it may be necessary for MNCs to duplicate data files and databases in foreign countries and to pay for extra hardware and software and the personnel to maintain them.

5. TDF regulation will inhibit the development of international data-processing networks that would offer services at the lowest possible cost.

6. It may be possible to transmit data only to those countries with reciprocal data protection laws, as there may be conflicts of jurisdiction between the MNC's home country and the subsidiary's host country.

7. United States time-sharing suppliers (TELNET, for example) may be effectively priced out of the market by PTT tariff escalations. Many U.S. MNCs rely on these suppliers to a significant degree.[10]

International organizations developed a common approach to privacy protection at an early stage to prevent the development of different concepts and national regulations that would impede transborder data flow. The main work in this field was done by OECD, the Council of Europe, and the European Communities.

OECD offered the following eight principles:

1. **Collection limitations**. Limits should be placed on the collection of personal data. Any such data should be obtained by lawful and fair means and, where appropriate, with the knowledge or consent of the data subject.

2. **Data quality.** Data should be relevant to the purpose for which they are to be used and should be accurate, complete, and up-to-date.

3. **Purpose specification.** The purposes for which personal data are collected should be specified not later than at the time of data collection.

4. **Use limitation.** Personal data should not be disclosed, made available, or otherwise used for purposes other than those previously specified except with the consent of the data subject or by the authority of the law.

5. **Security safeguards.** Reasonable safety measures should be taken.

6. **Openness.** A general policy of openness about developments, practices, and policies with respect to personal data should be maintained, as well as the means to establish the existence and nature of personal data.

7. **Individual participation.** The data subject has a right to access and control the data.

8. **Accountability.** Accountability rests with a data controller for complying with measures giving effect to all these principles.[60]

PRIVACY CONCERNS

Every time you make a telephone call, purchase goods using a credit card, subscribe to a magazine, or pay your taxes, that information goes into a database somewhere . . . All these records can be linked so that they constitute . . . a single dossier on your life—not only your medical and financial history but also what you buy, where you travel, and whom you communicate with. It is almost impossible to learn the full extent of the files that various organizations keep on you, much less to assure their accuracy or to control who may gain access to them.[11]

Privacy is now a very public issue. A 1990 Harris poll, conducted for consumer-data giant Equifax, showed that 79 percent of respondents were concerned with threats to their personal privacy—up from 47 percent in 1977.[35]

Congress is attempting to bring some order to the variety of privacy and technology laws, and the U.S. Office of Consumer Affairs has targeted privacy as one of its main concerns. Various consumer advocacy groups are now turning privacy into the issue of the 1990s. Among these groups are the Consumer Federation of America and the American Civil Liberties Union.

Some of the places where a private citizen's privacy is being invaded include the following:

- **On the phone.** Cellular and cordless phones become small radio stations. With the right receiver (a high-tech scanner, a child's walkie-talkie, or a baby monitor), it is possible for a neighbor to tap into personal calls.

- **At the supermarket.** Several companies are testing systems for supermarkets that will record every purchase. The supermarket could then offer these listings to organizations who could, in turn, offer discount coupons, encourage additional purchases, and target a competitor's customers, among other efforts.

- **On the job.** Many computer systems now allow supervisors full access to an employee's computer activities (including reading E-mail messages and monitoring keystrokes as a way of measuring productivity).

- **In your mailbox.** Mailing lists can be compiled that are used to target likely prospects. The direct mail marketers know where you live, your age, income, credit history, and personal habits.[35]

Another source of data that is then sold originates from information given freely by consumers to get credit or insurance. This information is then sold for other uses without the consent or knowledge of the individual.

Where does this information end up? Some of the companies who purchase this information include junk mail distributors, mail order companies, banks wishing to target credit-card customers, and telemarketing companies.

Buying Information

Companies can buy the results of the entire 1990 census linked to a street-by-street map of the United States on several CD-ROM disks. Based on the Privacy Consumer Consent, companies would have to notify consumers each time they intend to pass along personal information. The European Community is passing tough laws that will take effect after 1992.[35]

Here in the U.S., the direct-market industry claims that these regulations would be prohibitively expensive and, in turn, would restrain fair trade. The present foundation of privacy laws in the U.S. are stated in the U.S. Constitution.

Although the word 'privacy' does not appear in the document, the Supreme Court has interpreted the Constitution to grant individuals a right of privacy based on the First, Fourth, Fifth, Ninth, and Fourteenth Amendments. In addition to the constitutional basis, several other government regulations address privacy:

- Privacy Act of 1974
- Fair Credit Reporting Act
- Right to Financial Privacy Act of 1978

The newest efforts to regulate privacy range from simple fixes to a full-fledged constitutional amendment.

Cashless Society

Consumers are welcoming the advent of the "cashless society" because of the convenience and speed of transfer, but this paperless concept introduces with it several issues to be considered:

- There is simply no practical way to cope with the amount of paper needed to record and verify the millions of personal and business transactions taking place every day.
- Transactions are not only increasing in number but involving much larger amounts of money, and essential control over transactions is lacking. As almost all of us use credit cards to make purchases, we have already made a radical departure from our traditional cash-oriented society.

The dilemma we now face is the public's penchant for convenience and credit cards—and the resultant "sale" of the information to others. George Orwell's predictions in *1984* that Big Brother (the police state) is watching you may be taking on a new identity. "Perhaps the realization of Orwell's predictions was prevented by the very fact that someone did foresee the extreme possibilities."[46] How we are to control the monster we've created may present one of the biggest challenges to face technology.

ERGONOMICS

As we consider the protection of computers and information and develop procedures to build secure systems, we must also address the health, safety, and protection of the individuals who will operate this equipment. If the continuity of workers is not considered, a very critical "missing link" becomes evident.

The introduction of video-display terminals (VDTs) into the workplace has also generated a list of health-related complaints and safety infringements. Some of the noted complaints include:

- Eyestrain
- Eye irritation
- Blurred vision
- Neck, shoulder, and back pain

■ Increased stress

■ Evidence of anxiety, irritability, and fatigue

The VDT dilemma has two major facets to be addressed:

■ Health and safety of the worker

■ Slackening productivity

Once of the first steps to follow to lessen or alleviate some of these problems is ergonomic awareness. **Ergonomics** is the study of making machines fit people, not the other way around. Ergonomics is more a matter of education than training.

> Ergonomics gained prominence around World War II when it was discovered that pilots were starting 5,000 foot dives from an altitude of only 1,000 feet. They were misreading the dials on their altimeters. Dials that worked fine on clocks were confusing pilots.[53]

The implications of fatigue, strain, and decreasing productivity (judgment) experienced by these pilots may have a counterpart in the computer arena as employees perform repetitive tasks that require alertness and accuracy.

Pregnancy Considerations

In addition to the strain and fatigue factors, a greater issue has emerged concerning the danger of electromagnetic radiation from computer terminals—particularly to the unborn fetuses of women who sit in front of a terminal for long periods.

A Swedish group has found more evidence that electromagnetic radiation from computer terminals can have harmful effects. This group detected statistically significant effects on pregnant mice.[64]

There is some strong evidence to support the claims of miscarriage, still-births, birth defects, premature births, and other pregnancy-related problems of mothers who are overexposed to VDTs (8–10 hours a day). The VDTs currently being used in offices present an environment that needs careful evaluation and control in order to protect workers.

Legislation

Ergonomic legislation has been introduced in many states to protect the health and safety of computer-industry workers. The content of VDT bills has taken three forms:

1. Ordering a study on the issue

2. Directing a department to develop guidelines for VDT use

3. Mandating a set of requirements for VDT use in the private and public sectors.

Some of the standards to be included in ergonomic regulations include furniture and lighting, allowing workers a 15-minute break every two hours to get away from the screen, and eliminating any other known hazards. (See Figures 14.1 and 14.2.) The National Institute of Occupational Safety and Health (NIOSH) is actively involved in monitoring the ergonomic issue.

FIGURE 14.1
Ergonomic Factors

> ☑ Adjustable chairs should be used that provide the proper angle when viewing the screen.
> ☑ The top of the VDT screen should be 10 degrees and the center 20 degrees below the user's straight-ahead seeing position.
> ☑ The screen should be approximately 14 to 20 inches from a user's eyes.
> ☑ Screens should be kept away from windows and other harsh lighting sources to avoid harsh glare.
> ☑ Overall lighting should be maintained at 30- to 50-foot candles, which is less than half of the customary office lighting to reduce glare.
> ☑ The screen should be approximately three to four times greater than room lighting.

NEW TECHNOLOGIES

Many new technologies are being introduced into the Age of Information that present security challenges. At the same time, new uses of existent technologies are growing rapidly—adding to the already over-burdened security profession. How we deal with this new "high tech" will become our ultimate challenge.

Artificial Intelligence

Artificial intelligence (AI) is the ability to mimic or duplicate the functions of the human brain. An AI system includes all the people, procedures, hardware, software, data, and knowledge needed for computer systems and machines that demonstrate characteristics of intelligence. An **expert system** is an information system that can make suggestions and reach conclusions in the same way a human expert can. Can computers really think? Most experts would say no. They can be programmed to make it appear that they are, in fact, thinking, but all parameters and decision rules must be built into the program. The science fiction movie *2001: A Space Odyssey* showed audiences a futuristic glance at what could happen with "thinking computers."

We are now using computers to make medical diagnoses, explore for oil, determine what is wrong with mechanical devices, and assist in designing and developing new computer systems.

With past computer systems, including sophisticated decision-support approaches, programs were developed to produce reports, documents, graphics, and other outputs. Computers did not display "intelligent" behavior. AI now involves developing computer programs and approaches that give computer systems "thinking" ability.

FIGURE 14.2

Ergonomic Workstation

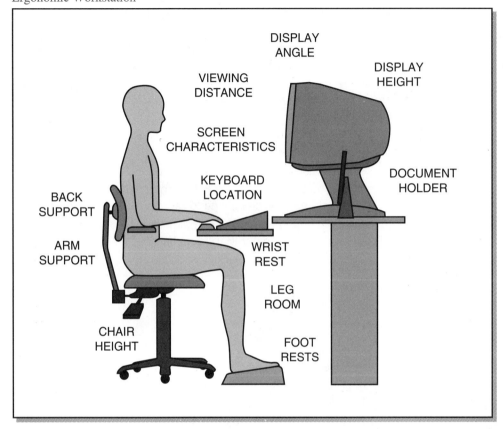

AI is a broad field that includes several key components, such as expert systems, robotics, vision systems, and natural language. Many of these components are interrelated, and advances in one can result in advances in others. Research in AI in the past 20 years has resulted in expert systems that explore new business possibilities, increase overall profitability, reduce costs, and provide superior service to customers and clients.

Joseph F. Coates offers this scenario to demonstrate the misuse of an AI system:

> Dobbs vs. State of New York has been accepted by the Supreme Court for hearing this year. Dobbs had elected to have his guilt adjudicated and his punishment determined by Big Judge Mark III expert system. After Dobbs received a five-year suspended sentence, it was discovered that the expert system had been tampered with. Furthermore, there was evidence that Dobb's son-in-law's brother, who holds a Ph.D. in computer science from M.I.T. and is associated with the vendor of the hardware, may have been involved with the tampering. Dobbs argued double jeopardy, that proper operation of court equipment is the responsibility of the court, and that a tenuous family connection is irrelevant and mention of it slanderous.[15]

Artificial intelligence (and expert systems), as with any new technology, can be misused if proper security procedures are not set in place at the inception. The potential damage and ramifications resulting from unsecured AI are monumental.

Electronic Data Interchange (EDI)

In a broad sense, **EDI** is the computer-to-computer exchange of intercompany business documents in a public standard format. EDI can also be used for the electronic exchange of information between organizations engaged in ordering, shipping, and payment of goods and services. "The electronic access systems already used between banks and their trade service customers are EDI applications in function if not in name."[64]

> Entire industries are beginning to agree on communications standards supporting vast networks for the electronic interchange of business data. Such industries— including automobile, grocery, hardware and housewares, pharmaceutical, transportation, and warehouse—have established industrywide networks or agreed to adopt an EDI vendor's central electronic mailbox for depositing and retrieving that data. Companies with major market shares must decide whether to build proprietary EDI systems or work with competitors, suppliers, and customers to develop public systems.[50]

Raymond Strecher, Vice-President and Product Manager in the international trade services consulting practice at American Management Systems in New York, predicts that "by 1995 a small number of banks will invest in creative EDI strategies as a competitive advantage in global banking. In response, other banks will begin to offer EDI products so as not to lose global trading customers or be left out of the global market."[64]

Today, it is estimated that approximately 10,000 companies are actively engaged in EDI communications with suppliers and/or customers.

Some of the initial problems that use of EDI has presented are as follows:

- **Legal implications.** "Paperless" contracts may not be enforceable. The contract can be enforced if it can be determined that it was the intent of the parties to enter into the agreement.

- **Legal ambiguity.** Applications of the Uniform Commerce Code (UCC) are not clear. The UCC states that business contracts are to be executed as "signed writings," but it is not clear how this applies to EDI.

- **Fraud.** Because there are no supporting documents on paper, electronic fraud is a primary concern. Using a call-back system is one possible solution.

Electronic Imaging

An **image processing system** is a system that is able to enter data from documents using a scanner and to manipulate the data. This device can be used to enhance photographic prints, pictures, and graphical or non-textual data.

A new wave of the future may soon be affecting the way we travel, shop, or even bank. A new scanning service produced by International Imaging Systems in

Milpitas, California, may soon be able to scan crowds and recognize particular people. How will this device aid the United States, as well as the rest of the world? The main foreseen purpose of such a device is its use to monitor crowds in airports for terrorists. Through the use of computers, this new technology uses brain-like qualities to obtain a three-dimensional picture of people. After vigorous preparation, the computer can obtain pertinent facial facts and refuse strange or unneeded facts to obtain a positive identification of people.

Original pictures are scanned to obtain a mathematical likeness, then the computer can scan a crowd and detect likenesses to reveal a positive identification. The programmer decides on how wide or narrow a search is conducted, meaning whether all features are used for comparison. These systems could also be used to help prevent robberies or even assassinations of prominent figures.

Another application of image processing involves applying computer technology to news photography, which promises faster delivery and easier processing of more and better photos. The downside of this new process is that "it also threatens deception by sophisticated, undetectable, and relatively easy alteration of component image data and the news information conveyed by the retouched photograph."[56] The central ethical question from a professional point of view is how to use and not use technology that already has been employed to deceive magazine and newspaper readers. "Though deception is seldom the word used, the fact remains that deliberate changes have been made to the content or appearance of news, feature, and advertising photography."[56]

CARD-BASED FINANCIAL TRANSACTION SYSTEMS

Card-based financial transaction systems enable customers to do their banking any day, anytime, and almost anywhere. They also enable card-issuing banks to provide banking services at nonbank locations and to reduce the number of branch locations as well as the size and capital demands of individual branches.

Several different types of card-based or computerized systems are gaining a wide acceptance and are used by consumers and banking institutions:

- **Electronic funds transfer (EFT)** involves the use of an automated device to electronically transfer funds. Consumers generally use bank cards (debit cards) in their transactions.

- **Automated teller machines (ATM)** are input/output devices used to conduct transactions with a bank.

- **Point-of-sale terminals (POS)** are computerized pieces of equipment for checking out items at grocery or department stores; they usually are a combination of optical scanners for input, computer for processing and transmitting data, and printer for output.

- **Financial transaction card (FTC)** is a wallet-sized plastic card that is used as the primary access device for automated operations that do not directly involve bank employees.

All of these automated and remote devices also "invite fraudulent misuse and physical abuse by thieves and vandals, as well as bank employees."[66]

With most of these systems, a **PIN (personal identification number)** is used. The PIN is a secret number entered into an accepting device (e.g., an ATM) by the cardholder.

Computers have become so advanced and so important in society that, soon, there will be no use for cash or checks. The banking industry has become almost completely automated. Home banking will be the trend of the future. Electronic funds transfer (EFT) systems handle more than 80% of all bank transactions, and electronic payment systems have been growing at a rate of about 20% annually. The use of automated teller machines (ATMs) exceeds the use of credit cards. Consumers can pay bills by dialing their bank's computer. Although EFT systems look promising for the future, there is evidence that these systems are victims of multimillion dollar crimes. The U.S. Bureau of Justice Statistics in February 1984, warned that "EFT systems may create the potential or opportunity for new types of criminal activity." EFT systems move more than $100 trillion annually, but how safe are the systems?

In an EFT transfer transaction, electronic blips replace paper. Financial information is initiated, transmitted, and recorded electronically. The elements of most EFT systems are as follows:

1. **Access devices**—plastic cards and account numbers that allow entry into the system
2. **Personal identification numbers (PINs)**—passwords that identify the user to the system
3. **Terminals**—the means by which the user can contact the computer and direct it to carry out specific transactions
4. **Computers**—the devices that record and transmit information on financial transactions

More than a dozen EFT systems are in use. The five most common are:

1. Wire transfers
2. Automated teller machines
3. Point-of-sale terminals
4. Telephone bill paying
5. Home banking

Wire transfer (WT), the oldest EFT system, is generally used by government and businesses to carry funds and information at very high speeds. The most widely used WTs are Fed Wire, Clearing House Interbank Service, and Bank Wire.

ATMs enable bank customers to deposit, withdraw, and transfer funds to or from various accounts by using an access card and a secret PIN.

With point-of-sale terminals, customers are given POS cards that they insert into a terminal at a retail store. The terminals are hooked up to a bank's computer. By using a PIN number, the customer conducts a sale by transferring funds from his or her account to that of the store.

Telephone bill-paying systems (TBP) allow a consumer to transmit bill-paying information directly to the bank's computers.

Home banking systems (HB) allow for a number of transactions, such as transferring funds, making account balance inquiries, and rescheduling payment.

Security experts are becoming increasingly concerned with the potential of misuse with all types of EFTs. Annual EFT crimes are estimated to range from $100 million to several billion dollars.

The four basic vulnerabilities of EFTs that are identified by security experts are the following:

1. **Unauthorized use of access devices.** The theft of EFT cards and codes has become more popular.
2. **Frauds by authorized users**. The user falsely claims that a third party used his access device to make withdrawals from his account.
3. **Internal manipulations by dishonest employees**. Employees use a customer's card and PIN to make unauthorized withdrawals and transfers.
4. **Sabotage**. Electronic interceptions are often directed at a system's communication lines.

Jerome Svigals, president and founder of Jerome Svigals, Inc., of Redwood City, California, a consulting firm that specializes in financial services and security device implementation, offers these appropriate countermeasures.

ATMs. The following events (with suggested responses or preventive measures) can signal fraudulent financial transactions:

- **Cardholders deny transactions that appear on the billing statement.** Firm and determined risk management focused on the learning and response process must be implemented. Coordination with the legal department before such an event prepares an appropriate course of action as a rational decision process for investigating cardholder claims.
- **Overstated ATM deposits.** A timely process of checking deposits and reversing incorrect entries is the first line of defense.
- **Claims of incorrectly revoked cards.** An after-hours, secure recovery procedure, coupled with an examination of the activity log, is the safest approach. This method allows adequate time to return the card by the next business day, after confirming that the card should not have been revoked.

POS Units. The following events and transactions associated with POS units must be addressed:

- **Complaints about PIN requirements.** A careful examination of a customer selection option is needed.
- **Double transactions by clerks.** Vigorous pursuit and timely follow-up of customers' concerns are needed.
- **Cash back at the POS.** This approach offers an alternative to ATM transactions but requires the same level of account control, including PIN validation.

■ **Large transactions.** These must be handled by a highly secure process. For example, any transaction over $1,000 requires a verbal exchange of preestablished identification codes.[66]

As the computer industry continues to expand, so will the ability to commit computer-related crimes, unless security measures can be installed and enforced.

Supporters of the "cashless" society say it will reduce traditional crime, but a new type of crime is growing. These new crimes are difficult to detect and guard against. It is important for society to realize the threat of these "computer criminals" to the financial sector. Security is the first step in protecting EFT systems.

PBX Fraud

PBX (private branch exchange) is a customer-operated, computerized telephone switching system. It provides internal telephone communications between stations located at a given facility, as well as telephone communications between the company and other public or private telephone networks. "Abuse of PBX long-distance codes can be expensive and dramatic. It is consistently referred to as a multimillion dollar a year problem, and sometimes viewed as the fastest-growing segment of network abuse."[17]

Many organizations with PBX phone systems use remote access codes that allow authorized employees to call in from outside the system and access an outgoing line or a voice mailbox. The abuse of such remote access codes, often referred to as PBX long-distance codes or extender codes, can be costly and far-reaching. "In one case, computer hackers involved in a 'call sell' operation generated over $1.4 million in telephone calls against one PBX owner's extender code over a four-day holiday period."[17] Hackers have stated that they attack PBX and other "hot" codes en masse deliberately so as to make detection and prosecution of hackers less likely.

Figures 14.3 and 14.4 offer some further details on how PBX fraud is conducted.

FIGURE 14.3
PBX Abuse

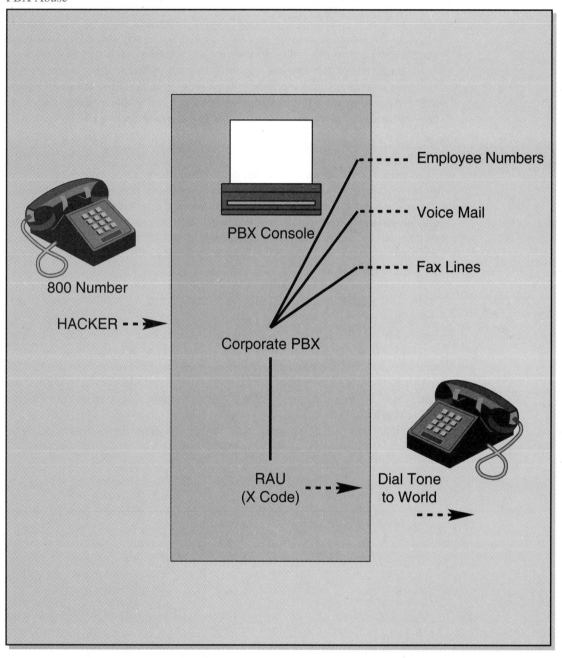

☑ **Dumpster diving.** Obtaining information by searching through trash, especially forms, printouts, and manuals that have been discarded.

☑ **Extender code.** A numeric password required by a remote access unit to permit an outside connection.

☑ **Finger hacking.** Trying groups of numbers on the telephone keypad until a valid access code number is found.

☑ **Remote access unit.** A device attached to a PBX that allows calls (which may originate inside the PBX or from an outside call) to obtain a dial tone that enables the caller to make long-distance telephone calls through the PBX.

☑ **Social engineering.** The practice of obtaining information by calling a company and pretending to be someone else, either in the company itself or connected to it (such as a customer). The effectiveness of social engineering depends on the hacker's ability to develop a trust relationship over the telephone through the use of the business's own jargon.

Source: William J. Cook, "Paying the Bill for Hostile Technology: PBX Fraud in 1991," *ISP News,* September/October 1991, p. 34

SUMMARY

The key questions to be considered in assessing the future of information systems (and the resultant security issues) are: Where is the computer industry headed in the next decade? What privacy issues are becoming most paramount? What effect do computers have on the nature of work performed?

In the next century, the focus on team management, demographic changes in the work force, and an increase in educational demands will challenge the flexibility of security professionals.

Some of the computer crimes that are growing worldwide and threaten information systems are manipulating data, altering of data order, counterfeiting, false propaganda, falsifying data, holding computers hostage, victimizing POS terminals, and hacking.

The international issues of paramount concern are terrorism by computer, software infringement, commodity transfers and export licensing, and **transborder data flow**.

Privacy concerns are now a very public issue. These issues involve phone use, marketing data, job-related monitoring, and mailing lists.

Ergonomics, the matching of humans to machines, is of vital concern as more and more workers are using computers for extended periods and experiencing some health problems. Legislation has been introduced in many states to protect the computer industry workers.

New technologies that are introducing significant security concerns as they become widely adopted are **artificial intelligence, electronic data interchange, electronic imaging, card-based financial transaction systems,** and fraudulent use of **PBX** systems.

REFERENCES

1. "Are VDT's Health Hazards?" *Newsweek*, October 29, 1984, pp. 122-123a.

2. Bellis, Paula S., "Friend or Foe? The VDT Controversy Continues," *The Office*, February 1986, pp. 40-42.

3. Benbow, Gary, Jason Masters, and Barry Cooper, "Computer Security in Australia," *Computer Control Quarterly*, Spring 1986, p. 32.

4. Bequai, August, *Technocrimes* (Lexington, MA: Lexington Books, 1987).

5. Blume, P., "Legal Information Economics," *Information Age*, 12(3), pp. 176-180.

6. Blume, P., "Problems Concerning Transborder Data Flow," *International Computer Law Advisor*, January 1991, pp. 23-27.

7. Burnham, David, "Flaws Are Cited in Treasury Computers," *The New York Times*, February 13, 1986, p. D5.

8. Burnson, P., "The Perils of Going Global," *Infoworld*, 11(33), 1989, pp. 34-40.

9. Campbell, Douglas, "Computer Sites: Targets for Destruction," *Security Management*, July 1988, p. 57-60.

10. Chandran, Phatak, and Sambharya, "Transborder Data Flows: Implications for Multinational Corporations," *Business Horizons*, November/December 1987, pp. 73-82.

11. Chaum, David, "Achieving Electronic Privacy," *Scientific American*, August 1992, pp. 96-101.

12. Chaum, David, "Privacy Protected Payments: Unconditional Payer and/or Payee Untraceability," *Smart Card 2000: The Future of IC Cards*, North-Holland Press, 1989.

13. Chaum, David, "Security Without Identification: Transaction Systems to Make Big Brother Obsolete," *Communications of the ACM*, Vol. 28, No. 10, October 1985, pp. 1030-1044.

14. Clement, John, "Rethinking Security," *Government Data Systems*, March/April 1987, pp. 45-46.

15. Coates, Joseph F., "Computer Crime Patterns: The Last 20 Years (1990-2010)," *Computer Security Issues and Trends*, Computer Security Institute, 1987. (Advertising Supplement in *Datamation*), September 15, 1987.

16. "Computer Matching and Privacy Addressed in Senate Bill," *Data Management*, December 1986, pp. 13-14.

17. Cook, William J., "Paying the Bill for Hostile Technology: PBX Fraud in 1991," *ISP News*, September/October 1991, pp. 34-37.

18. Corson, Richard G., "VDTs—New Evidence Indicates Helpfulness Over Harmfulness," *Data Management*, December 1986, pp. 24-41.

19. "DPMA Active on VDT Issues," *Data Management*, December 1986, p. 8.

20. Davies, Donald, "Confidentiality, Integrity, and Continuity," *Computer Control Quarterly*, Spring 1986, pp. 28-31.

21. deButts, John, "Securing Our Future Through Technology," *Computer Security Issues and Trends*, Computer Security Institute, 1987. (Advertising Supplement in *Datamation*, September 15, 1987).

22. Dickman, Steven, "'KGB Hackers Get Off Lightly," *Nature*, March 1, 1990, p. 6.

23. "Editor's Comments," *MIS Quarterly*, March 1988, p. iii-vi.

24. Elmer-Dewitt, Philip, "Peddling Big Brother," *Time*, June 24, 1991, p. 62.

25. "Export License by Phone Proposed: U.S. Policy Criticized," *Computer Law and Tax Report*, December 1986, p. 7-9.

26. Fiderio, J., "Information Must Conform to a World Without Borders," *Computerworld*, 24(40), pp. 91-95.

27. Fitzgerald, Kevin, "Computer Crime Detection," *Computer Control Quarterly*, Vol. 9, No. 3, 1991, pp. 41-49.

28. Forcht, Karen, "International Trade and Electronic Data Interchange: Legal Issues," *IBSCUG Quarterly*, Vol. 2, No. 4, Winter 1991, pp. 5-6.

29. Fox, Barry, "Making Money from Radiation Fears," *New Scientist*, August 1987.

30. Freedman, W., *The Right to Privacy in the Computer Age* (New York: Quorum Books, 1987).

31. Freidfeld, Karen, "The VDTs," *Health*, March 1985, p. 75.

32. Greguras, Fred, Gary Reback, and Joel Riff, "Software's Legal Protection Around the World," *Information Strategy: The Executive's Journal*, Vol. 7, No. 1, Fall 1990, pp. 23-29.

33. Guynes, Jan L., C. Stephen Guynes, and Ron G. Thorn, "Countering the National Barriers to Transborder Flow of Financial Data," *Financial and Accounting Systems*, Spring 1991, pp. 36-41.

34. Hembree, Diana, "Warning: Computing Can Be Hazardous to Your Health," *Macworld*, January 1990, pp. 150-157.

35. "How Did They Get My Name?," *Newsweek*, June 3, 1991, pp. 40-42.

36. Hughes, Gordon, "The Copyright Infringement Penalty in Australia," *Computer Control Quarterly*, Vol. 10, No. 1, 1992, pp. 30-32.

37. Icove, David, "Keeping Computers Safe," *Security Management*, December 1991, pp. 30-32.

38. Kaverman, Steven C., "2000 and Beyond," *Security Management*, September 1990, pp. 53-57.

39. King, W. R., and V. Sethi, "A Framework for Transnational Systems," S. Palvia, P. Palvia, and R. Zigli (Eds.), *The Global Issues of Information Technology Management* (Harrisburg, PA: Idea Group Publishing, 1991).

40. King, William R., and Vikram Sethi, "An Analysis of International Information Regimes," *International Information Systems*, January 1992, pp. 1-37.

41. Kirby, M., "Legal Aspects of Transborder Data Flows," *International Computer Law Advisor*, February 1991, pp. 4-10.

42. Lamb, John, and James Etheridge, "DP: The Terror Target," *Datamation*, February 1, 1986, pp. 44-46.

43. Lewis, Mike, "Wrapping VDTs in Red Tape," *Nation's Business*, June 1984, p. 60.

44. Madsen, W., "Effect of Transborder Data Flow Upon Information Security and Integrity," *Information Age*, 11 (3), pp. 131-137.

45. Manheim, Marvin L., "Global Information Technology: Issues and Strategic Opportunities," *International Information Systems*, January 1992, pp. 38-67.

46. Markoff, John, "Arrests in Computer Break-Ins Show a Global Peril," *The New York Times*, April 4, 1990, pp. A1, A16.

47. Morrow, George, "A Computerized Cashless Society," *Creative Computing*, Vol. 10, No. 11, November 1984, pp. 271, 272, 274.

48. "Mugshot Maven," *Discover*, April 1990, pp. 14-16.

49. "New Software Protection Laws in Korea and Other Asian Nations," *Computer Law and Tax Report*, December 1986, pp. 9-10.

50. Norris, Richard C., "The Industrial Politics of EDI Communications Standards," *Information Strategy: The Executive's Journal*, Vol. 7, No. 1, Fall 1990, pp. 18-22.

51. O'Connor, Kevin, "The Privacy Act in Operation: Some Issues Relevant to Fraud Control," *Computer Control Quarterly*, Vol. 9, No. 3, 1991, pp. 16-21.

52. Ploman, E. W., *International Law Governing Communication and Information: A Collection of Basic Documents* (Westport, CT: Greenwood Press, 1982).

53. Reidenberg, J., "Information Property: Some Intellectual Property Aspects of the Global Economy," *Information Age*, 10 (1), 1988, pp. 3-12.

54. Reynolds, Ern, "Responding to the Ergonomics Challenge," *Government Data Systems*, September/October 1986, pp. 64-65.

55. Ritsema, H.A., "Information Technology and the Law," *International Journal of Technology Management*, 4 (4,5), 1989, pp. 551-562.

56. Rosenbery, Jim, "Computers, Photography, and Ethics," *Editor and Publisher*, March 25, 1989, pp. 40, 42, 44, 54.

57. Rozen, Arnon, and John Musacchio, "Computer Sites: Assessing the Threat," *Security Management*, July 1988, pp. 41-51.

58. Sauvant, K.P., *International Transactions in Services: The Politics of Transborder Data Flows* (Boulder, CO: Westview Press, 1986).

59. Schwartau, Winn, "Seven Weapons for the Well-Armed Computer Terrorist," *ISP News*, September/October 1991, pp. 38-39.

60. Sieber, Ulrich, *The International Handbook on Computer Crime* (New York: John Wiley and Sons, 1986), p. 95.

61. Solomon, Elinor Harris, *EFTs and Payments: The Public Policy Issues* (Boston, MA: Kluwer-Nihoff Publishing, 1987).

62. Southard, C.D., IV, "Individual Privacy and Governmental Efficiency: Technology's Effect on the Government's Ability to Gather, Store, and Distribute Information," *Computer/Law Journal*, Vol. IX, 1989, pp. 359-374.

63. Stanulis, Gary E., "Benefiting from EDI," *New Accountant*, November 1991, pp. 28, 29, 51, 52.

64. Strecker, Raymond E., "Automation Is Still an Advantage in Global Banking Services," *Information Strategy: The Executive Journal*, Vol. 7, No. 1, Fall 1990, pp. 5-8.

65. "Studies Underline Hazards of Computer Terminals," *New Scientist*, August 1987, p. 33.

66. Svigals, Jerome, "Countering Fraud in Card-Based Banking Systems," *Financial and Accounting Systems*, Spring 1991, pp. 56-62.

67. Tate, Paul, "Risk! The Third Factor," *Datamation*, April 15, 1988, pp. 58-64.

68. "Terrorists Turn to Hacking," *Computer Control Quarterly*, Vol. 9, No. 3, 1991, p. 49.

69. The, Lee, "Easy on the Eyes," *Personal Computing*, October 1985, p. 90.

70. Ward, Stewart R., O.D., "Making VDTs Easier on Eyes," *Nations Business*, April 1984, p. 73.

71. Weiss, Kenneth P., "Dick Tracy in Cyberspace," *Computer Control Quarterly*, Vol. 9, No. 3, 1991, pp. 8-9.

72. Yasin, Mahmoud, and Ronald F. Green, "Why Risk Government Regulation of HR Databases?" *Information Strategy: The Executive's Journal*, Vol. 7, No. 1, Fall 1990, pp. 37-41.

REVIEW QUESTIONS

1. What are some of the methods currently being used to perpetrate computer crime and abuse?

2. How is terrorism by computer conducted?

3. What action is currently being taken to curb the global problem of software infringement?

4. What are the security considerations of commodity transfer and export licensing?

5. What is transborder data flow? What are the security concerns?

6. What are some of the privacy concerns of collection of information in databases?

7. What is ergonomics? How does this term relate to computer-related jobs?

8. What is artificial intelligence? What are some of the security concerns?

9. What is electronic data interchange? What are some of the security concerns?

10. What are some of the current card-based financial transaction systems?

DISCUSSION QUESTIONS

1. The information age revolution has brought with it innumerable benefits, but it also has created problems and risks that didn't exist before its onset. Explain the concept of information age vulnerability.

2. Most information systems are designed to improve the quantity of the output rather than the quality of the input. As we enter the next century, what does this statement tell us?

3. The most fertile and unpredictable source of computer crime is not always greed or social conscience but intellectual challenge. What procedures should be put into place to thwart the contemporary hackers of the 1990s?

4. Though deception is seldom the word used, the fact remains that deliberate changes have been made to the content or appearance of news, feature, and advertising photography. How do these acts violate the public's right to be informed and bring journalistic freedom into a new light?

5. A growing number of studies link intensive computer use to a host of painful and even debilitating maladies. Meanwhile, scientists argue the possible long-term effects of computer radiation. What steps should be taken to make work places safer?

EXERCISES

1. Choose a computer location on campus and interview users about eyestrain, backaches, and any other effects of long-term computer use.

2. Interview a banker or financial officer to ascertain his or her use of EDI and TDF.

3. Locate an article in a periodical or newspaper that discusses copyright infringement.

4. Observe an ATM sight and report on any possible security breaches that you may observe.

5. Interview a retail sales clerk about his or her use of POS terminals.

PROBLEM-SOLVING EXERCISES

1. The team concept that is predicted to become the primary way of doing business in the future takes on many aspects. What are some of the misleading conclusions (and security considerations) that could be drawn from each of the following statements?

   a. Productivity increases as workers are given responsibility for their own product or service.

   b. Gains in productivity of 30 percent are not uncommon.

   c. Having teams do the whole job rather than just parts of it creates pride of ownership.

   d. Teamwork makes it possible for each person to do several or all of the tasks necessary to produce one product or provide a single service.

2. The pace of technological progress is not a matter to be decided by the technologists, but by the entire body of our citizens. Ultimately, technological progress depends upon public acceptance, and that depends finally on public understanding. What are the implications of this statement? (Note: consider security and control issues.)

3. Speculation is growing in the United States that a brokerage service is in operation that puts terrorist groups in touch with suitably skilled hackers. Candidates for this brokerage include the various espionage agencies of the former Eastern Bloc countries, including those of East Germany and Romania. Even the KGB has now taken to placing advertisements in the Russian editions of Western newspapers, offering "intelligence services" in return for hard currency. How can this activity be eliminated to prevent hackers from "hitting" the global markets?

4. Although legal protection for software is increasing rapidly around the world, the scope and effectiveness of that protection varies significantly from country to country. These multilateral efforts are intended to promote effective protection of intellectual property rights while ensuring that individual countries' laws enforcing such rights do not themselves become barriers to trade. Are the current protection mechanisms sufficient to thwart the illegal distribution of software? What additional activities are needed?

5. As databases grew in the 1980s, many of the controls were removed. President Reagan said, "We're going to get government off the backs of the American people." Many people feel that statement meant, "We're going to get the government regulators off the backs of business." What effect did this "de-regulation" have on the use of people's private information?

CASES

E-Mail Snooping

Elaina Stubbins, a former administrator for a large U.S. company, was fired after trying to make her supervisor stop reading co-workers' E-mail. The company said she "got the ax" for insubordination. She countered that the evidence used against her was her own E-mail—and it was misinterpreted.

1. Is it ethical and legal for her supervisor to monitor E-mail messages?
2. What action should she take to "right the wrong"?

The Key Is the SSN

The growing amounts of information that different organizations collect about a person can be linked because all of them use the same key—your social security number. This identifier-based approach trades off security against individual liberties. The more information that organizations have (whether the intent is to protect them from fraud or simply to target marketing efforts), the less privacy and control people retain.

1. In your opinion, is the SSN making the best use of a common identifier?
2. Is there an economical alternative to the mass use of SSNs?
3. How many databases can you identify that contain your personal information and are identified by the SSN?

INDEX